JOINT SPECIAL OPERATIONS

JSOU
UNIVERSITY

RESISTANCE
OPERATING CONCEPT (ROC)

Otto C. Fiala

Foreword by
Major General Kirk Smith
Brigadier General Anders Löfberg

The JSOU
MacDill Air Force Base, Florida
2020

Contents

Figures

Tables

Foreword: U.S. Air Force Major General Kirk Smith

Nearly thirty years after the Cold War ended, European nations once again face threatening neighbors with potentially expansive intent. Many threatened nations belong to military alliances, but even those with strong alliances and friends do not necessarily have the power to prevent incursion. This could make those states appear to be easy targets. This book is intended to prevent that.

This multi-year project developed by a diverse group of people, combines myriad instruments of national power into a focused guide on a difficult topic and makes it accessible to all. This publication applies a sharp focus on one aspect of unconventional warfare: resistance.

The concept of resistance is a familiar one. Historically, nations have resisted invasions using conventional forces (CF). Some of those events are included as case studies in this book. The concept of resistance is not limited to Europe, but applies when sovereign nations are threatened by other nations. Today's threats are not limited to conventional incursions by uniformed military forces. As witnessed recently in Crimea and other parts of the Ukraine, proxy forces can be used by nations to threaten sovereignty and territorial integrity, challenging our traditional understanding of invasion by a foreign power. This is hybrid warfare, especially when the hostile power tries to mask its intentions. When a nation loses control of territory, resistance is applicable.

Nations under clear threat from neighbors must build resilience into their societies. Resilience is a nation's will and ability to withstand external pressure, influence, and possible incursion. A united society with a vibrant sense of national identity can more easily withstand external malign influence. Resilience allows nations to recover and resume their self-determination. Resilience is the fundamental foundation of resistance.

Nations typically plan to protect critical infrastructure during crises. This book acknowledges these efforts, but also explains how to increase national resilience against incursion by planning for, establishing, and developing organized national resistance capabilities. Starting now, threatened nations must formulate transparent national legal frameworks and policies which establish organized and controlled resistance capability. Resistance planning must involve not only the military, but government agencies and ministries at the national and local levels. It must also involve nongovernmental elements of society; the nation as a whole must plan for and be involved in various aspects of its defense. Resistance planning must also include a framework for reestablishing sovereignty when occupiers have been repelled. Potential adversaries must be put on notice that they will not succeed; they will be ousted.

When national resistance planning is integrated with allies and partners committed to the ideals of national sovereignty, territorial integrity, and self-determination, it can become a powerful message against a potential adversary. It places a potential adversary on notice that it cannot violate a nation's territorial integrity and attempt to establish a new status quo. The nation encroached upon will not accept defeat. Rule of law will prevail. National independence and autonomy will prevail.

That is the message of this book.

Kirk Smith
Major General, U.S. Air Force
Commander, Special Operations Command Europe

Foreword: Swedish Special Operations Command Commander, Brigadier General Anders Löfberg

This publication is unprecedented in many ways. It is a unique collaborative, comprehensive, combined, and joint effort that addresses an understudied area, namely resistance and resilience. In recent years, most of the contemporary literature and studies have focused on insurgency and counterinsurgency operations concerning the wars and conflicts since 2001. In contrast, this important publication fills a knowledge void (resistance and resilience) in order to complement existing understanding, thoughts, and ideas on insurgencies, conventional warfare, and asymmetric conflict that is apparent today. It is unique because the Resistance Operating Concept (ROC) is an innovative, complementary tool in the modern military arsenal that speaks not only to the military power side of statecraft, but also to the civil society, the broader population, and the civilian defense as a whole. It could be called not just the Joint Force but a "Total Defense Force." Resistance—resilience together with legitimacy—will undoubtedly evolve into an important component of a nation-state's defense sovereignty and security in the near future. The underpinning principles are nothing new but will certainly play a vital part in broadening and building the concept of total defense for the time to come. Legitimacy, for example, can be seen as a key factor in how a social unit, structure, or formation is held together and therefore plays a critical role in the cohesion and appeal of a resistance movement, as well as how the conduct of its operations are viewed from the perspective of the all-important population. Resilience and its close companion—perseverance, are vital for a society to overcome adversity and setbacks and generate a unifying fidelity to the mission in the face of arduous circumstances to continue to learn and adapt to achieve its aims.

Moreover, this project is unique in the ways and methods used to develop this publication. It is a commendable effort. There are many contributors that include countries, organizations, partners, prominent academics, expert practitioners, as well as centers of learning who have contributed to the making of the publication. This unique effort speaks volumes of the network of people, who willingly share ideas, hard-won lessons and experiences, as well as valuable insights, and who are committed to appreciating and seeking to understand this particular field of war and conflict. I am certain this publication will serve its purpose for strategists, policymakers, researchers, academics, and practitioners to mention a few that will find the content in this work both informative and interesting as a building block and foundation for further studies, projects, education, and doctrine development. It is a document for the future.

Anders Löfberg
Brigadier General
Commander, Swedish Special Operations Command

Acknowledgements

The Resistance Operating Concept (ROC) originated as an initial effort under the guidance of Major General Michael Repass, Commander, Special Operations Command Europe (SOCEUR). Mr. Byron Harper took that guidance and began the Resistance Seminar Series in 2014, supported by the next SOCEUR Commander, Major General Gregory Lengyel. They had the foresight to realize that it was necessary for the United States and our allies to have a common understanding of national resistance, and cooperate in planning for such an eventuality, based on Russian actions that continue today. Though developed with European partners, this concept has worldwide application.

We are grateful to all participants of the Resistance Seminar Series and Writing Workshops. The knowledge from these seminars and workshops coalesced into the first edition of the ROC, finalized in January 2017. This concept is a result of many contributors. Major General Repass planted the seed and supported its germination. Early supporters were Brigadier General Urban Molin, Chief, Swedish Special Forces Command, Mr. Lars Hedström, Dr. Richard Schultz, Major Matt Dreher, Mr. Mark Grdovic, Dr. Doowan Lee, Mr. Chris Donnelly, Ms. Linda Robinson, Dr. Sebastian Gorka, and Mr. Derek Jones. Professor Michal Matyasik went above and beyond to get us into Jagiellonian University. Major Andrejs Zaburdajevs, Latvian Special Operations Forces (SOF); Colonel Riho Uhtegi, Estonian SOF; Colonel Modestas Petrauskas, Lithuanian SOF; and Dr. Ulrica Pettersson of the Swedish Defence University, helped to counterbalance American-centric ideas. Colonel Eugene Becker, Ms. Heather Moxon, Mr. Christopher Blaylock, Dr. Daniel Troy, Mr. Mark Stottlemyre, Colonel Mark Vertuli, Sergeant Major William Dickinson, Ms. Elizabeth Kuhl, Mr. Piotr Hlebowicz, Mr. Sergei French, Mr. Daniel Riggs, and Mr. Robert Yates provided either broad or specific ideas, and efforts used to produce the first edition of the ROC.

Dr. Otto C. Fiala and Colonel Kevin D. Stringer, Ph.D., the managing editors of the first edition of the ROC, distributed it informally to partners for review in January of 2017. Later, as the ROC's primary researcher, organizer, and writer, Dr. Fiala built on the original version which was based on seminar after action reports and workshop written outputs, by buttressing it with further professional and academic literature, supported with U.S. Army and U.S. joint doctrine, and further operationalized the concept in the follow-on, limited distribution versions. After additional workshops, tabletop exercises, and further research, Dr. Fiala produced this final version. Supporting this effort in editorial support were: Lieutenant Colonel Jedediah Medlin, Lieutenant Colonel Randy Martin, Major H. Gavin Rice, Ms. Glennis F. Napier, Ms. Molly MacCalman, Mr. Georges Egli, Mr. Jim Worrall, Mr. Serge French, Mr. Aristotle Kestner, Mr. William McKern, Major Michael Weisman, Mr. James Del Castillo, Ms. Mila Johns, and Ms. Diane Le-Farnham.

The authors thank the doctrine writing team of the U.S. Army John F. Kennedy Special Warfare Center and School and the United States Army Special Operations Command (USASOC). The Johns Hopkins University Applied Physics Laboratory National Security Analysis Department recognizes the influence of their Assessing Revolutionary and Insurgent Strategies (ARIS) within the ROC. Mr. Paul J. Tomkins Jr., is the USASOC project lead for ARIS, with Mr. Robert Leonhard as the acting editor.

We also thank Mrs. Tatjana Havel for her generosity in allowing the use of "The Meeting" as the primary cover artwork, Ms. Cathey Shelton for her immeasurable graphic designs support, and Sergeant Andrew S. Donnan, Specialist Monique O'Neill, and Private First Class John R. Cruz for their admirable design of the front and back covers.

In bringing this document to publication, we are extremely grateful to the Joint Special Operations University for their support to this project.

Preface

This Resistance Operating Concept (ROC) explores actions that a sovereign state can take to broaden its national defense strategy and prepare to defend itself against a partial or full loss of national sovereignty. This document is a result of inputs to the Resistance Seminar Series, initiated in 2014, as a succession of seminars dedicated to studying resistance as a means of national defense.

The Resistance Seminar Series assembled a multinational academic and practitioner network to foster a broader intellectual perspective and build a common understanding among the disparate groups and individuals needed to support a resistance. Our partner participants were primarily from northern and north-central Europe, with common concerns. The seminars provided a structured forum for stakeholders to critically evaluate and develop "pre-crisis" activities, including preparation, deterrence, and other activities, while allowing for discussion and exchange of ideas on theoretical, historical, and practical elements of resistance. The seminars promoted critical thinking on resistance themes, creating a basis for collaboration and mutual understanding at strategic and operational levels.

As a culmination of these efforts, the ROC:

- Defines *resilience* as: The will and ability to withstand external pressure and influences and/or recover from the effects of those pressures or influences.
- Defines *resistance* as: A nation's organized, whole-of-society effort, encompassing the full range of activities from nonviolent to violent, led by a legally established government (potentially exiled/displaced or shadow) to reestablish independence and autonomy within its sovereign territory that has been wholly or partially occupied by a foreign power.
- Provides a common understanding of terms defined within previous seminars and by reference to publicly available U.S. military doctrine. "Adversary" and "enemy" are used to describe an aggressor state. The term adversary is used to describe the aggressor state prior to conflict, while enemy is used after that adversary becomes the foreign occupier and national resistance becomes necessary to restore national sovereignty.

The ROC employs many examples of historic insurgencies within the main body in order to examine the similar tactics, though we distinguish resistance from insurgencies. Resistance, as used here, specifically describes a national resistance, its organization, and activities against a foreign occupier to restore national sovereignty, and not other political grievances which serve as impetus for insurgencies. The examples herein explain historical ideas and tactics and do not endorse the groups, movements, or tactics identified. The ROC also provides common terminology to continue to explore resistance concepts concerning the integration of resistance planning, preparation, and procedures.

The Resistance Seminar Series originated under the guidance of Major General Michael Repass, Commander, Special Operations Command-Europe (SOCEUR). The initial multinational writing workshops were then begun under his successor Major General Lengyel. The writing of this ROC, based on that work, was begun under his successor, Major General Schwartz, in 2016. His support for the concurrent continuation of the seminar series allowed for refinement of the ideas contained in this book. It also allowed for expansion to include participants from outside the special operations communities and defense ministries. After the initial writing workshops, SOCEUR continued the development and expansion of this concept and added additional chapters and case studies to illustrate the points made in this concept while further expanding on the efforts of the workshops, resulting in this volume. Under his successor, Major General Smith, participation in the seminar series grew even further by bringing in more participants from outside of the special operations and defense communities. This extremely valuable exchange sped both intranational planning cooperation, as well as planning cooperation among the participating nations.

National Resistance is not a new concept, nor is it limited to any particular region of the world. Nations have resisted more powerful foreign occupiers throughout history. In today's interconnected world, communication and rule of law are critical to success. Throughout this work, we continuously stress strategic communication and the necessary establishment of legal frameworks. Governments must be prepared to effectively communicate to their populations, both occupied and unoccupied, their allies, and even the adversary's or enemy's public for success in the information environment, which contributes to success in the physical environment. The necessity and criticality of national legal frameworks to support the organization, development, and potential specifically authorized use of this form of warfare cannot be overemphasized. A legal framework offers internal legitimacy to the actions of the threatened nation and facilitates allied and partner support by communicating legitimacy on the international stage to secure support to restore sovereignty.

The foundation for the preparation and possible conduct of national resistance rests firmly on national resiliency. A nation must have a citizenry that identify as members of that nation and must have a desire to remain a sovereign and independent nation. This necessity is outlined here before even discussing resistance. The abovementioned legal framework supporting the development of a resistance capability is an integral part of national resiliency. The population must see this capability as a legitimate form of warfare, grounded in law, which is acceptable and suitable. It must then be willing to support resistance. That is the foundation of resiliency against a foreign threat.

This concept relies heavily on U.S. doctrinal terminology as an accessible basis for common understanding. U.S. doctrinal terms such as "underground," "guerrilla," "auxiliary," and "shadow government" are taken from U.S. unconventional warfare literature and used here to achieve this commonality of understanding. The intent is for allies and partners to be able to find further literature regarding those components through use of those terms. The terms also function to provide some political neutrality to allies and partners since the terms are from U.S. doctrinal literature.

Though this written concept was developed in cooperation with several specific allies and partners for use as a common planning guide against a particular threat, its applicability is worldwide. There exist many nations physically located very near much larger, threatening neighbors, with historical records of expansion and subjugation. This Resistance Operating Concept is a brief (re-) introduction to national resistance and a general planning guide.

Otto C. Fiala, Ph.D., J.D.
Colonel, U.S. Army Reserve (Ret.)
Chief Editor and Author

Introduction

> We shall fight on the beaches, we shall fight on the landing grounds, we shall fight in the fields and in the streets, we shall fight in the hills, we shall never surrender. - Winston Churchill, 4 June 1940

1. Purpose

A. The Resistance Operating Concept (ROC) encourages governments to foster pre-crisis resiliency through Total Defense (also known as Comprehensive Defense), a "whole-of-government" and "whole-of-society" approach, which include interoperability among its forces and those of its allies and partners. This establishes a common operational understanding and lexicon for resistance planning and its potential execution in Total Defense, incorporated within National Defense Plans. Several states in Europe and other parts of the world employ such a Total Defense posture. The ROC seeks to identify resistance principles, requirements, and potential challenges that may inform doctrine, plans, capabilities, and force development. It also identifies opportunities for intra-and intergovernmental support and collaboration, while promoting overall allied and partner nation (PN) interoperability. This document is the result of a cooperative effort to understand this topic.[1]

B. The government and military, with popular support, take action against an enemy in a traditional conventional defensive environment. The population is the primary actor in a resistance or Total Defense situation. Allies and partners play a very significant role in supporting the resistance effort. The most significant difference between traditional and Total Defense is the preparation required to ensure that the population is ready to fulfill its Total Defense role (see fig.1). In a Total Defense construct, the population has an increased/greater, more significant role. This preparation is part of the resilience concept. Enhancing and institutionalizing collaboration among government ministries, civic organizations, and the larger public is critical to success. This collaboration helps build a more resilient society and strengthens resistance networks established in the event resistance is required.

2. Scope

The primary focus of the ROC is developing a nationally authorized, organized resistance capability prior to an invasion and full or partial occupation resulting in a loss of territory and sovereignty. Resistance, as a form of warfare, can be conceived as part of a layered, in-depth national defense. Toward this end, the ROC first seeks to delineate the concept of national resilience in a pre-crisis environment. The ROC describes resilience as withstanding and recovering from external pressures. National resilience is enhanced with the formation of a national resistance capability. Second, the ROC seeks to develop resistance requirements, and support planning and operations in the event that an adversary compromises or violates the sovereignty and independence of an allied or PN. The ROC attempts to demonstrate both the significance of national resilience and the criticality of maintaining legitimacy during the conduct of resistance operations during the struggle to restore and resume national sovereignty.

3. Total Defense or Comprehensive Defense[2]

A. **General.** The concept of Total Defense has been adopted by several nations, particularly those bordering hegemonic powers. It includes all activities necessary to prepare a nation for conflict in defense of its independence, sovereignty, and territorial integrity; it consists of both civil and military defense. This concept recognizes that attacks can come not only in the form of traditional conventional military actions, but also attacks against the country's economy and society, designed to weaken its cohesion and resolve to defend against threats to its independence, sovereignty, and territorial integrity. This defense concept includes not only governmental agencies and functions from national to municipal level, but also private and commercial enterprises, voluntary organizations, and individuals. Total Defense encompasses all societal functions. Depending on national culture and historic experience, a legal framework may be constructed to mandate participation in planning, as well as to legally mandate certain civil activities (e.g., energy and food distribution; ground, air, sea, and rail transportation priorities; and communications) during a declared crisis and during an actual conflict. Regardless of its extent, a legal framework is critical.

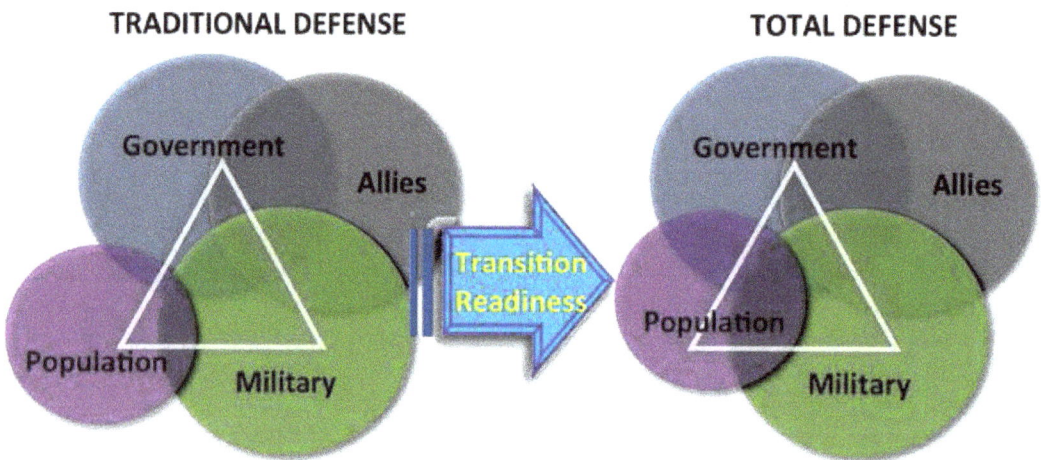

Figure 1. Comparing Traditional Defense to Total Defense.
SOURCE: CREATED BY PARTICIPANTS AND BASED ON AFTER ACTION REPORT, RESISTANCE SEMINAR, RIGA, LATVIA, 23-25 NOVEMBER 2015, 6

B. **Strategic Purpose.** Total Defense is integral to a National Defense Strategy.[3] The intent of Total Defense is to mobilize all support necessary to defend the nation and its territorial integrity against armed attack. This includes cooperation with other states, on home territory or abroad, to protect and promote security, freedom, and autonomy. This focus is designed to afford maximum time and space for military maneuver and political decision-making. Additionally, it calls for the mobilization of a prepared, firm, resolute, and perseverant resistance capability against an enemy in any of its occupied territory. The nation's resistance capability must be able to demonstrate to a potential aggressor that an attack will be extremely costly. This concept, applied domestically, must be joined with diplomatic, political, and economic measures taken by the state and its international partners, bilaterally and multilaterally,[4] to deter a potential aggressor. This deterrence should include agreements for those other nations to act against the aggressor.

C. **Lead Agency.** A nation must establish a governmental office of primary responsibility in order to advance, coordinate, and synchronize this effort. Such office is likely best placed inside of the ministry of defense (MOD) as its primary task is to make recommendations that will support the armed forces and national defense. That office can oversee and coordinate both military and civil defense planning and preparation as an all-inclusive, mutually supporting, and synchronized effort. It can also make recommendations for agreements with allies and partners, internal legal and policy framework adjustments, and interagency agreements to support this effort. Recommendations then require the political assent of the nation's chief executive and legislature to create the necessary supporting legal and policy frameworks.

D. **Civil Defense.** The objectives of establishing a civil defense are to protect the civilian population, safeguard the continuing functioning of necessary public services, attenuate the accompanying harmful effects of conflict, and contribute to the operational capabilities of the armed forces during conflict.[5] Civil defense increases the capability of the people to prepare themselves through personal preparedness (i.e., storing canned food, bottled water, medicines, and batteries) to alleviate supply, transportation, and distribution assets directed toward national defense while increasing the willingness of the individual to defend his or her country. Individuals and organizations inside and outside of government may be conscripted for roles in support of wartime organization. This may involve the establishment of municipal or regional reinforcement pools for rescue services, medical services (e.g., the designation of a hospital in a designated civil defense region as a war hospital), and home protection organizations, as well as the creation and maintenance of shelters and evacuation capabilities. Based upon these assignments, basic training and education for people in their wartime roles may be necessary.[6] Voluntary defensive organizations such as home guards or defense leagues[7] have a large role to play in the civil defense activities within the Total Defense concept. Voluntary organizations often involve significant numbers of people and seek to develop and strengthen the will to defend their country.

E. **Military Defense.** In cooperation with internal civil partners and jointly with external partners, the objective of military defense is to promote the nation's security, maintain a force with a high-readiness capability to deter a potential aggressor, be able to defend the nation against attack, and secure a military victory to maintain or restore national sovereignty and territorial integrity. Defensive preparations should include the planning for stay-behind military forces and the establishment of a national resistance capability while under enemy occupation. The military must be prepared to protect the nation's sovereignty, prevent and win conflicts if they arise, and support the civil authorities during crisis or war.

F. **Relationship to Resilience.** Success of the Total Defense concept requires a population to be willing to commit to their specific individual and group defensive roles during peacetime and during national emergencies, and be willing to defend their country during conflict. Training and education can prepare individuals and larger groups for resistance activities and strengthen the population's will to resist an aggressor. A strong civil defense requires social cohesion and the maintenance of a common culture and national traditions while ensuring that all citizens have access to state services and benefits and are treated equally under the law; this is sometimes expressed as social defense. Developing and maintaining the will, resolve, and determination to overcome a national crisis and defend the nation is an element of national pride and mutual respect which contributes to the strength and commitment to overcome a crisis and the faith that it will be overcome; this can also be referred to as psychological defense. The planning and preparation (e.g., storage of foodstuffs and fuel) undertaken during peacetime to prepare the society directly contributes to national resilience during a crisis or war and can form a significant aspect of deterrence. Additionally, nations should build resiliency in their economies by protecting critical infrastructure against cyber and physical attacks.

G. **Relationship to Resistance.** Under the Total Defense concept, military and civil defense authorities are jointly responsible for establishing and maintaining a capability to conduct military and civilian resistance activities in their national territory. This can range from activities conducted by military stay-behind forces and government-recognized or sponsored voluntary military units, to organized nonviolent resistance efforts by and within the general population. These activities are to be conducted within a legal and policy framework established by the government and will be responsive to the legitimate national government, whether displaced or exiled. The goal of resistance activity is the restoration of the status quo antebellum, the restoration of the legitimate government with all previously attendant powers and national sovereignty over all national territory.

4. U.S. Department of Defense (DOD) Doctrinal Terminology

A. For the sake of clarity, resistance must be delineated from similar and complementary terms describing operations and activities within irregular warfare (IW), defined in U.S. doctrine as, "[a] violent struggle among state and non-state actors for legitimacy and influence over the relevant population(s)."[8] Though they are related terms, resistance must also be distinguished from U.S. security cooperation (SC), security force assistance (SFA), unconventional warfare (UW), and foreign internal defense (FID). We note the following U.S. doctrinal definitions to distinguish these U.S. doctrinal activities from resistance. These terms encompass United States DOD activities to support a PN preparing to conduct resistance.

1. **SC** is the U.S. DOD doctrinal term that encompasses U.S. support to resistance planning to bolster partner capability and legitimacy. This activity is defined as "[a]ll Department of Defense interactions with foreign security establishments to build security relationships that promote specific United States security interests, develop allied and PN military and security capabilities for self-defense and multinational operations, and provide United States forces with peacetime and contingency access to allied and partner nations."[9]

   a. **SFA** falls under SC. It is DOD activities that support the development of the capacity and capability of foreign security forces (FSF) and their supporting institutions. It is the set of DOD SC activities that contribute to unified action by the United States Government to support the development of the capacity and capabilities of FSF and their supporting institutions, whether of a PN or an international organization (e.g., regional security organization). [10]

2. **UW** is a U.S. DOD term, doctrinally defined as, "activities conducted to enable a resistance movement or insurgency to coerce, disrupt, or overthrow a government or occupying power by operating through or with an underground, auxiliary, and guerrilla force in a denied area."[11] It can be either a supporting effort within a larger campaign or the strategic main effort. In the context of supporting a resistance, UW is conducted in the territory in which the ally or partner has lost sovereignty, to assist a resistance organization in its efforts to remove an aggressor and restore its national sovereignty.

3. **FID** is a U.S. DOD term, doctrinally defined as, "the participation by civilian and military agencies of a government in any of the action programs taken by another government or other designated organization, to free and protect its society from subversion, lawlessness, insurgency, terrorism, and other threats to their security."[12]

   a. The focus of FID is to support the host nation's (HN) internal defense and development (IDAD). IDAD is the full range of measures taken by a nation to promote its growth and protect itself from subversion, lawlessness, insurgency, terrorism, and other threats to their security. It focuses on building viable institutions that respond to the needs of society.

b. FID encompasses U.S. activities that support a HN IDAD strategy designed to protect against subversion, lawlessness, insurgency, terrorism, and other security threats, consistent with U.S. national security objectives and policies.

B. Within these U.S. DOD definitions, the U.S. engages in SC and SFA when supporting the PNs development of an organized resistance capability. If that PN loses full or partial sovereignty over its territory to a hostile actor, then the U.S. can engage in UW to assist resistance forces. If that PN is under pre- or post-crisis threat from a foreign actor interfering in the PN domestically, then the U.S. engages in FID to help free and protect the partner from foreign subversion or insurgency.

5. Understanding, Defining, and Differentiating Resilience and Resistance

A. **Resilience.** A society's resilience contributes to deterrence and supports national defense planning to include resistance to regain national sovereignty, as well as the final post-resistance restoration of sovereignty. Generally, the survivability and durability of a society may also accurately describe this term. Essentially, resilience is the will of the people to maintain what they have; the will and ability to withstand external pressure and influences and/or recover from the effects of those pressures or influences.[13]

B. **Resistance.** Resistance is the natural response of a sovereign government and its people when faced with a threat to their sovereignty and independence. Government proactive preparation and planning across its organizations and the whole-of-society is vital to ensure appropriate mechanisms are in place to conduct organized resistance against an occupier. Resistance factors include the geographic and historical relationship between the adversarial/enemy governing authority and the population resisting it. In its objective of seeking the restoration of the pre-conflict status quo, resistance (unarmed or armed, nonviolent or violent) is distinguishable from terrorism, insurgency, or revolution. The methods and intensity of resistance are determined by the degree of coerciveness of the occupier or its proxy government being resisted. Resistance is a nation's organized, whole-of-society effort, encompassing the full range of activities from nonviolent to violent, led by a legally established government (potentially exiled/displaced or shadow) to reestablish independence and autonomy within its sovereign territory that has been wholly or partially occupied by a foreign power.[14]

6. Resilience/Resistance Conceptual Model

A synergistic relationship exists between ongoing government planning and preparation activities that foster national resilience and defend sovereignty, and how these activities contribute to resistance conducted to regain national sovereignty. This relationship of defending and regaining national sovereignty is depicted in figure 2. Government planning and preparation activities enhance national resiliency and set favorable conditions for the resumption of sovereignty.

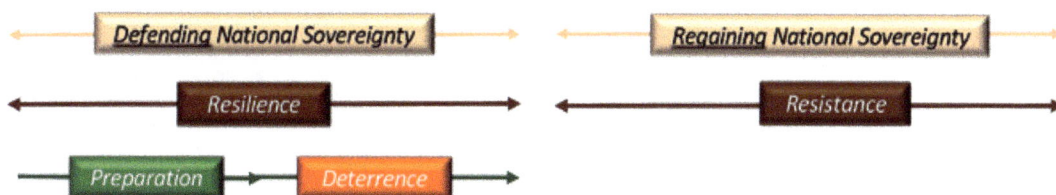

Figure 2. Resilience and Resistance in National Defense.

SOURCE: CREATED BY PARTICIPANTS AND BASED ON AFTER ACTION REPORT, UNCONVENTIONAL WARFARE/ RESISTANCE SEMINAR, BALTIC DEFENCE COLLEGE, TARTU, ESTONIA, 4-6 NOVEMBER 2014, 4

Chapter 1. Resilience As A Foundation For Resistance

Deterrence today is significantly more complex to achieve than during the Cold War.[15] - National Security Strategy of the United States of America, December 2017

1.1. Operating Environment

Governments responsible for fostering resilience face a highly dynamic and complex contemporary operating environment. Self-organizing human networks engage in multifaceted, nonlinear behaviors, with an absence of centrally controlled responses, complicating power dynamics, and consistency of power distribution. Foreign governments, networks of governments, and influential non-state actors cooperate and compete for influence within populations in ways that can undermine resilience. Yet, despite the growth of non-state actor influence, governments overseeing state apparatus retain the primary responsibility for fostering resilience and organizing national resistance.

1.2. Structural Elements of the Operating Environment

The operating environment has structural elements that are consistent, but may vary in degree. Nations cooperate or compete in all elements. A government must understand, through self-assessment, the relative position of strength or weakness its society possesses in each element. Potentially exploitable weaknesses must be recognized and addressed to increase resilience. These elements are best described—and the strengths and weaknesses assessed—through familiar assessment tools:

- National means/resources typically available to governments are expressed through the elements of national power: diplomacy, information, military, economic, financial, intelligence, and law/law enforcement (DIMEFIL).
- An assessment of the systems of an operating environment such as a country or state can be expressed as: political, military, economic, social, information, infrastructure, physical environment, and time (PMESII-PT).
- An analysis of the civil considerations of an operating environment is expressed in the six categories: area, structures, capabilities, organizations, people, and events (ASCOPE).
- Other key social drivers include industry, diasporas, key leaders, history, demographics, climate, technology, regional considerations, and other stakeholders. It is also necessary to analyze the way various factors interrelate or are weighted in importance.

1.3. Prerequisites for a Successful Resistance

A. A strong foundation of resilience is necessary to engage in a successful resistance against an aggressor. A highly resilient population can be created through the development of a strong national identity accompanied by preparation to overcome crisis, which strengthens a nation's will to resist. The government can also engage in practical psychological measures to strengthen popular identification with its national identity, emphasizing the homeland rather than the government in power. The government must proactively assess and identify its society's vulnerabilities and use a comprehensive approach to reduce these vulnerabilities. It is also the government's responsibility to identify potential external threats to the nation and prepare against these potential threats through military and civil preparedness. The government must also strengthen relevant allied and PN relationships and increase interoperability with external support as a method of deterrence and increasing the cost of aggression to an adversary. National and local emergency plans for natural and manmade disasters are also a part of national resilience. Finally, the government should communicate the existence of potential external threats to its own population, along with its plans for the population and military to counter or mitigate those threats through preparedness, training, and the necessary institutional and legal structures and policies to develop, establish, and conduct resistance if and when necessary.

B. **National Identity.** A strong national identity and values, also known as national cohesiveness, are prerequisites for a national resistance as they maintain/strengthen the population's resilience and motivation to resist. A national identity is obtained through promoting measures such as historical and patriotic education consistent with identified cultural values, transparent communication with minority populations to ensure their inclusion in civic and governmental life, separation of politics from national defense policymaking to the greatest degree possible, and national unity messaging to encourage patriotic and civic-minded activity (e.g., youth scouting, camping, sports leagues, and clubs) through nongovernmental organizations (NGOs) or other associations at the local level. Official and public encouragement of young people to participate in activities like scouts or a national defense or civic support league can reinforce patriotic sentiment. Emphasizing a national identity instead of support for a government in power may help overcome some citizens' reluctance to support a resistance because of dislike for the current government.

C. **Psychological Preparation.** Psychological preparation of the population should begin long before conflict. It should include patriotic education that stresses good citizenship or affinity for the nation or land and should be incorporated into the education system at lower levels to help children build immunity to an adversary's propaganda.

1. Psychological preparation does not end with the onset of hostilities. The resistance will need to maintain popular support. Therefore, many of the activities that build resilience will continue throughout the resistance itself, if perhaps in modified form. In addition, the organized resistance will engage in actions to maintain and increase popular morale in order to continue popular focus on regaining independence and national sovereignty over all its territory.

2. Psychological preparation includes political mobilization and must be planned around a set of easily understood political objectives. Ongoing influence activities directed against the occupier can be used to unify the population in support of the state, the resistance, and in opposition to the enemy. This includes a narrative that fits the population's psychological needs and supports the strategic goal of restoring national sovereignty.

D. **Knowledge of Vulnerabilities.** Governments must identify and mitigate internal vulnerabilities/weaknesses. Analyzing the internal operational environment and identifying areas of potential vulnerability that an adversary can exploit is the first step in preparing the state and nation. In any given operational environment, there may be particular relevant elements or sub-elements that drive power dynamics and grievances. These can be exploited by external powers as openings to gain leverage into a society in order to influence it. An adversary may use these elements or drivers as tools to create or increase divisions within populations. Examples of such drivers include identity, religion, economics, perceptions of repression, corruption, exploitation, and lack of essential services. These sub-elements of the operational environment, depending on their strengths as tools of exploitation, may require additional attention during resiliency building.

E. **Vulnerability Reduction.** Reducing vulnerabilities requires whole-of-government and whole-of-society approaches that will cover all the elements of the operating environment. Examples include proactively countering adversary messaging, diversifying (to the extent possible) and protecting the national economy and critical industries/infrastructure, facilitating a common operating picture[16] among relevant organizations, protecting basic standards of living, securing borders, promoting national unity, adopting data and cyber protection and information assurance measures, reducing vulnerabilities of key populations, and maintaining existing military advantages. The drivers must be identified early and assessed as potential vulnerable points of political leverage (influence) by an adversary, followed by actions to reduce those vulnerabilities. There are a number of efforts a nation can undertake, depending on the nature of its vulnerabilities; for sample assessments to identify vulnerabilities, see appendix G.

F. **Potential External Threat Identification.** The government's ability to recognize and define the threat will drive the governmental actions necessary to mitigate the threat and prepare for resistance, if deemed necessary. Governments must understand and recognize the manner in which the external threat can exploit its advantages in the domestic operational environment. Governments should inform and educate the population about the threat, particularly those elements of the population that are vulnerable to adversary activities and influence. Communication themes and messaging efforts must also consider audiences outside the nation's borders such as friendly governments and their populations, potential aggressor governments and their populations, and the national diaspora, and should include counter propaganda efforts.

G. **Preparation Against the Threat**

1. **International Preparation.** Strong relationships with allies and partners support deterrence and help guarantee external support during resistance. Stakeholders with whom governments may make agreements to coordinate preparation include but are not limited to: international organizations and alliances; ministries of defense, intelligence, and foreign affairs; finance and law enforcement organizations; civil institutions; and community groups, particularly among the diaspora. Formal public agreements should promote capabilities and can create a strong deterrent for an adversary. International coordination and agreements should include intelligence sharing, law enforcement information sharing, and conducting exercises. Planning should include obtaining international recognition for the government's position, contingency operations with and without external support, and post-conflict stabilization upon restoration of national sovereignty. Establishing a clearly defined end state, such as the restoration of pre-conflict national sovereignty, helps achieve unity of purpose among the legitimate government, its resistance organization, the occupied and unoccupied national population, allies, and partners. Governments should negotiate, codify, and sometimes exercise roles and responsibilities with allies and partners prior to crisis. This facilitates the development of capabilities and international legitimacy and support.

2. **Interoperability with External Supporters**

   a. Agreements with allies and partners in place prior to commencement of hostilities can facilitate timely cooperation, while enabling and sustaining resistance efforts. These agreements may delineate types of support requested from each ally or partner during different phases. Agreements also help ensure legal recognition of resistance networks to address the potential concerns of allies and partners regarding cooperating with such legally designated resistance networks and distinguishing them from illicit networks. When building external support, the government must recognize that becoming too reliant on, or even appearing too reliant on, external support can undermine the credibility of the government and possibly undermine the credibility of an indigenous domestic organization representing the people. At the same time, the guarantee of outside support, particularly in the form of early material support followed by eventual combat troops to help oust the occupier, may improve the capacity and willingness to conduct resistance operations.[17] Most of all, governments and their resistance leaders should coordinate their plans with allied and PNs to ensure interoperability during either overt external support or clandestine support.

b. Agreements and exercises with partners and allies facilitate interoperability and may help to discover important issues that must be addressed prior to the provision of external support. For example, an exercise may discover that many countries require 14 days for customs clearance and customs checks of certain equipment and material, while others require as little as seven days. In such a case, the receiving allied or PN may need to establish an allowance or waiver within its legal framework for certain customs procedures, country clearances, and overflights to enable rapid infiltration of forces and/or material during times of near or actual crisis.

3. **Domestic Preparation**

   a. Governments can engage in proactive strategic communication to distribute essential information and instructions to the nation's citizenry, while countering the information operations (IO) of potential adversaries. Pre-crisis IO should be prepared, have clear narratives and synchronized actions and information, and occur in sequence with resistance acts that can be initiated as crisis nears and continue through the first weeks of enemy occupation, even from a displaced location. Policymakers should develop the necessary laws and legal structures to enable flexibility and initiative, while still managing risk in a peacetime environment (see appendix A). These legal structures must account for resistance measures as well as account for continuity of government during resistance and upon resumption of national sovereignty. Planning can include establishing operational contingency resistance plans for ministerial and interagency involvement for a whole-of-society approach, rules of engagement, command and control, physical and personal security, secure communication, interoperability, and counterintelligence (CI). Many of these efforts can be combined with civil defense efforts in preparation for a natural or man-made disaster. In addition, implementing resistance plans during exercises, well in advance of a crisis, can help ensure readiness down to the local level. Compartmentalized resistance plans add an extra layer of security. As part of Total Defense, these are tangible demonstrations of the national will to resist. Two examples of communication through distribution of information to the populace to enhance resiliency and boost the ability to resist, follow:

      01. In 2015, the Lithuanian Ministry of National Defense issued the third edition of its citizen's guide, *Prepare to Survive Emergencies and War: A Cheerful Take on Serious Recommendations*.[18] The 75-page manual offers survival techniques, focusing on an invasion scenario. The manual states "It is important that the civilians are aware and have a will to resist–when these elements are strong, an aggressor has difficulties in creating an environment for military invasion." The manual tells its citizens to pay attention to the actions of neighboring Russia, even noting that Russia may use "denial and ambiguity" at the beginning of an invasion. This edition explains how Lithuanian citizens can observe and inform on the enemy if Russia succeeds in occupying part of the country. It does this with detailed images of Russian-made tanks, guns, grenades, and mines to assist citizens in easily identifying equipment and allowing each citizen to be an observer who can then report. It also covers basic first aid and surviving in the wilderness. The manual reminds its citizens that defending the country is the "right and duty of every citizen." It recognizes that citizens are a crucial part of the country's early warning system.

Issuance of this manual also falls into the realm of remarkably transparent strategic communication to deter aggression. Further, as part of its early warning system, the government has even established a telephone hotline for Lithuanian citizens to report suspected foreign spies.[19]

02. In May 2018, the Swedish Government's Civil Contingencies Agency *Myndigheten för samhällsskydd och beredskap* (MSB), distributed *If Crisis or War Comes*, a 20-page pamphlet (see appendix I) with pictures to all 4.8 million homes to show how its population can prepare in event of an attack and contribute toward the country's "Total Defense." The pamphlet detailed tips on home preparation, explained how people can secure food, water, and heat, how to understand the warning signals, where to find bomb shelters, and to be prepared to sustain themselves without government help for at least a week. This was the first time in more than half a century that the Swedish government issued such public information guidelines. Additional highlights of this pamphlet: In the event of armed conflict, "Everyone is obliged to contribute and everyone is needed" for Sweden's "Total Defense." And if Sweden is attacked, "we will never give up. All information to the effect that resistance is to cease is false."

b. **Cyberspace Considerations.** Resiliency and defending national sovereignty mean that the government must also be able to defend in cyberspace. This is also the operating space in which an adversary would likely launch attacks as part of a hybrid campaign. Therefore, understanding and detecting attacks in this environment is critical. This environment consists of the interdependent network of information technology infrastructure and data, including the internet, telecommunications networks, computer systems, and embedded processors and controllers. Detecting and defending against attacks in cyberspace supports the nation's ability to defend its national sovereignty and security. This requires the combined collaborative and integrated efforts of the military, other government agencies, and civilian organization stakeholders.

1.4. Planning for Resistance

A. **Strategic Planning.** The strategic planning process should stem from the government's understanding of an adversary's mechanisms of control, and in turn identify ways to undermine the adversary's efforts. It should also account for the political, strategic, operational, and tactical levels, which the government should prepare for in advance, through exercises to test its plans and ensure readiness down to the local level (see chapter 3: Interagency Planning and Preparation). The strategic planning process should address the establishment and development of underground and auxiliary networks and specify how they will be activated. It should identify roles and responsibilities, requirements for securing external support, and operating with external support. The plan should also achieve the appropriate balance of control between national and local levels to ensure effectiveness and a common purpose for a potentially diverse set of actors. Planning must be comprehensive, proactive, and incorporate all relevant stakeholders. Strong continuity of government planning is essential and should include such aspects as communication, organization, security, and oversight.

B. **Planning for Exiled or Displaced Government.** As part of the continuity of government, with the goal of reestablishment of national sovereignty over any occupied territory, the government must plan for internal displacement or foreign exile. The concept of internal displacement is significant because an adversarial state may seek to displace the government from its capital location to make it more difficult for the legitimate government to govern. This could also induce domestic confusion as to legitimate state authority and control of state functions and reduce popular faith and confidence in the ability of that government to govern. This contingency of exile or displacement should be established within the national legal framework, and made publicly known without necessarily revealing details, to ensure the credibility and legitimacy of such displaced or exiled government. The possibility of exile, to remove the most senior government leaders from an impending threat must be planned with the ally or partner who will agree to receive and host the government. This pre-crisis plan, likely secret, should include critical officeholders (e.g., prime minister or president, defense minister, minister of foreign affairs, etc.) as well as certain staff members, and their families to avoid occupier threats to those families in an attempt to affect governmental decision making. The plan must include office location as well as plans to obtain the necessary equipment and communications to function and ably represent the nation and its sovereignty from the external location. The physical space can be one of its foreign embassies in a friendly state. Embassies on foreign friendly soil are good places to store critical documents, including plans for continuity of government and resistance. This exile plan must also include secure communication and transportation requirements for those persons designated to be removed from the country when the threat reaches a predesignated threshold. If the state does not have adequately secure capability to remove these persons when necessary, then a plan must be made with a capable ally to assist in securely removing those persons when necessary. To ensure that legitimacy remains with the pre-conflict government, the legal framework can state that any political decision made by a public legislative or executive authority or body that is physically within occupied territory is null and void. This legal or constitutional framework can also establish and delegate limited legislative and executive powers to a War Delegation or War Council to last through the time of emergency (see Swedish Constitution of 1974, amended through 2012, chapter 15, "War and Danger of War.")

C. **Pre-Crisis Resistance Component Organization and Core Cadre.** In case of displacement or exile, the government must also plan for a stay-behind leadership structure, operating among the people, to assist in the conduct of resistance operations and provide governance (shadow government) to compete with the enemy's occupation regime. This shadow government, in the occupied territory, will coordinate underground and guerrilla networks. Therefore, in addition to predesignated shadow government leadership, underground and guerrilla components must also be established prior to a crisis. Each of these components need not be completely filled with all anticipated members, but a core cadre of leaders and certain trained specialists, who must be ready to be activated and used as crisis clearly approaches, should exist. This cadre should receive the training and education, during peacetime, necessary to fulfill positional duties and, if necessary due to occupation, to later guide and then further develop the organizations (see fig. 3). Many cadre members should not have military or government records, to prevent the enemy occupier from searching for resistance leadership among the records of active, reserve/auxiliary, or retired members. Cadre leadership should also have a role in locating potential additional members who will be vetted by the government during peacetime and contacted to become resistance members when necessary (see appendix E, case study 2, Switzerland). Component leadership must understand how to flow communications, personnel, intelligence, and supplies into, out of, and throughout the territory under occupation. This preparation also allows leaders to build connections with the population and develop active support that will benefit the resistance in a conflict. As part of network development, the leadership will have to develop a CI system to assess the loyalty of each resistance member or auxiliary participant. To preserve communications security, the resistance leadership must establish methods for resistance elements to communicate internally while also maintaining security through compartmentalization.

1. **Centralized Resistance Planning.** One possible course of action is a civilian-led national crisis management center within the MOD or ministry of interior (MOI) to organize, oversee, and lead government resistance planning and preparation, including activating resistance plans and essential resistance elements upon government order. Planning for the resistance phase includes identifying

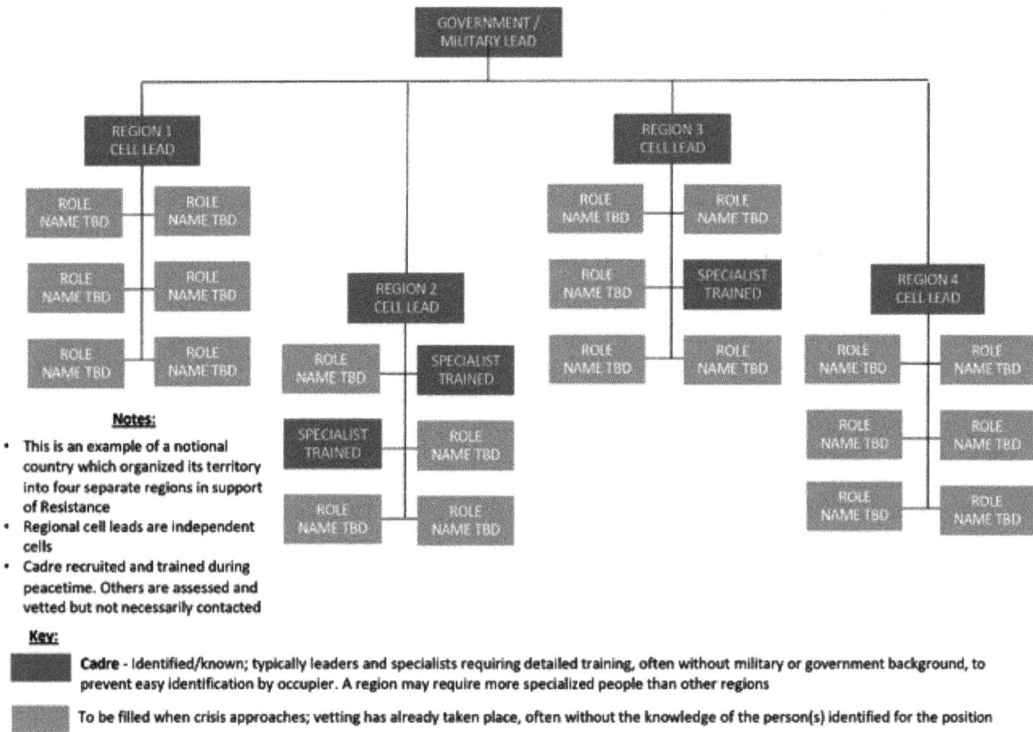

Notes:

- This is an example of a notional country which organized its territory into four separate regions in support of Resistance
- Regional cell leads are independent cells
- Cadre recruited and trained during peacetime. Others are assessed and vetted but not necessarily contacted

Key:

Cadre - Identified/known; typically leaders and specialists requiring detailed training, often without military or government background, to prevent easy identification by occupier. A region may require more specialized people than other regions

To be filled when crisis approaches; vetting has already taken place, often without the knowledge of the person(s) identified for the position

Figure 3. Pre-Conflict, Cadre-led, Partial, Resistance Organization Sample.

SOURCE: OTTO C. FIALA, CAPTAIN JONATHAN JANOS, DIANE LE-FARNHAM, AND JAMES "JIM" WORRALL

key operational positions and the personnel to fill those positions. The government should identify population elements that should remain in place, in order to utilize their expertise to assist the resistance in an auxiliary role, to include collecting and disseminating information or disinformation on adversary activities.

2. **Resistance Organization Structure.** The appropriate resistance organizational structure will depend on specific attributes of the country in question, though will likely adhere closely to the general components of underground, guerrillas, auxiliary, and shadow government. The political, physical, sociocultural, and other landscapes will determine the size, shape, activities, and scope of the resistance. It is critical that the government understand these attributes and how they affect the establishment, organization, and development of a resistance capability in regard to the strategic objective of regaining national sovereignty by removing the occupier.

3. **Identification of Key Personnel.** It is important to place the most suitable people in key decision-making positions, ensure they have access to information that best supports their decisions and operations, and hold them accountable. Experience and roles in society may determine which network some individuals are best qualified for e.g., underground, auxiliary, guerrilla force, or shadow government. Not everyone, however, will be suitable for active participation in the resistance. Many members of society such as government, civic and business leaders, and even popular figures will likely be too well known to participate actively and regularly in the underground, and while they may be appropriate for intermittent auxiliary functions, they should be compartmentalized from other resistance operations

since the enemy will almost inevitably be monitoring their actions. Operational security should always be considered in assigning roles and responsibilities.

4. **Passive Resistance.** The population at large that are not members of the resistance organization has many opportunities even outside of organized events to passively resist. Individuals from all layers of society can take individual actions to weaken enemy morale or disrupt the daily operations of the enemy, through nonviolent, clandestine, or passive resistance. Prior to a crisis, the government should ensure that the population is aware of how they can contribute to a possible resistance against an occupation, such as passive methods for the majority who will not be active members of the organization. During an occupation, the responsibility to communicate these options falls particularly to the underground within the occupied territories.[20] Passive resistance encompasses slow performance of work, ignoring certain procedures or rules while claiming to be unaware, miscounting or not accounting for goods required by the enemy, and many other activities or lack of activity that can be excused or disguised as acting out of ignorance, fear, or wrong information (see appendix D, case study 2, Poland, for examples).

5. **Resistance Components Ratio.** Resistance organizations have traditionally required both a rural element and an urban element to be successful. In countries that provide sufficient terrain in which larger guerrilla forces could operate, the resistance's organization and strategic plan would place greater emphasis on these forces first taking control of large swaths of hinterland. It would then turn its attention to the cities, where the government would already be significantly weakened due to the loss of its base of support in the rest of the country.[21] The ratio of guerrilla forces to underground and auxiliary members in this environment would be greater than in more urban environments, where overt guerrilla forces could not move or operate as freely. With recent significant increases in urbanization around the world, the balance between the urban and rural elements of a resistance organization have shifted, with greater importance placed on the urban operations undertaken by the underground, auxiliary, shadow government, and public component if allowed to exist.[22] Additionally, improvements in technology that make it easier to identify guerrilla safe havens in rural terrain have increased the importance of urban networks, particularly the traditional undergrounds, that benefit from the anonymity of crowded spaces and can operate among urban populations.[23]

6. **Factors of Ethical Organizational Behavior**

    a. Organizational structure must support ethical behavior and control violence. Based upon the quality of individual resistance member performance, leaders can ensure adequate compensation or take corrective action to enforce ethical behavior; therefore, it is vitally important to attract willing resistance members instead of attempting to rely on conscripts. Compensation can be intrinsic rewards or recognition in the form of inputs into decisions, payments or support to families, or a promotion within the organizational structure. Implementing measures to evaluate resistance unit leadership, members, and operational performance can inform overall leadership of shortfalls requiring correction. This process helps ensure the resistance meets strategic objectives that achieve the desired end state.

    b. Regardless of structure, resistance elements that are isolated and cut off from legitimate government control and support may be more likely to disregard legal and ethical codes in an effort to survive. Additionally, resistance leadership should be very wary of working with criminal networks for short-term gain because the enemy will exploit these types of relationships to turn the population and international opinion against the resistance. These two matters, of isolated resistance forces and criminal networks, should be addressed legally and organizationally during pre-crisis planning.

7. **Centralization versus Decentralization.** An effective organizational structure can help resistance leaders maintain operational control and potentially manage escalation of tension or violence. The resistance must balance competing impulses for centralization versus decentralization. Centralization answers a national-level need to coordinate activities across sectors and allows for increased operational compliance within legal frameworks and ethical mores but increases the risk of the adversary detecting resistance members and deconstructing the organization. Decentralization places authority and responsibility for day-to-day operations at the level of the local resistance elements that best understand the security requirements of their area of operations. Decentralization also permits more rapid adaptation to events that do not unfold according to plan. Decentralization requires knowledgeable and competent leaders to understand and execute the intent of their higher-level command (in U.S. Army doctrine; Mission Command.)[24]

D. **Resistance Physical Infrastructure.** Supplies such as cash, weapons and ammunition, medical equipment, and communication equipment necessary for resistance operations should be obtained prior to a crisis. These materials can be stored and maintained in pre-planned locations or cached centrally under the authority of a designated government agency awaiting distribution orders. Cache sites should be surveyed, designated, prepared, and supplied as and when needed. Material can be placed at the cache sites on a permanent basis or used as a combination of pre-storage and distribution to caches on order. The government should also ensure sufficient funding and logistics to support all other preparatory activities and must incorporate this aspect into their legal framework. The government can authorize caches of certain materials (e.g., weapons and ammunition) outside of government or military facilities during peacetime or only allow sites to be stocked during certain pre-crisis events or upon the onset of a designated crisis (see Cold War Resistance Case Studies, appendix E).

E. **Required Resistance Networks.** Required networks include (but are not limited to) logistics, medical, information/messaging, finance, education/training, transportation, recruiting, communication, intelligence/CI, security, and sabotage and subversion.

F. **Validate/Exercise the Infrastructure and Plans.** Once the personnel are identified, material is obtained, and networks are built, the government should take the opportunity to exercise the capability in order to identify its strengths and weaknesses. Exercises are a highly effective means of identifying strengths and weaknesses, while promoting coordination between agencies, organizations, allies, and PNs. Government organized resistance elements must train, rehearse, practice, and develop knowledge of their resistance capacities prior to a crisis. This activity is also an opportunity to engage with trusted allies and partners to ensure interoperability and to identify requirements for any non-standard items required for resistance activities. Coordination and synchronization among key stakeholders are essential. Such exercises can not only assess capabilities, but can also be part of the strategic communication deterrence message:

1. In June 2018, Sweden activated all 40 battalions of its Home Guards (approximately 22,000 volunteer soldiers) in Sweden's biggest surprise preparedness exercise since 1975 to strengthen military deterrence by improving operational capabilities. In an event of a crisis, the Home Guard is responsible for protecting the core functions of the Swedish state to free the professional army for other front-line duties.[25]

1.5. Resilience as an Aspect of Deterrence

The first step in deterrence is to determine the actions a nation wants to discourage the adversary from taking, based on an understanding of what the adversary hopes to accomplish and how that adversary hopes to accomplish it. Increasing the expected costs of occupation through credible resistance preparations can influence adversary decision making in favorable ways. Deterrence in the information environment is a more complex challenge for Western democracies due to many types of freedoms and protections. Asymmetry between democracies and authoritarian regimes can limit a democracy's counter-propaganda messaging activities. Coordinated, collective defense is preferable and likely to have a greater deterrent effect than preparations and actions by individual states alone. Thus, publicizing resistance preparation and exercising portions of it can itself contribute to deterrence e.g., Lithuania's 2015 *Prepare to Survive Emergencies and War* or Sweden's 2018 *If Crisis or War Comes* (appendix I).

1.6. Resistance within Defense Planning

A. **Planning and Frameworks.** A resistance campaign plan should be a part of national defense planning and should also be considered to be a component of building national resilience. A resistance campaign needs to be planned prior to the requirement to resist an occupier. The government must establish a transparent legal framework authorizing the organization, provide for the training and equipping of such an organization, and establish a policy framework to authorize its mobilization, use, and demobilization.

B. **Communication and Deterrence.** Information concerning legal and policy frameworks must be part of the public domain to assure domestic legitimacy and thus will also be available and transparent to a potential aggressor. The frameworks should also be part of the national strategic communication strategy and support the overall national deterrence strategy.

C. **Limited Role of Resistance Organization.** A resistance organization is designed for use against an occupier and should not be used prior to conflict to supplement or enhance law enforcement activity against clandestine aggressor actions (see appendix E). Specific resistance-designated personnel, such as selected cadre, should not only receive necessary training but also participate in defense preparation exercises, within strict security measures to protect details of resistance plans and identities of individuals. These exercises should also function to educate key resistance personnel regarding resistance operations as well as incorporating aspects of the legal and policy framework governing the resistance campaign and its activities.

D. **Interagency and Population.** Resistance as part of defense planning should also be expanded to include additional government and civilian organizations and agencies that are not a direct part of the defense structure. Other government ministries, as well as civilian security and law enforcement organizations must understand the concept of resistance and their potential roles in resistance and contribute their expertise and resources to the planning effort (see chapter 3 and appendix J). The government has a responsibility to include the general population through aspects of the Total Defense concept, with its elements of civil defense. Within this concept, the citizenry can be made aware of methods of nonviolent and passive resistance. A historical analysis of non-military civilian participation can guide these holistic planning efforts. These above aspects are captured in figure 4. See also appendices D and E.

E. **Risk and Confrontation Progression** (see fig. 4). Engaging in resistance planning is a low-risk activity for a nation that may be threatened by an aggressor. A threatened nation should engage in such planning as a deterrent, a means of national resiliency building, and to defend its people in case of conflict. Publicizing certain aspects of planning and preparation of conventional military and resistance capabilities can act as a deterrent, combined with conventional military exercises and force development. Additionally, the nation should assess its own vulnerabilities and engage in national resilience building and psychological preparation of the population for potential conflict and resistance to occupation. When the aggressor initiates conflict, the nation's conventional forces (CF) are mobilized and the resistance organization is activated. Under occupation, the population can engage in nonviolent and passive resistance activity, IO begin against the enemy, and the nation engages in national resistance warfare. During this period, once enemy occupation commences and friendly CF have been defeated or otherwise neutralized by the enemy or alternatively moved to external friendly territory, the nation continues its resistance, helping to create the conditions for the entry of friendly CF (fig. 4, green bar). Upon defeat of the enemy and re-assertion of national sovereignty over all formerly occupied territory, forces are demobilized and reintegrated, and deterrence resumes. The next chapter focuses on resistance planning to further detail this progression.

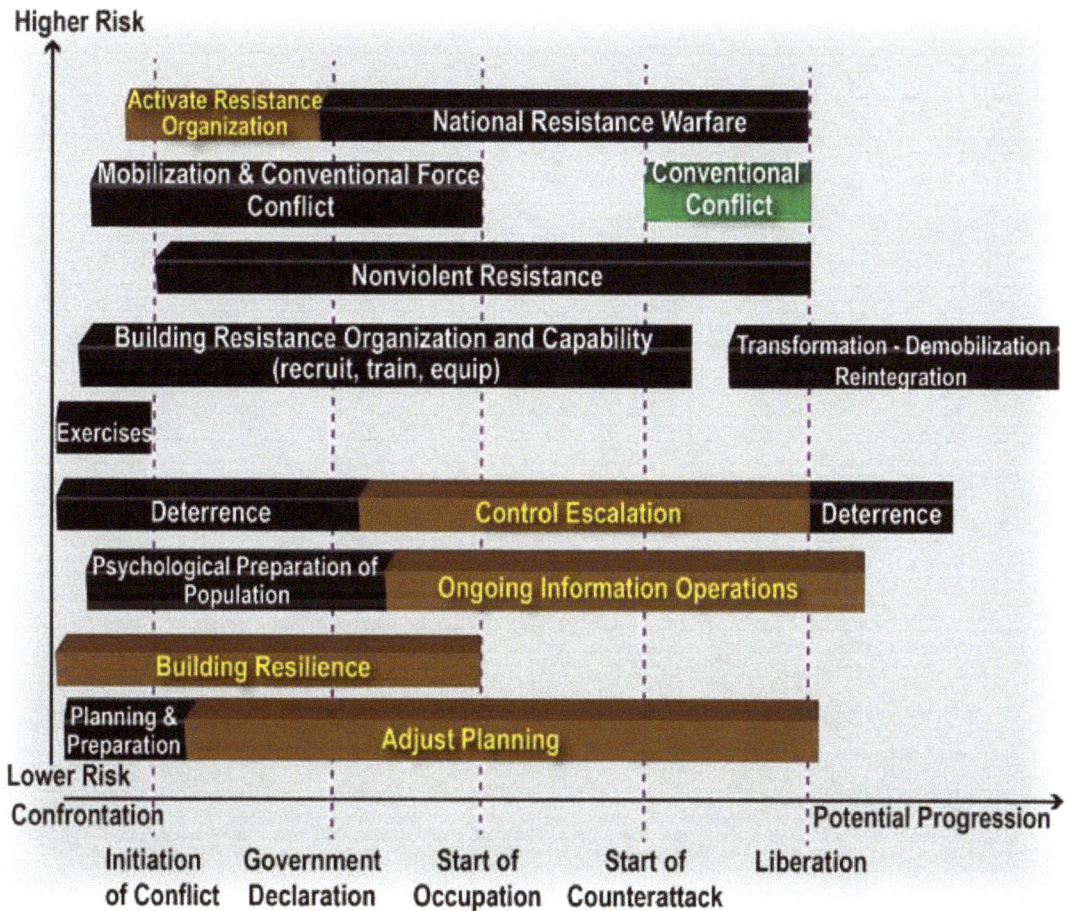

Figure 4. Resistance Activities and Risk Levels.

CREATED BY PARTICIPANTS, MODIFIED FROM ORIGINAL GRAPHIC CONTAINED IN AFTER ACTION REPORT, RESISTANCE SEMINAR, SWEDISH DEFENCE UNIVERSITY, STOCKHOLM, SWEDEN, 26-27 JULY 2016

Chapter 2. Resistance

A general uprising, as we see it, should be nebulous and elusive; its resistance should never material-ize as a concrete body, otherwise the enemy can direct sufficient force at its core, crush it, and take many prisoners. When that happens, the people will lose heart and, believing that the issue has been decided and further efforts would be useless, drop their weapons. On the other hand, there must be some concentration at certain points: the fog must thicken and form a dark and menacing cloud out of which a bolt of lightning may strike at any time. - Carl Von Clausewitz, *On War*[26]

2.1. Operating Environment

National resistance against an occupier is triggered when a foreign power occupies sovereign national territory directly or by proxy. However, the resistance spectrum of activity can be ambiguous. There may be elevated risk levels associated with activities during pre-conflict which may themselves act as triggers to aggressor actions. The boundaries separating pre-crisis resilience building and crisis resistance opera-tions and activities can be ambiguous if the adversary engages in asymmetric warfare methods. Therefore, activities conducted during resilience and resistance can overlap (see fig.5). It is critical to understand and delineate indicators of adversarial activity and conduct during advanced planning and preparation prior to a crisis. Depending upon various aspects of resilience and deterrence, the adversary's actions can bring about one of three conditions, creating a new operating environment: partial loss of sovereignty, full loss of sovereignty, or sovereignty infringements.

Figure 5. Resilience as the Foundation of Resolute Resistance in National Defense.
SOURCE: MODEL WAS CONSTRUCTED BY OTTO C. FIALA, ASSISTED BY CATHEY SHELTON

2.2. Aggressor Actions and National Resistance

Figure 5 describes aggressor actions that bring about national resistance through a change in the operational environment. It does not account for the actions of the nation's CF, except that they would be engaging in defensive military actions during the periods of "National Military Defense" and "Defend/Resist." During pre-crisis peacetime, in recognition of potential aggression against it by a larger state, the nation engages in resilience building (see chapter 1) and resistance planning. Figure 4 assumes that the aggressor is likely to begin with lower level forms of aggression, which may later combine with a conventional force incursion, which is a form of hybrid warfare (see appendix C). Once the aggressor's intent becomes clearer based on indications and warnings, the nation begins activation of its resistance organization, almost concurrently with its conventional force mobilization. Yet, the extent of the threat may not necessarily be clear, possibly due to the use of proxies, and so a period of crisis sets in based on aggressor activities. Once the aggressor clearly infringes upon national sovereignty of its target state, it begins occupation of physical territory and national resistance in the occupied territory begins soon after. As the aggressor's intent becomes clearer, there will likely be an ebb and flow of resistance activities. If resistance meets moderate success against the enemy by frustrating enemy goals, then that enemy is likely to partake in coercive population control measures. The effect of these measures is to likely reduce resistance activity against the enemy. After the situation stabilizes further, the resistance will seek and find opportunities for actions against the enemy and to maintain popular morale and hope of restoration of independence. Eventually, the resistance will assist in setting conditions for entry of friendly CF and/ or terms of settlement to reinstate national sovereignty in all occupied territory (see fig.5).

2.3. Enemy Effects on National Defense Resources

Prior to enemy actions, the targeted state will rely upon national capabilities such as intelligence, law enforcement, conventional military forces, Special Operations Forces (SOF), and if necessary based on the nation's assessment of its own defense capability against a specific aggressor, an organized resistance (see fig. 6). The nation's intelligence organization(s) should be the first elements to become aware of aggressor activities and pass this information through existing channels to the nation's leadership. An adversary could conceivably develop asymmetric capabilities within a target state over the course of years, while testing the target state's intelligence and law enforcement capabilities. We will limit our treatment of asymmetric attacks as part of hybrid warfare to the actual execution of these attacks immediately preceding an overt (direct or proxy) incursion effort (see fig. 6). During hybrid attacks, several facets of the nation's capabilities likely react to the adversary. Law enforcement, supported by intelligence, is likely the first security asset to come into contact with elements of a hybrid attack, such as proxy forces. When the adversary engages in overt incursions on the nation's sovereign territory, the CF and SOF have larger roles to play. As the enemy begins to consolidate its occupation, ordinary law enforcement efforts, such as targeting typical criminal activity, should continue. If eventually defeated, CF are demobilized by the occupier but some (possibly predesignated) can blend into the stay-behind forces and the resistance, some can even be shifted to a PN to continue the fight later when joined by greater CF, as was done by Poland in World War II. The nation's intelligence services function underground in occupied territory and historically gain much intelligence from law enforcement; the SOF that remain as stay-behind forces blend into the population and conduct activities, which are not always distinguishable by the enemy from the activities conducted by the nation's resistance organization (see fig. 6).

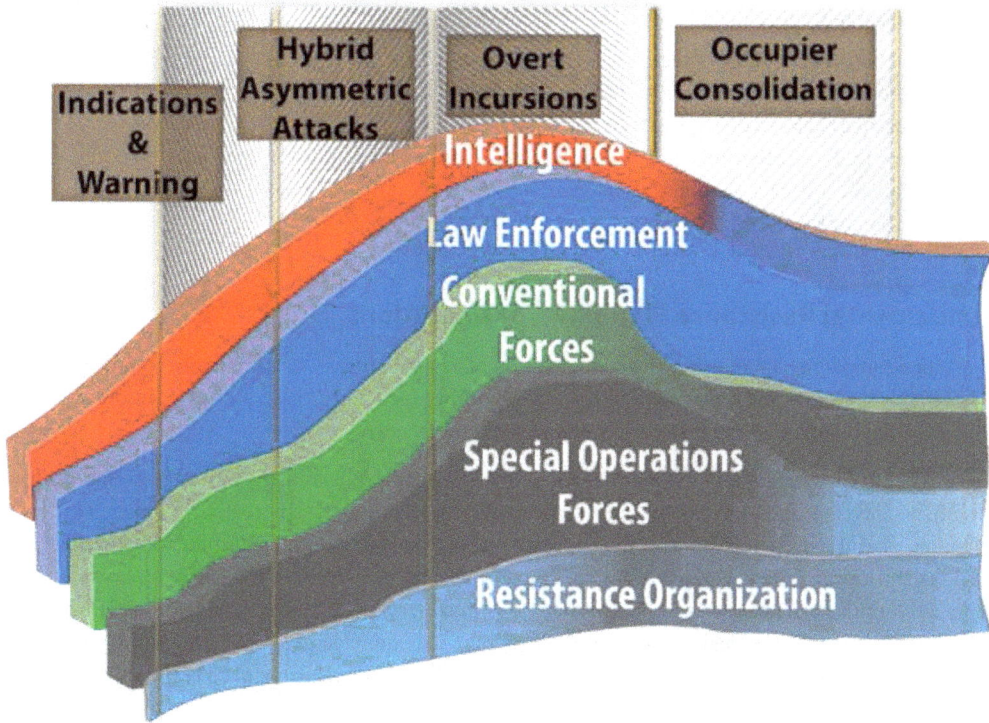

Figure 6. Effects of Enemy on Nation's Intelligence, Law Enforcement, and Military.
SOURCE: MODEL WAS CONSTRUCTED BY OTTO C. FIALA, ASSISTED BY CATHEY SHELTON

2.4. Adversary Imposed Conditions onto the Operating Environment

A. The adversary will assess the targeted state using variants of assessment tools (e.g., DIMEFIL, PMESII-PT, ASCOPE, and social drivers) and on their own ends, ways and means, and risk analysis of objectives. Based on that assessment, the adversary may impose one of three conditions:

1. **Partial Loss of National Sovereignty.** A hostile force places a portion of the targeted state's territory under its authority. Partial occupation applies only to the territory where hostile authority has been established.[27] The legitimate national government maintains control over only a portion of its previous territory.

2. **Full Loss of National Sovereignty.** The entire targeted country is under the authority and effective control of a foreign occupier, likely ruling through proxies supported by the presence of an armed force, and is no longer administered by the country's legitimate national government. Under this condition, a government-in-exile may direct resistance operations from outside the occupied territory.

3. **Sovereignty Infringements.** Adversary subversion of national power through coercive IW, targeting one or more vulnerable means of national power, will result in a reduced effectiveness of the legitimate sovereign government's ability to govern. This subversion may be localized within the targeted state or occur throughout the entire state. This results in eventual de facto loss of territory and/or sovereign ability to act.

B. Techniques used to impose the above conditions can include:

- Cyberwarfare/disruption/sabotage,
- Exploitation of vulnerable populations (e.g., ethnic, religious, racial),
- Key industry attacks to disrupt economic/financial sectors,
- Hybrid warfare and asymmetric tactics identified in professional military journals,
- Conventional/traditional military attacks,
- Combination of the above.

2.5. Fundamental Resistance Planning Considerations

A. A resistance campaign plan requires the consideration of several critical fundamentals to guide such planning. These considerations must be factored into all planning, tactical through strategic, in order to bring about the intended outcome of the restoration of national sovereignty.

1. **Organizing Entity.** Ideally, prior to a crisis, the government establishes an organizing entity to establish, develop, and guide a resistance organization for use against a potential occupier. This entity, suitably placed within the MOD or MOI, is responsible for pre-execution planning and organization of the resistance campaign. The government's intelligence services have a large role to play in organizing and recruiting for the resistance effort and can make good use of retired personnel. However, for security purposes, much resistance leadership and key personnel should not have military or government records. The organizing entity should recruit effective leadership for each resistance component to assist in resistance organization and preparation, and who will form the core of the cadre. The organizing entity develops, rehearses, and tests the resistance plan.

   a. This organizing entity can also be a body of higher-level decision makers from throughout the ministries of the central government whose identities would be secret, as well as the possibility of the body itself being secret. These individuals can meet to recommend actions to the head of the government (i.e., prime minister or president) who can then choose to direct the recommended actions in whole or in part or send the recommendations back for further consideration and development. The advantage of such a secret body is to protect as much knowledge of the resistance plan as possible from an adversary or invading enemy. Specific planning such as details of networks, caches, support agreements with other states, and cadre personnel can be accomplished in a compartmented or cellular type structure. This limits aspects of resistance planning to certain compartmented planners or cells so that no one cell conducts all the planning. This can limit knowledge of the detailed planning so that no one person knows all the details of all planned activities, thus reducing risk of compromise of this knowledge to an adversary or enemy. Though the members and activities can be secret, as with other government secrets, the establishment and functioning of this body must comport with an authorized legal framework in order for it to be legitimate.

2. **Guiding Narrative.** A well-crafted resistance narrative will unify government and societal functions to integrate public and private sector efforts with minimal friction. Established well before a crisis, a nation's strategy to address national security threats provides the foundation for its narrative by outlining how a government builds resilience and prepares for contingencies that may necessitate resistance. A guiding narrative should be carefully crafted to ensure it resonates with the population. The narrative will provide the basis for communication between the government, its population, and the international community. Additionally, a well-crafted resiliency narrative can also serve as a strategic deterrent to adversaries.

3. **Unifying Purpose and Intent.** The success of a national resistance is dependent on a population's support, involvement, contributions, and necessary resources to accomplish resistance campaign objectives. Every individual in society can take part in the resistance campaign to varying degrees and in a wide variety of tasks, though it is the historical norm that the majority of the population will be passive or neutral. However, for maximum effectiveness, though it may not involve a majority of the people within the entire nation, all districts (jurisdictional subdivisions) in the country must contain elements of the resistance organization for an effective national resistance campaign. Governments must make known practical ways for people to resist without harm to them or their families. As one British general noted when the Indian population demanded independence in 1946-7, "You cannot collect taxes indefinitely at the point of a bayonet." Without consent, a government must rely on a high degree of coercion to force popular compliance, thus losing legitimacy domestically and internationally. Therefore, a common purpose and governmental ability to motivate the population to take action is fundamental to a national resistance (see appendix B: Methods of Nonviolent Resistance, and table 2: 198 Nonviolent Actions).

4. **Maintenance of Rule of Law and Political Legitimacy.** Adherence to the rule of law is intrinsic to the concept of legitimacy. These two concepts are integral to each other. The application of the rule of law maintains political legitimacy.

    a. **Rule of Law.** Adherence to the rule of law refers to programs conducted to ensure all individuals and institutions, public and private, and the state itself are held accountable to the law, which is supreme. It limits arbitrariness by confining the exercise of power within legal constraints. The rule of law in a country is characterized by legal frameworks, public order, accountability to the law, access to justice, and a culture of lawfulness. Rule of law requires laws that are publicly promulgated, equally enforced, and independently adjudicated. Laws must also be consistent with international human rights principles. It also requires measures to ensure adherence to the principles of supremacy of law, equality before the law, accountability to the law, and fairness in applying the law. Resistance activities must adhere to the rule of law at all times (see appendix A).

    b. **Political Legitimacy.** Political legitimacy is based on an understanding of the state as a political organization formed through a social contract. In the social contract, legitimate political authority originates in the consent of those being governed. To demonstrate its support for the rule of law and bolster its case for legitimacy, any violent or nonviolent measures the resistance conducts should fall within the framework of applicable national and international law. Overt violent measures include use of guerrilla raids, ambushes, sabotage, and other activities. If the resistance controls any territory, it will retain a large measure of legitimacy be adhering to the rule of law.

5. **Control and Oversight of Resistance Activities.** Effective planning between government ministries, interagency organizations, military, and other societal elements assists in establishing structure, processes, and expectations that help control and mold resistance behavior. Resistance leadership must maintain command and control of all resistance activities to ensure compliance with legal standards and ethical mores inherent in the resistance narrative. Elements conducting resistance activities outside the command and control of the regional and national resistance leadership risk discrediting resistance claims of adherence to the rule of law, which can allow the enemy to challenge the resistance organization's claim to leadership over a unified, organic movement. However, in practice it is very rare for resistance organizations or governments to control all activities. Indeed, it may even be detrimental to security for them to try to do so since such a level of control must use communication methods very frequently, which may be or may eventually become compromised, thus jeopardizing all those

along the communication chain. Establishing the intent and using mission command to guide activities can be both effective and secure. Spontaneous and serendipitous ideas should be encouraged within the population. Humor, for example, is a great weapon to satirize the enemy and maintain popular morale. Resistance leadership should anticipate the enemy carrying out false flag activities to turn the population against resistance elements. Additionally, the enemy may simply create and distribute false information concerning resistance activities to discredit the resistance. Any hint of disunity and lack of control within the national resistance can lend credibility to these false attacks or charges. Maintaining control of the resistance also allows the leadership to enforce a code of ethics in its operations, which helps maintain legitimacy.

6. **Whole-of-Government and Whole-of-Society Collaboration (Total Defense).** Government resistance planning prior to crisis outbreak is central to establishing the components and capabilities for resistance activities. To this end, the government should enhance and institutionalize collaboration among and across governmental organizations, civic organizations, and the larger public in order to prepare the society for resistance as part of building societal resilience. Many governmental organizations can capitalize upon existing connections with civic organizations, trade or professional associations, or NGOs. Greater integration and collaboration among government organizations is necessary to reach a broader audience and effectively plan for resistance.

7. **Agility and Adaptability.** A resistance campaign can be influenced by both supporting actors and adversarial actors. It is essential that resistance campaigns possess the ability to adapt to changes. Supporting actors may adjust the degree, type, or timing of support and the adversary may adjust the methods used to consolidate control over occupied territory and its tactics against resistance. If unable to adjust, resistance organizations, processes, and goals ultimately will not survive.

8. **Post-Crisis Continuity of Government.** Intragovernmental collaboration on resistance planning is necessary for resistance to be a viable course of action within national defense planning. Engaging in intragovernmental and international planning further increases the viability of a resistance plan. Additionally, continuity of government plans for a displaced or exiled government must be made and must address certain requirements, in addition to plans for a shadow government in occupied territory.

   a. **Intergovernmental and International Coordination.** This activity helps ensure resistance roles and responsibilities are well understood and increases the credibility and legitimacy of an exiled or displaced government. Intergovernmental planning engages other governments and international planning, including leveraging the diaspora and engaging other international organizations and entities. This understanding is important particularly with regard to continuity of government, messaging, intelligence collection, and information sharing. Exercises are an effective means of promoting coordination and testing plans.

   b. **Legal and Policy.** A legal framework must be established and recognized to enable preparation and execution of resistance plans and activities. The legal and policy frameworks should also enable interministerial and intergovernmental strategic communication and information sharing agreements that respect operational security requirements, support successful resistance operations, and enhance legitimacy.

   c. **International Agreements.** During pre-crisis, formal agreements with allies, partners, and international/multinational organizations (as applicable), in support of a resistance campaign facilitate legitimacy, increase resiliency, can deter an adversary, and streamline actions necessary during resistance.

d. **Resistance Networks.** Required resistance networks that should be planned and possibly staffed with core cadre during pre-crisis as part of the resilience effort include, but are not limited to logistics, medical, information/messaging, finance, education/training, transportation, recruiting, communication, intelligence/CI, security, and sabotage and subversion.

2.6. Resistance Organization Components

A. **Four Primary Components: Underground, Auxiliary, Guerrilla Forces, Public Component.** Resistance organizations are typically composed of four traditional primary components: the underground, auxiliary, guerrilla forces, and public components. The goals, objectives, and success of the resistance effort will determine the level of development of each and the relationships among them. The underground and guerrillas are politico-military entities that conduct both political and military activities. Typically, the underground is the first element of the resistance to be formed based on factors of leadership and popular willingness to oust the occupier from their midst. The part-time members of the auxiliary component provide clandestine support to both the underground and guerrillas. The public component is an overt, political entity. If the existence of the public component is tolerated by the occupier, then it may negotiate with the occupying power or the government installed by the occupier on behalf of the resistance, to include negotiating on behalf of a shadow, displaced, or exiled government. The four traditional primary components—underground, guerrillas, auxiliary, and overt public entity—exist in a dynamic and evolving relationship, changing in response to internal and external drivers.

B. **Additional Governance Components: Exiled Government, Internally Displaced Government/ or Government Remains in Unoccupied Capital, or Shadow Government.** In addition to the primary components, the government should also plan, organize, and prepare for an externally exiled government, an internally displaced government outside of occupied territory if the national capital is occupied, and a shadow government within the occupied territory. The existence of this continuing governance capability will enhance the legitimacy of the resistance. A government-in-exile may exist in case of total occupation. A displaced government may exist within the original state borders but outside the occupied territory in case of partial occupation where the capital was occupied, or the government may continue functioning in its original location if partial occupation excluded the capital. The shadow government exists in the occupied territory as a clandestine governance activity of the resistance. It may operate in functional parallel with the governing structure of the occupying regime in the area under occupation. It acts clandestinely within the occupied area on behalf and in support of the displaced or exiled government with the primary purpose of supporting and influencing the behavior of the population. The shadow government seeks to maintain or increase popular resistance to the occupier.

C. **Popular Participation.** Though not a primary component or an additional governing entity, popular support for the resistance is crucial. The resistance components are from and embedded within the population and thus rely on popular support for survival. Most of the population, even if they are opponents of the occupation, will not be active or even auxiliary members of the resistance. However, they can partake in passive methods of resistance to hinder enemy efforts of power consolidation, with minimal risk to themselves. Prior to a crisis, the government can make known many examples of such methods to inform the population, increase its resiliency, deter a potential aggressor, and assist a resistance while under occupation, if necessary (see appendix I). Passive resistance can range from small, isolated challenges to specific laws to complete disregard for governmental authority. Passive resistance can wield tremendous power, challenging even formidable military opponents, or at minimum, bringing them to the negotiating table. Planning, strategizing, and management are essential to integrating individual, isolated acts of protest or defiance into activities with greater combined effects against the enemy. The population can also engage in many overt nonviolent methods of resistance, if possible. These actions can be categorized into the following three categories: protest and persuasion, noncooperation, and intervention (see appendix B).

D. **Resistance Distinguished.** National resistance to a foreign occupier as a form of warfare is our primary focus. Such an organization uses nonviolent and violent means to undermine, weaken, or overthrow the occupying power. Popular resistance generally begins with the desire of individuals to remove an occupying power or its installed government. Resistance organizations and movements have taken many forms over time. Here we distinguish resistance from other conflicts such as the uprising of ethnic or religious groups against their rivals or insurgencies against long established governments that may possess several types of legitimacy.

E. **Leadership.** Resistance leadership typically draws from natural leaders such as former military personnel, religious leaders, college professors, local office holders, and neighborhood representatives. Leaders historically emerge from these groups during resistance and insurgencies, but they can also be vetted and designated for various levels of resistance leadership within the primary components during pre-crisis organization. The resistance leadership must at least maintain, but should also grow, the population's desire to resist and remove the occupier. The population must be convinced that regaining its sovereignty is feasible. As with similar forms of conflict, this is essential in maintaining support for resistance. As the movement grows, so should the primary components of the resistance organization.[28]

F. **Primary Components of the Resistance Organization**

1. **Underground.** The underground is, traditionally, a cellular organization that conducts operations in urban areas that are under the control of the occupiers' security forces. It is composed of politico-military entities designed to conduct political and military actions. Examples of underground functions include:
   - Intelligence and CI networks,
   - Subversive radio stations,
   - Media networks that control newspaper or leaflet print shops, social media, satellite television, and/or web pages to disseminate information to the population and outside world,
   - Special materiel fabrication such as false identification, explosives, weapons, and munitions,
   - Control of networks for moving personnel, hiding personnel sought by the enemy, handling logistics, and generating funding,
   - Conducting acts of sabotage in urban centers, and
   - Clandestine medical facilities.[29]

2. **Auxiliary.** The auxiliary refers to that portion of the population that provides active clandestine support to the guerrilla force or the underground. These are persons who may be asked to perform one activity one time, or called upon several times to perform certain activities. Their participation is typically intermittent. The auxiliary is not a separate organization; specifically, it is a component to an urban underground or guerrilla network that it supports. Auxiliary members are part-time volunteers whose value is based on their normal and accepted positions in the community. The auxiliary is not a separate organization, but consists of different types of individuals providing specific functions as a component within an urban underground network or guerrilla force's network. These functions can take the form of logistics, labor, or intelligence collection. Auxiliary members may not know any more than how to perform their specific function or service that supports the network or a component of the organization. Due to their typical physical proximity to the occupier and the occupier's knowledge of their daily work, in many ways auxiliary personnel assume the greatest risk. Due to their lack of

detailed knowledge of the organization and methods of the resistance, and their individual small roles, they also pose the lowest risk to the resistance in case of capture or if they otherwise are unavailable for use, and typically are easily replaced. While under occupation, resistance leaders have often used auxiliary functions to test a recruit's loyalty before exposing him to other parts of the organization. Auxiliary functions are like the connective tissue of the human body. They act as the innumerable fibers connecting the resistance organization to state and societal entities and support the capabilities and activities of the underground and guerrilla components. Such persons may also provide early warning of occupier activities based on their placement. Specific functions include:

- Logistics procurement and distribution (all classes of supply),
- Labor for special materiel fabrication,
- Security and early warning for underground facilities and guerrilla bases,
- Intelligence collection,
- Recruitment,
- Communications network staff such as couriers and messengers,
- Media distribution,
- Safe house management, and
- Logistics and personnel transport.[30]

3. **Guerrillas.** Guerrilla warfare is distinguishable from the conventional warfare of regular, well-armed and trained state forces operating in large units. Persons conducting this form of warfare are traditionally termed "guerrillas." Guerrillas are basically irregular soldiers who, in our context, comprise the armed or military component of an organized resistance. They can be small units of stay-behind military forces or selected and trained members of the civilian population, or a combination of both. They typically have significant disadvantages in terms of equipment and capability, and sometimes also in training if they are civilians with no military experience, as compared to the regular and specialized forces of the occupier. Guerrilla warfare techniques traditionally include raids, ambushes, sabotage, and harassment techniques to interdict enemy movements, subvert morale, and degrade material strength. If popular support to the resistance grows, then the guerrilla component can grow and even better sustain itself while also conducting more aggressive activities against the enemy.

   a. It is important to use the term guerrilla accurately in order to distinguish between other similar types of irregular forces such as mercenaries and criminal gangs. As used here, guerrillas are under the lawful control of a legitimate, possibly exiled, displaced, or shadow government. A guerrilla is an irregular and predominantly indigenous member of a guerrilla force, organized along military lines to conduct military and paramilitary operations within the occupied territory. Mercenaries are individuals or groups of soldiers whose allegiance is secured through payment. Criminal gangs are local, regional, or international organizations conducting illegal activities for profit. Each can be co-opted by all parties. In resistance warfare, the activities of mercenaries and criminal gangs can provide complex challenges to maintaining the legitimacy of the resistance organization. The occupying power or its installed government can use these groups for operational support or propaganda advantage against the resistance.

   b. As military type units, guerrilla units also have command relationships similar to conventional force units. Guerrilla commanders have subordinate staffs and units—though typically on a small scale in occupied territory—responsible for personnel, communications, medical, and logistical support, plans, operations, etc. The commanders also report up to the resistance leadership. An example of these relationships is graphically displayed above (see fig. 7).

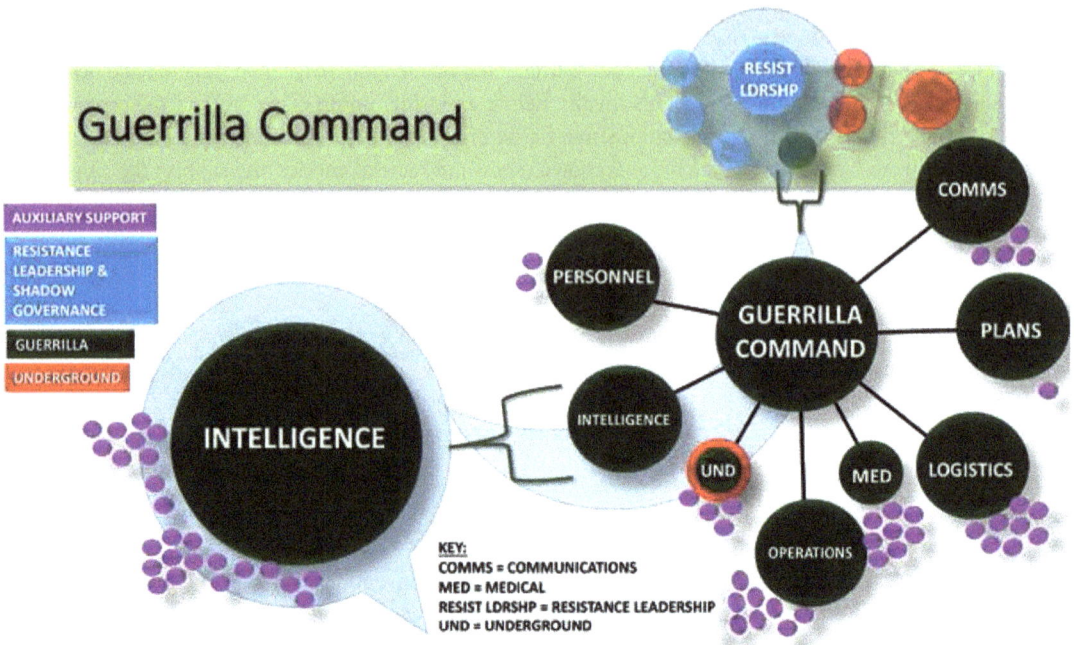

Figure 7. Guerrilla Command Relationships Sample.

SOURCE: CREATED BY U.S. ARMY SERGEANT MAJOR WILLIAM DICKINSON, SENIOR OPERATIONS SERGEANT, SOJ3, SOCEUR/USED WITH PERMISSION

4. **Public Component/Overt Political Arm.** The public component is an overt political expression of resistance within the occupied territory, if tolerated and allowed to exist by the occupier or its installed government. It is not a shadow government or government-in-exile. This component of resistance can be an opposition political party, allowing the resistance to simultaneously engage in both nonviolent and violent opposition.[31] Public resistance components may negotiate directly with the installed government or occupying power. They can also make overt appeals for support from domestic and international sympathizers while also organizing such support. In a resistance with no exiled, displaced, or shadow government, they may be the leadership of the resistance. In a well-prepared resistance, they are overt entities wielded by the leadership of a shadow, displaced, or exiled government. The scale of this component may be a large organization or a single spokesperson.

a. The occupier may gradually lose tolerance for such activity. As the occupying power or its proxies apply all manner of pressure to these public political adversaries, this public component will need to modify its overt political stance in order to be allowed to operate openly. At that point, its support for the resistance will have to be clandestine or presented indirectly in such a manner as to not threaten its ability to operate openly within occupied territory. The four resistance components exist in a dynamic and evolving relationship, changing in response to internal and external drivers.[32]

b. If existing public components are completely suppressed by the occupying power, then the resistance may then operate secretly or indirectly through other openly existing organizations to continue pursuing the movements' objectives. The public components may also relocate to another state where they can operate openly in support of the resistance.

2.7. Methods of Warfare

A. Within this concept, resistance is defined as "a nation's organized, whole-of-society effort, encompassing the full range of activities from nonviolent to violent, led by a legally established government (potentially exiled/displaced or shadow) to reestablish independence and autonomy within its sovereign territory that has been wholly or partially occupied by a foreign power." Delineation of methods of resistance is based on ways and means to achieve the end of reestablishment of national sovereignty. The means are the nation's population and the components of the resistance, as well as any external assistance. Excluding passive and non-violent resistance, the other methods, or ways, fall within the realm of armed conflict. Primarily, these methods are subversion and sabotage, to be executed by the underground and the traditional methods of raids and ambushes executed as guerrilla warfare by the guerrilla component. The other members of society who do not directly participate as members of one of the four resistance components, have many methods of passive and nonviolent resistance available to them.

Association of Components with Methods		
Underground	Guerrillas	Auxiliary
Subversion	Ambushes	Non-Violent Clandestine
Sabotage	Raids	Resistance

Table 1. Association of Components with Methods.
SOURCE: CONSTRUCTED BY OTTO C. FIALA AND DIANE LE-FARNHAM

B. The association in table 1 is not intended to be all encompassing or to restrict the activities of each component. There can be cross-over based on such factors as mission, accessibility, and opportunity. The following sections deal primarily with methods associated with the underground and auxiliary components. The methods associated with guerrillas (ambushes, raids) can be found in many military manuals concerning guerrilla or IW and thus will not be detailed here.

1. **Underground–Subversion**

    a. Subversion is defined in U.S. joint doctrine as "actions designed to undermine the military, economic, psychological, or political strength or morale of a governing authority."[33] This definition covers a broad list of subversive activities, including propaganda, IO, selective sabotage, and the targeted removal or killing of key individuals.[34]

    b. Subversion is designed to degrade the influence and power of the ruling authorities while simultaneously generating support for resistance among the population. A common, historically-based, and easily understood example of this is when a movement creates civil unrest against the ruling authorities and those authorities overreact with coercive measures or repressive violence. Any abuses, or even questionable abuses, committed by those authorities, can be documented and highlighted by the resistance to undermine the new regime among the people, the international community, and the aggressor state's or occupier's own domestic audience. An example of removal of persons potentially dangerous to a resistance comes from the Polish underground during World War II. In an attempt to remove a portion of the local Volksdeutschen population (citizens of German descent and sympathies), the underground attempted to have them transferred onto active duty with the German army by forging their signatures on letters to Berlin requesting the honor to serve in the military.[35]

    c. Subversion during resistance to an occupation should achieve one or more of the following objectives: undermine the authority and reputation of occupier military and security services, infiltrate and subvert key government organizations run by occupation forces and their supporters, conduct psychological operations against the occupier, and undermine the political authority of the occupation government.

2. **Underground-Sabotage.** Sabotage can be defined as "the injury, destruction, or knowingly defective production of materials, premises, or utilities used for war or national defense."[36] Sabotage can be lethal or nonlethal, and can be divided into selective and general sabotage.

   a. **Selective Sabotage**

      01. An underground can use selective sabotage to incapacitate installations that cannot be easily replaced or repaired in time to meet the occupier's critical needs. An example is a tactical target such as a bridge that is essential to transporting enemy troops and supplies. Another form of selective sabotage is a non-kinetic strike such as a cyberattack on a supervisory control and data acquisition system or transportation center. Undergrounds typically excel at selective sabotage because they can fine tune the timing of the destruction as opposed to the coordination of a larger attack by CF. An example is the timing of the destruction of a bridge. If such physical infrastructure is destroyed or severely damaged without regard to a necessary or critical effect on the enemy, such as the tactical use of that bridge for movement of supplies or troops, then destruction or damage to the bridge could simply place an unnecessary hardship on the population.[37]

      02. Resistance members handling explosives are typically not demolition experts and, therefore, require explosives that are relatively safe, with accessible materials, are simple to produce, and easy to use. However, safety, access, and simplicity of production and use are not enough because some instruction is still required in the effective use of explosives. Such training is an important function of the underground and often requires support from external entities. Alternatively, knowledge can be disseminated via literature or online references. For example, during World War II, the Soviets broadcast a ten-minute program called "Course for Partisans" twice a day and published/distributed the *Soviet Handbook for Partisans*.[38]

   b. **General Sabotage.** Underground sabotage operations not only hamper the enemy's war effort but can also encourage the populace to engage in general acts of destruction. Information disseminated to the population can increase the use of sabotage from the selective to the general by encouraging the populace to engage in general acts of destruction. These actions can be divided into active and passive. Though these acts may not have significant material effect, inducing people to perform minor acts of sabotage links them more firmly to the cause. These more general instructions for active sabotage can include incendiary devices such as "Molotov cocktails," or the assembly of fragmentary hand grenades from evaporated milk cans, blasting caps, and metal filler for shrapnel. Fires can be started by deliberately overloading machines. Some machinery can be destroyed or seriously damaged by placing emery dust or sand in delicate bearings. Enemy transportation can be disrupted by putting bleach into gasoline tanks, strewing nails onto roads, blocking roads with stalled trucks or felled trees, or misdirecting convoys by changing road signs. The resistance can also encourage people to practice passive sabotage by failing to lubricate machines, misplacing spare parts, slowing production, or practicing absenteeism.[39]

3. **Auxiliary-Clandestine Resistance.** The auxiliary component of the resistance organization functions in a supporting role. The actions of its members are clandestine in nature because the members do not openly indicate their sympathy to or involvement with the organized resistance. Auxiliary personnel typically have full-time jobs, often in positions that can be advantageous to the resistance (e.g., government office clerks, government associated communications and transportation specialists, medical personnel). Their participation in resistance activities is occasional, based on the requirements of the resistance organization. They may be contacted by the underground for specific assistance, such

as the services of a surgeon, or they may voluntarily render such assistance based on information in their possession, such as enemy troop movements.

2.8. Communication Synchronization/Strategic Communication

A. During peacetime, the government must have the ability to form a narrative, themes, and messages and be able to effectively synchronize and communicate them. During occupation, the displaced or exiled government must also have an effective communication capability. It must synchronize this with appropriate resistance leaders and networks, allies and partners, its occupied population, and the international community to deliver its narrative, themes, and messages. At the same time, it must also ensure that targeted audiences receive and understand the delivered communication. This capability can influence opinions and perceptions toward the opponent and mitigate opponent information activities and effects. Communication capabilities are the ways and means that influence opinions and perceptions of targeted audiences. They must support the legitimate sovereign national government opposing the occupation and support its resistance. These communications must also undermine or mitigate the potential effects and capabilities of adversary or enemy information activities.

B. Technological advances in methods of media distribution have resulted in a continuous and rapid flow of communication. Communication has also become fragmented with the proliferation of media platforms, allowing the coexistence of multiple, and sometimes conflicting, narratives. This has created a situation where it is very difficult to defeat the opponent's narrative. Therefore, communicating the government's narrative and message while countering the adversary's or enemy's narrative or message requires proactive, varied, and agile processes and capabilities. This results in the battle of the narrative. The goal of this battle is to gain superiority over the adversary's narrative, to diminish its appeal and following, and when possible, to supplant it, or make it irrelevant. This battle is fought in the information environment, and its success or failure is measured in the cognitive dimension (perceptions/attitudes) and physical domains (behavioral changes). The goal is to gain superiority over the opponent's narrative.[40]

C. National communication varies widely based on technology and established cultural practices for sharing information. This variety spans the spectrum from word of mouth to technologically facilitated social media. Further, contextual understanding of a message or narrative varies across cultures, different demographics, and local conditions. Often, people filter media communicated information through social non-media systems to establish information trustworthiness. Synchronizing communications in the information environment during resistance requires a comprehensive process and should include the following considerations.

1. **Narrative, Themes, and Messages**
   a. **Narrative:** An overarching expression of the context and desired results. A short story used to underpin operations and to provide greater understanding and context to an operation or situation. A key component of the narrative is establishing the reasons for, and desired outcomes of, the conflict in terms understandable to relevant publics.
   b. **Theme:** Unifying ideas or intentions that supports the narrative and are designed to provide guidance and continuity for messaging and related products.
   c. **Message:** A narrowly focused communication directed at a specific audience to support a specific theme. A tailored communication directed at a specific public, aligned with a specific theme, in support of a specific objective.[41]

d. Established well before a crisis, a nation's strategy to address national security threats provides the foundation for its narrative by outlining how a government builds resiliency and prepares for contingencies. The narrative is the overarching expression of the context, reason, and desired results associated with the resistance campaign. A psychologically unifying, crafted narrative, meeting strategic and operational objectives, should also resonate with the population to enable control, unity of purpose, and encourage ethical behavior. Themes are formed at the strategic, operational, and tactical levels. The themes at each level (each level should have several themes) must be nested under the themes of the higher level. Messages are subordinate to themes and support the themes and narrative. They deliver precise information to a specific target audience to create desired effects while supporting a specific theme. Messages are tailored for a specific time, place, delivery mechanism, and target audience. During resistance, messages can be tailored to specific events so that the enemy and the population believe that things that go wrong for the occupier are somehow being orchestrated by the resistance. Typically, plausibility is adequate to sustain the resistance's legitimacy. The two themes that should be consistent from the beginning of the crisis until restoration of national sovereignty are legitimacy and sovereignty.[42]

01. **Legitimacy**

    i. The concept of political legitimacy is based on an understanding of the state as a political organization formed through a social contract with its citizens. In the social contract, legitimate political authority originates in the consent of the governed, while outlining a reciprocal relationship of mutual obligations and rights between the governed and the government.

    ii. In the case of a nation resisting encroachment by an aggressor state, the new proxy ruling state authority emplaced by the aggressor state will have no legitimacy among most of the national population it seeks to govern. Similarly, the aggressor state will have little success in gaining any recognition for this new state authority in the international community. The legitimate national government must constantly and consistently emphasize its legitimacy in its strategic themes, even if exiled or shadow. At the same time, it must also prevent the occupier from achieving acceptance and a degree of legitimacy in the information environment within or outside of the occupied territory.

02. **Sovereignty: Juridical versus Empirical**

    i. "Juridical sovereignty" is the international recognition of state sovereignty; that is, the international recognition of the right of the state to rule within a certain geographical area. In this case, the new state authority of the occupier will achieve very little such recognition in the international community, except perhaps for the aggressor state and a few of its partners. Therefore, the displaced national government will be at the advantage of holding technical juridical sovereignty over the population within the contested state.

    ii. "Empirical sovereignty" is normally associated with the state's right to use force within its borders and the provision of services to the population. The government authorities of the occupier will continue services in an attempt to minimize the backlash of public discontent and to attempt to behave as the sovereign authority in order to eventually gain internal and international recognition. It will also claim the right of the use of force and will have no basis on which to legitimize that right internally or internationally; however, that will most likely not prevent it from applying force.[43]

    iii. The legitimate displaced or exiled government must always emphasize its juridical sovereignty in the international arena while also communicating this to its population within the occupied territory. For example, the occupier may send a government representative from the occupied state to an international forum and claim legitimacy for that representative. The legitimate government must challenge each instance of such occurrences to emphasize to the international community that it is the sole sovereign and representative of the nation. It must also continuously challenge the occupier's assumption of empirical sovereignty by challenging and discrediting any use of force by the enemy against the occupied population.

2. **Message Coordination and Integration.** Resistance leadership develops communications synchronization to support resistance efforts in influencing target audiences. Communications synchronization provides intent, objectives, theme, and message guidance, information coordination and integration among all partners, to include other governments, international organizations, and NGOs. This synchronization ensures consistency of messages and actions.

3. **Continuous Communication of Unity of Purpose.** Communicating unity of purpose and coordination of roles and responsibilities can improve collaboration effectiveness and demonstrate unity of effort. Continuous and consistent communication of unity of purpose can attract external supporters to the nation's cause while disharmony can attract malign actors intent on exploiting potential schisms and also make it more difficult for external supporters to continue support.

4. **Disrupt Adversary Narrative.** In the pre-crisis environment, the aggressor will use its own unifying narrative to leverage the support of a potentially diverse set of minority groups (e.g., ethnic or religious) opposed to the target state or to set one segment of the population against another. It will continue and adjust this narrative in the crisis environment it creates. After occupation, the resistance must use IO to deny support to the occupier from the populations targeted by the occupier as well as the domestic population of the occupier state. The resistance should understand the occupier's narrative and identify ways to disrupt its attempts to create support among the population.[44]

5. **Controlling Information Flow.** The unifying narrative is indispensable to the success of the resistance. The ability to control the flow of information to support that narrative is often even more important than offensive actions—which must always support the narrative. This requirement necessitates that networks be in place before a resistance is activated to facilitate communication between different elements of the resistance. It also requires channels for continuous dissemination of the unifying narrative to the population. These channels will allow the resistance to respond quickly and accurately to any events or other developments that could affect the allegiance of the population. Controlling the flow of information also requires the resistance to identify ways to disrupt the adversary's ability to communicate among themselves and to the population, without contradicting the resistance narrative. The resistance can inhibit the enemy from disseminating its message to the population through sabotage of its facilities, subversion of personnel, and counter messaging.

6. **Communication Strategy.** Governments need to adjust messaging and targets according to phase when considering communication strategy between the resilience and resistance phases. In the resilience phase, the government will communicate the national narrative within its borders and to allies, partners, the international community, and the potential adversary. However, in the resistance phase, a critical qualitative aspect of the communication strategy is ensuring the legitimacy and credibility of the potentially displaced or exiled government. Should the legitimacy and credibility of such government come into question, the ability of that government to represent the interests of the nation to the international community, and orchestrate resistance efforts, will be at risk. During resistance, an exiled government must tailor messages to different internal populations: supporters of resistance,

supporters of the adversary, and those who are neutral. The individuals and organizations transmitting the messages must have credibility with the target audience (intended receiver of message) to ensure the messages are well received. During resistance, the diaspora is an audience that takes on new significance and importance and should be included in the communication strategy as a base of support. Further, communication is two-way, and a feedback assessment capability must exist to ensure the message was correctly received to allow adjustment and refinement. A communication strategy will usually have a high correlation with other engagements, activities, and actions taken for or against a particular audience (i.e., armed resistance/guerrilla actions can themselves be messages but must support the communicated narrative). Figure 8 depicts these communication lines in both phases.

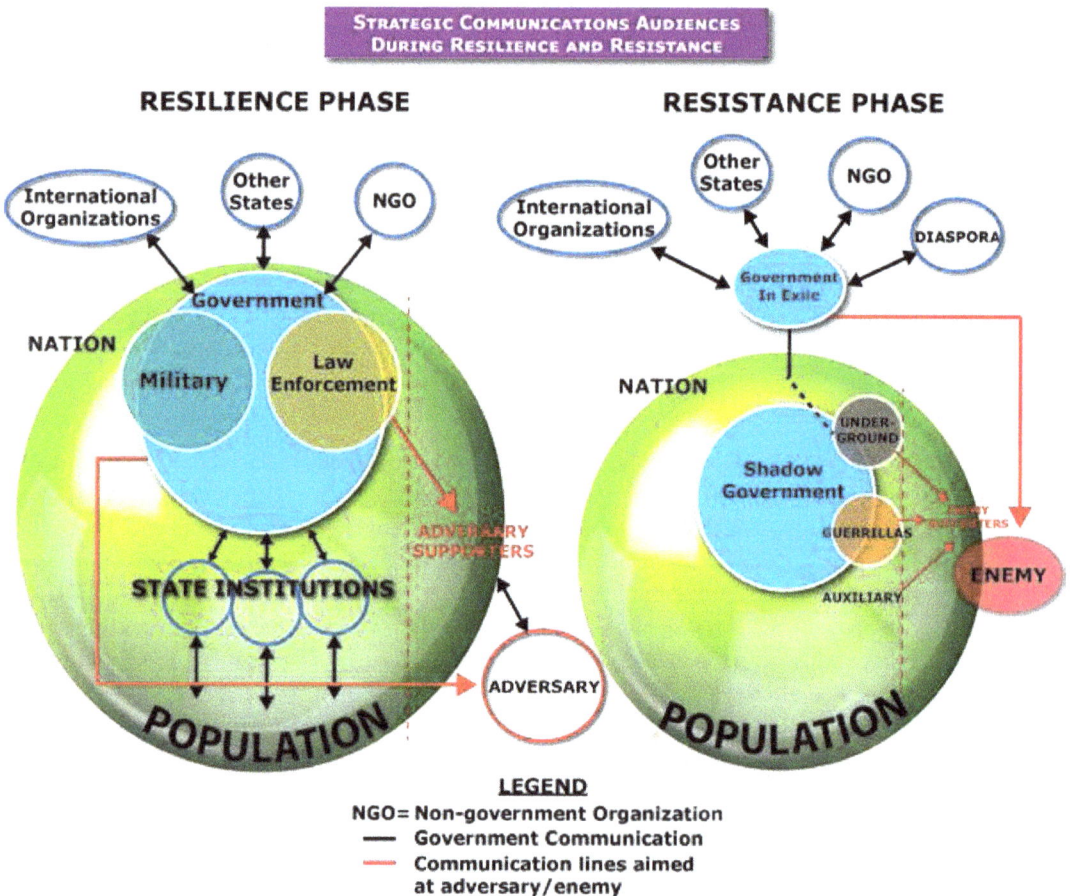

Figure 8. Communication Lines during Resilience and Resistance.

SOURCE: CREATED BY PARTICIPANTS, MODIFIED FROM ORIGINAL GRAPHIC IN AFTER ACTION REPORT, RESISTANCE SEMINAR, RIGA, LATVIA, 23-25 NOVEMBER 2015, 5

2.9. Organizing the Underground for Resistance

A. The leadership and organization of the resistance is perhaps the most critical feature of a resistance organization. The most successful resistance organizations are resilient organizations whose leaders possess the flexibility and vision to adjust operations, organization, and methods of command and control in pursuit of the end state of restoring national sovereignty.

B. Unlike the auxiliary component, which often has ad hoc membership based on placement and opportunity and thus lacks firm organization, the numbers and criticality of the underground's functions in maintaining national resistance require special attention to developing its most effective organization. An effective organization, in turn, requires the underground to perform certain essential "housekeeping" administrative functions, as well as operational functions. A sample organization is graphically presented in figure 9.

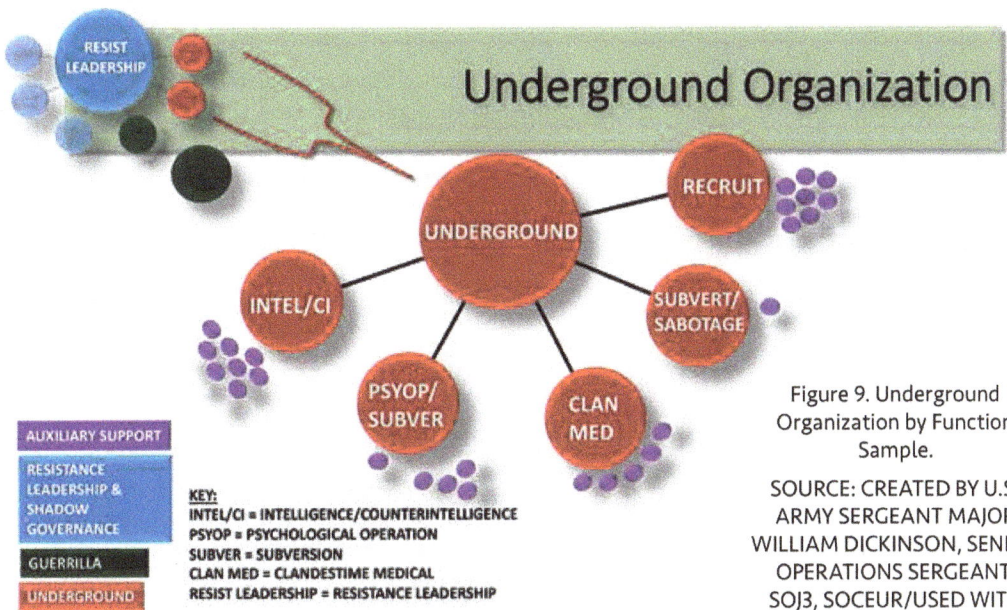

Figure 9. Underground Organization by Function Sample.

SOURCE: CREATED BY U.S. ARMY SERGEANT MAJOR WILLIAM DICKINSON, SENIOR OPERATIONS SERGEANT, SOJ3, SOCEUR/USED WITH PERMISSION

C. There are no standardized ways to accomplish a given network function. The choice and effectiveness of a given technique depends largely upon the resources available and the countermeasures used by the security forces. As undergrounds adopt certain practices, security forces invariably develop countermeasures that undermine the effectiveness of those practices once they become aware of the practices. Consequently, both governments and undergrounds often change techniques and develop new ones. Moreover, the availability of resources governs the underground's choice of scope and depth of organization, organizing and development techniques, and determines how flexible it can be in responding to occupier actions. There are several factors that determine the organizational structure.[45]

1. **Strategy**

   a. A resistance strategy should be part of national defense plans long before a crisis occurs. Resistance organizations operate on both political and military fronts. The choice of which front an underground will emphasize depends on the likelihood of success of one over the other, opportunities, and enemy strengths and weaknesses. It is also subject to modification when unexpected events compel the underground to revise its strategy. When it is impossible to seize the state instruments of control from occupation forces by force, then the underground predominantly chooses the political strategy of trying to weaken the occupying force's governing effectiveness, typically by subversion. Depending upon availability of resources, support for the occupier from a small section of the population, and the extent and effectiveness of enemy coercion, the resistance should have developed contingency plans available for a variety of enemy courses of action.

b. An underground may shift from a predominantly political strategy to one that is predominantly military if the expression of popular support is curtailed by the enemy. Such a shift depends on resources available, intended effect on popular perception and behavior, intended effect on external audiences, intended effect on the enemy, its own domestic political support, and the extent of external support. The underground may conduct military activity to disrupt enemy operations and maintain popular morale and popular hope for the restoration of sovereignty. It should also retain significant capacity for use against the enemy at a later time, especially if external allies and partners are preparing for military action to restore national sovereignty. Underground military activity may be most useful and critical when it is available to coordinate with and assist allied and partner forces when those forces are ready to enter the country and restore national sovereignty. A rapid shift to a predominantly military strategy will likely be required at that time to assist incoming friendly forces.

c. After deciding to shift primary emphasis from either the political or military front to the other, the entire underground may require major reorganization. During the Malayan "emergency" in the late 1940s, communist leaders shifted from political efforts, through the organization of united fronts and infiltration of unions, to the militarization phase of terrorism and guerrilla warfare. After this failed, they turned back to political activity. With each shift in strategy, units had to be organized, disbanded, or reorganized at no small cost to the movement. These units included escape and evasion networks for both guerrillas and underground, intelligence networks in urban areas and throughout the countryside, secret supply depots and supply routes networks, and recruiting teams for both underground and military units.[46] Different strategies will give different emphasis to underground functions, and performance of these functions, in turn, requires different organizational structures.

2. Underground Origins and Leadership: Historic versus National Defense Planning

a. Undergrounds usually develop from within existing social, political, or military organizations. Such existing organizations can provide the necessary structure and relationships for recruiting members. In 1994, the Hutu leadership in Rwanda organized tens of thousands of recently unemployed young males into soccer clubs that would eventually become their operational elements against the Tutsis.[47] There exist many independent associations and quasi-governmental entities during peacetime and during resilience hardening that can provide part of the organizational base of a potential resistance as well as provide membership for the components of the resistance organization.

b. Historically, the organizational character of an underground has differed according to whether it had its beginning in political or military organizations and whether it was led by a political or military leader. Many of the European underground organizations of World War II developed from pre-war political parties, particularly communists, which were directed by political leaders. Consequently, the organization reflected the influence of the political leader. Some movements have both political and military leaders who function independently of each other. In this form of organization, the underground operates under direct political leadership while the guerrilla forces are led by a military leader. This occurred in the anti-fascist resistance in Italy during World War II and in Morocco during the fight for independence (1953-1955).[48]

c. When national defense planning includes a potential resistance organization, the unified political and military organization can be planned, and the leadership designated (core cadre), as part of an overall national defense plan. This increases reliability of a resistance organization in terms of training, arming, equipping, and loyalty to the cause of reinstituting the pre-war government and returning national sovereignty to the people in the occupied territory. It also assures domestic and international legitimacy and credibility for the resistance and its cause as well as retaining governmental control over it. Concurrently, it reduces the political and operational space for the development of organizations at odds with those goals by seizing the concepts of sovereignty and legitimacy, particularly when authorized and designated within a national legal framework.

d. During national defense planning, strategic, operational, and even some of the tactical level leadership of a national resistance can be predesignated to conduct this form of warfare. This cadre of persons can be associated with, previously associated with, or not at all associated with the government through employment (military, law enforcement, or other government workers). A large portion should come from nongovernment employment. Their roles and association with a potential resistance organization must be kept secret since secrecy under occupation will allow them to operate, whereas their revelation would subject them to arrest or targeted coercion by the occupying power to leverage them against their compatriots in the resistance organization.

3. Conflicting Requirements of Security and Expansion

a. An underground operates in a hostile environment in which governing forces attempt to seek it out and destroy it. In order to survive, an underground must adapt to two major factors, occupier countermeasures and its own successes and failures. If the enemy forces become increasingly effective, then the underground must emphasize security, which usually means smaller and fewer organizational elements.

b. Under occupation, a successful resistance underground might become deluged with new recruits. If it fails to expand its organization quickly, it may pass up an opportunity to expand and attain short and long-term objectives. On the other hand, recruiting non-fully vetted persons in order to quickly expand membership may provide the occupier's security forces with the opportunity to infiltrate the resistance. Therefore, to preserve its security, the organization should place potential new recruits in auxiliary roles, to test their loyalty if they cannot also be vetted via other procedures (see the Forest Brothers case study in appendix D).

c. Often, in order to achieve its objectives, the underground must be expansive and aggressive, but in order to survive, it must take precautions and emphasize security. The leadership must decide when to adjust the organization, its size, and security requirements to achieve an optimum balance between the possible need to expand, the need to replace members who leave the organization for whatever reason, and the need to maintain security.

4. Command and Control

a. An underground requires concerted action expressed through a centralized command. Without this, its activities can occur in a haphazard manner, may not support the strategic communication narrative, and may lose much of their cumulative effect (e.g., units may attack the same targets and may fail to attack a more operationally significant target). However, for security reasons, decentralization of activities is desirable. Therefore, leaders must strike a feasible, situationally determinative, balance between centralized and decentralized command and control.

b. Decentralized control methods cannot direct each subordinate unit and therefore must rely on mission orders issued by the central command describing the tactical objective and recommending activities to best accomplish that objective. This conduct of military operations through decentralized execution based on mission-type orders is also known as mission command.[49] Each of the subordinate units, which must place a premium on survival, can devise independent plans for carrying out the orders. Consequently, the subordinate units usually have the authority to make independent decisions on local issues and to operate autonomously with only general direction and guidance from the centralized command. This does not preclude special assignments or centrally directed activities.

c. The central command may not know precisely how many members belong to the organization's subordinate elements. It can test the strength of the organization by calling demonstrations, strikes, or other trial actions. Such tests provide the central command with some estimate of underground strength and ability to react without jeopardizing the identity of its members. These tests can also help the organization determine the length of time necessary to mobilize its units.

5. **Centralization of Administrative Functions**

a. Undergrounds should consolidate many activities in the central command to provide services to subordinate elements. These activities typically include:
- Production of false documents
- Collection of funds
- Purchase of supplies
- Analysis of intelligence information
- Security checks/vetting of new recruits

b. These centralized services are also typically best performed outside the contested physical space or in a place of sanctuary within an allied or PN. The Kosovo Liberation Army (KLA) relied heavily on administrative support from its diaspora in Germany and elsewhere.[50] During World War II, much of this centralized activity was conducted for continental European resistance organizations by their governments in exile, located in England.

6. **Decentralization of Units–Cell Structure**

a. Undergrounds usually organize into territorial units. Each territorial unit then subdivides into districts and then into cells. Within each district, and at each level of organization, there are different groups responsible for specific functions. Undergrounds can often organize units within existing occupational activities, such as railroad workers unions. Thus, the underground is typically organized by professional or occupational groups as well as territorial units such as provincial, city, and cell.

b. The basic underground unit is the cell. The cell usually has from three to seven members with one appointed as the leader who is responsible for making assignments and verifying their execution. As the underground recruits more people, the cells are not expanded, but new cells are created.

01. A cell may be composed of persons who live in a particular geographical area or work in the same occupation. Often, individual members do not know the places of residence or real names of their fellow members and only meet at prearranged times and locations. If the cell operates as an intelligence unit, its members may never come in contact with each other. An agent usually

gathers information and transmits it to the cell leader through a courier, a mail drop, or the clandestine use of computer networks (e.g., email, social media, and coded websites). The cell leader may have several agents, but the agents never contact each other and only contact the cell leader through intermediaries. Lateral communications and coordination with other cells or with guerrilla forces are also conducted in this manner. This way, if one unit is compromised, its members cannot inform their superiors or other lateral units.

02. To reduce the possibility of member detection, the underground disperses its cells over widely separated geographic areas and groups. This extends the occupier's security forces and prevents their concentration in one area. The underground attempts to gain as wide a representation as possible among various identity and interest groups. The Malayan Communist Party during the emergency of the 1950s was easy prey for security forces partly because it was almost entirely composed from a minority Chinese population and mostly restricted to a limited geographic area.

03. The underground organization and its activities are based upon a "failsafe" principle, meaning it is organized so that if one element fails, the operational effect will be minimal due to back-up elements. The underground should have a back-up element (i.e., alternate cell) that can perform the same duties as the primary element if the latter is compromised. Thus, the organizational expansion of undergrounds is usually in a lateral direction by duplicating units and functions.

    i.   An underground organization that provided an excellent example of these principles and served as a model for other movements was the Provisional Irish Republican Army (PIRA). In response to increasingly effective countermeasures by the British in 1972 and swelling numbers of volunteers, the PIRA moved away from a larger hierarchical military structure (e.g., battalions) and relied on a smaller cellular structure to enhance both security and operational effectiveness.

    ii.   The basic "cell" of the PIRA organization was the active service unit (ASU), which carried out the military operations of the PIRA. Each cell typically had four members and one operations commander. Each ASU was responsible for the bulk of the operational expenses as well as their safe-houses and transportation. The operations commander of an ASU usually knew the identity of only one higher commander in his organization. Part-time members were often men and women who held regular jobs in the community and would participate in PIRA operations on the weekends or after work hours. A few full-time members were paid by the PIRA with a weekly stipend and received additional support from the community in the form of donations, food, and clothing.

2.10. Underground Functions

A. An underground function is a necessary activity for the resistance to conduct to be successful and is integrated with other functions. The underground component is the component with the greatest and most varied responsibilities. Each function should be established and organized prior to a crisis. During and in the immediate aftermath of a crisis, though likely centered on the underground, aspects of these functions are typically shared between the components with specifics guided by the situation. These functions may increase or decrease in scope or emphasis but are present in all resistance campaigns. They are present even though the operating environment, the duration of the resistance campaign, the equilibrium between stakeholders/competitors, or tactics, techniques, and procedures may differ. Key functions are:

- Recruitment
- Intelligence
- Financing
- Logistics/Sustainment
- Training
- Communications
- Security

B. Recruitment

1. **Recruiting to Task.** Resistance organizations must accomplish a wide variety of tasks to be success-ful. These activities fall along a spectrum from nonviolent administrative record-keeping to IO, train-ing and equipping of guerrilla forces, executing targeted acts of violence, subversion, and sabotage. Therefore, recruitment under occupation can be a challenging endeavor. Resistance organizations are comprised of people; therefore, recruiting and retention of personnel is an essential function. This function is necessary even for a planned resistance of short duration.

2. **Continual Recruiting.** Resistance planning prior to a crisis will have identified many members and potential members for the various components of the resistance effort. However, during a crisis many members will be unavailable for many reasons, ranging from voluntary departure to arrest. There-fore, finding, investigating, contacting, and assimilating people throughout the life of the resistance campaign are crucial to its success.[51]

3. **Recruitment Considerations.** Core recruitment considerations include who, where, and how to effectively recruit, based upon resistance requirements for specific tasks and skills to meet the orga-nization's objectives and end state. Recruiting is linked to the appeal of the resistance cause, which is a factor of legitimacy and potential for success, but it is constrained by security risks.

4. **Recruiting Leaders.** Resistance senior leadership must be identified prior to a crisis challenging national sovereignty. A peacetime environment coupled with patriotic appeal against a potential threat to the nation allows for easier and more secure access to a greater pool of potential leaders. In this environ-ment, there is less risk both to the potential recruit and to the recruiter, as well as better means to vet the potential recruit. Leaders tend to join a resistance organization primarily because of the appeal of the cause, in this case, nationalism and the restoration of national sovereignty. During occupation, senior leaders are often not recruited but rather assimilated because they intellectually favor the cause and have independently developed their own conviction to support the cause, sometimes coupled with personal grievance caused by the actions of the occupier. As the resistance expands under occupation, the underground organization must also recruit and develop middle level leaders such as provincial or district leaders, influential agents within a university or government agency, and military leaders. During peacetime preparation and planning for a resistance organization, many of these civilian poten-tial members can be properly vetted by the government and be contacted for willingness to participate at a later time, closer to crisis. They need not be active organizational participants during a time of noncrisis. Only the core cadre of leaders, and some specialists, need to participate in training and exercises in peacetime. When resisting less oppressive aggressors, recruiting key figures who already wield great influence in society is of great advantage to the underground because it adds significant new influence capability to the movement.

5. **Recruiting in Rural Populations.** While under occupation, the resistance organization typically has better access to rural populations primarily because of their remoteness from centers of occupier control. The occupier usually focuses its power in urban areas due to greater population. This rural population often has larger numbers of unemployed or underemployed young people. Recruiters can offer money, excitement, and upward mobility. In southwest Asia, the Vietcong was very successful in recruiting in rural areas, primarily because the government had little influence there. This recruitment was also part of a larger objective to limit and disrupt the reach of the central government into rural areas, thus allowing physical space for guerrilla formations.

6. **Recruiting in Urban Populations.** The bulk of populations in advanced states reside in urban environments. However, once a resistance campaign is underway, urban recruiting carries increased security risks due to the proximity of the occupier's security forces, but it has the advantage of containing a very large population base from which to recruit. In the 1970s, the *Fuerzas Armadas Revolucionarias de Colombia* (FARC) in Colombia expanded its recruitment into the cities, based on an opportunity presented when poor urban workers began to protest their living conditions and economic stagnation. The FARC took up this grievance and represented their movement as a proletarian struggle against imperialism and government corruption. They established student groups and civic action programs within universities and schools and used these platforms to persuade people to vote for left-wing politicians and agitate for reforms of benefit to their insurgency. The FARC also drew upon the burgeoning urban population for recruitment into local militias and mobile guerrilla units. Whereas, rural recruits were typically drawn into their insurgency with promises of basic necessities, urban youth responded more to strong ideological propaganda.

7. **Recruiting Techniques**

   a. **Precursor Groups.** A consistently reliable technique to recruit people into a component of the resistance organization is to induce them to join benign and easily accessible groups as a precursor to later recruitment into the resistance. People can join student groups, protest events during occupation, or other activities at little risk because the activities are nonviolent, often legal, and typically carry lower risk. Such groups and events then serve as useful recruiting pools because they comprise large numbers of people already somewhat aligned with the cause of the resistance.

   b. **Selection and Vetting.** Whether during pre-crisis or while under occupation, this recruitment process gradually—sometimes over an extended period of time, especially under occupation—brings together an individual inclined to act on behalf of the cause and the underground, which is concerned about security. When underground operatives judge that the participant possesses the requisite enthusiasm and abilities and is an acceptable security risk, they approach the prospective member. Analysis of this technique showed that it took a while for many individuals to realize that they were being recruited. The prospective recruit might then spend a lengthy period in an entry level status, usually performing an auxiliary associated function, while his or her reliability and potential are assessed. The recruit may then advance in the organization into increasingly committed levels of activity. This may be a very quick process depending on cultural context and resistance requirements and resources.

   c. **Recruiting Through Personal Ties.** Using personal ties to contact potential recruits is very common in such recruiting contexts, particularly while under occupation. Relations between recruiters and potential members can emanate from family ties, religious affiliation, student groups, and other activities.

d. **Recruiting in Prisons.** During occupation, prison recruiting takes advantage of a literally captive audience to swell the ranks of the resistance organization. Regular criminals whose activities would reduce the legitimacy of the resistance are typically not the target. However, at a certain point after the crisis which resulted in a loss of sovereignty, the occupying regime installed by the enemy has likely imprisoned many people for various forms of resistance. Not all of these people were members of the resistance organization when imprisoned, but many of them are now ripe for recruitment into some level or component of the resistance organization.

e. **Subversion of Individuals.** As it relates to recruiting, in the context of resistance, subversion implies changing someone's loyalties. This form of recruiting targets individuals, typically within the occupier's government administration (e.g., officials, administrators, police, or military personnel). One of the best examples of the use of subversion as a recruiting technique was the Vietcong doctrine of *bihn van* (i.e., the promotion of desertion and defection from the government of Vietnam). Through agitation, persuasion, coercion, and threats, Vietcong operatives targeted key military and civilian officials to weaken the government's ability to rule and to swell the ranks of the insurgency. Subversion against the military was effective in diminishing the will to fight of many soldiers and in some cases, caused soldiers to provide intelligence on military operations or even to defect.

f. **Subversion of Unions and Critical Industries.** Undergrounds should have members placed within communication and transportation industries because agents therein can sabotage facilities needed by enemy military, and security forces can provide intelligence on occupier activities. They can also, to a degree, use their positions in communication or transportation to assist the resistance. Often, the labor in these industries is represented by unions or other associations. These organizations, if allowed to operate under the occupation regime, should have underground members within them, because controlling or influencing these groups enables an underground to call strikes and weaken occupier control. Additionally, funds in these organizations can be diverted to underground activities. Underground funds may also be concealed in union or association accounts by falsifying records. Strikes, demonstrations, and riots also diminish the effectiveness of the occupation forces because security forces or regular army troops may be required to control them, drawing manpower from units assigned to combating the underground. One classic example of labor union power is Poland's Lech Walesa-led Solidarity movement during the 1980s and the eventual transformation of that labor movement into a political party that undermined the authority of the Communist Party and placed Lech Walesa into office as president of Poland in 1990.[52]

C. Intelligence

1. **Shared Function.** One of the critical functions of an underground resisting a foreign occupier is intelligence. If external allies are engaging in military preparations to oust the occupier, then the underground must have networks and capabilities to assist that effort (see Types of Intelligence: Military Intelligence section). The underground typically shares responsibility for intelligence with the auxiliary. The underground establishes and controls the networks while auxiliary members collect most of the intelligence.

2. **Decision Input.** Leaders need relevant and timely information to make operational and tactical decisions. Therefore, underground operations typically include provisions for the systemized collection of intelligence on enemy forces and dispositions, political developments, lucrative targets for sabotage or guerrilla action, defectors, population dynamics, law enforcement activities, and a variety of other factors. Intelligence feeds senior level decision-making as well as small unit tactics and almost every

other activity of the underground, guerrilla force, shadow government, and public component if one exists.[53]

3. **Cellular Configuration and Training.** Intelligence gatherers should not contact each other or be able to identify each other. Cell members transmit gathered intelligence to the cell leader through remote methods to protect against identification. Intelligence cells and their members should not jeopardize their anonymity and thus effectiveness by conducting operations, such as sabotage, that could compromise them. Placement and access to essential information is a key factor when selecting intelligence gatherers. Training in collection, CI, operational security, etc., may be required in order for them to perform effectively. Therefore, conducting such training prior to a crisis is essential. Many of these persons should be trained in peacetime as part of the cadre. Such training can also be provided post-crisis if necessary, possibly with the assistance of a friendly state.

4. **Indigenous Knowledge.** The resistance has the advantage of knowing the terrain and the people, which assists its ability to gather intelligence against the enemy. Drawing on the strength of clandestine networks and friendships, resistance networks often have access to intelligence by virtue of having confederates operating throughout the area. Most PIRA members kept day jobs that often helped in some way to support the PIRA. If they worked in a government office, gaining access to official paperwork/forms or intelligence was part of their duty. Catholics in government administrative jobs provided rich information such as the home addresses of policeman or loyalist paramilitary members, to the Provisionals.[54] As another example, this fact assisted the Viet Minh because, pursuant to the 1954 Geneva Accords, large-scale regrouping of the population both north and south of the 17th parallel left underground operatives distributed throughout the country and connected to each other through clandestine networks. The insurgents in the South were perfectly positioned to provide military and political intelligence to the communist leadership in Hanoi. As in that example, effective resistance organizations must directly leverage popular support.[55]

5. **Security.** The advantage of knowing the people and terrain also benefits the resistance in its CI efforts to protect itself against enemy penetration. The PIRA was successful in preventing British security forces from infiltrating base areas and gaining intelligence by establishing close relations with their communities and using both coercion and incentives to encourage loyalty. Eventually, government forces intensified their efforts to gather intelligence through informers, surveillance, and interrogation. The struggle for control of intelligence continued throughout the 1970s and the increasing success of British efforts led to the PIRA changing its organization and practices to better secure itself. It began to organize in small cells, rather than the larger battalions of its earlier years, and increased instruction to recruits to refrain from discussing operations with anyone. This sophisticated response to British intelligence efforts included training members to avoid leaving forensic evidence after an operation and resisting interrogation after capture.[56]

6. **Methods and Functions.** Many factors frame and influence the intelligence process and cycle. Language, cultural affinity, and methods of communication can facilitate or inhibit intelligence efforts. The intelligence collection process is linked to a resistance's success or failure. Adversary forces are keenly interested in resistance intelligence for targeting. To accomplish these functions, a sample network and sub-source functions is represented in figures 10 and 11.

   a. **Human Intelligence (HUMINT).** HUMINT is a category of intelligence derived from information collected and provided by human sources. These sources serve as human eyes and ears. When well placed, they can provide early warning of adversary or enemy activity. Typically, these persons are

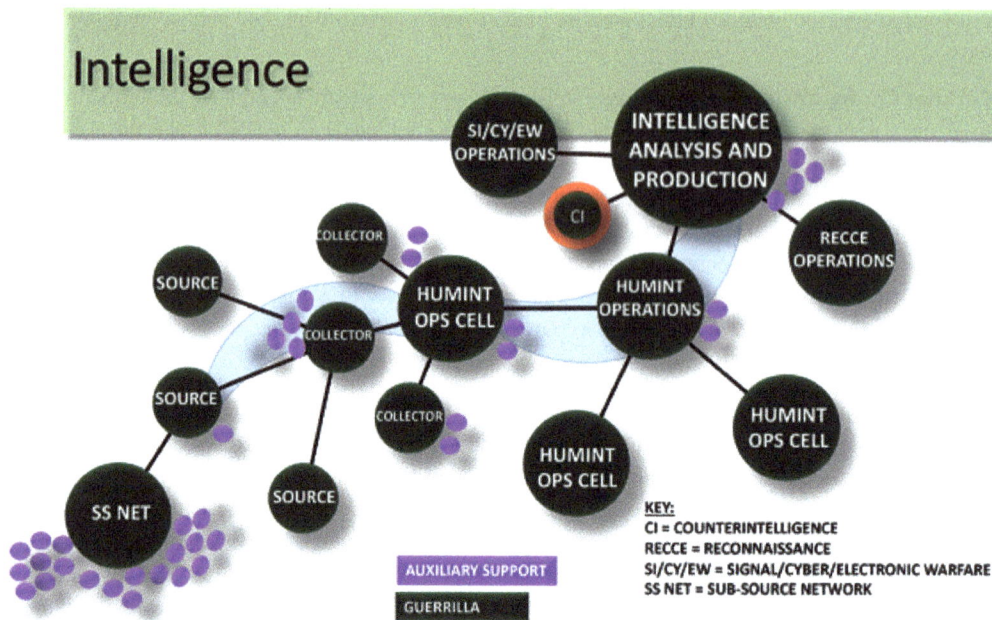

Figure 10. Intelligence Network Sample.

SOURCE: CREATED BY U.S. ARMY SERGEANT MAJOR WILLIAM DICKINSON, SENIOR OPERATIONS SERGEANT, SOJ3, SOCEUR/USED WITH PERMISSION

part of a source network that provides the equivalent of conventional reconnaissance and surveillance. Under occupation, assistance from friendly foreign intelligence services can provide identity intelligence (I2) on persons or groups through biometrics and forensics, signals intelligence (SIGINT), open-source intelligence (OSINT), CI, and document and media exploitation to assist in verification of information gathered.[57]

b. **CI.** CI encompasses activities conducted to identify, deceive, exploit, disrupt, or protect against espionage, other intelligence activities, sabotage, or assassinations by the adversary. CI is used offensively by penetrating and deceiving the opponent, and defensively to protect vital information and prevent it from being obtained by the opponent. In the pre-crisis environment, CI assets work closely with intelligence, security, and law enforcement to protect vital assets and information to prevent opponent infiltration of the resistance organization. During pre-crisis or under occupation, it includes all actions to identify, disrupt, or protect the resistance organization against espionage. The opponent will attempt to develop and use informants and double agents to attempt to gather critical information about the resistance. Therefore, operations security (OPSEC) is critical to the survival of the organization.[58]

c. **OPSEC.** OPSEC is a threat-based risk management process intended to prevent an adversary or potential adversary from obtaining awareness or early warning of resistance organization activities. OPSEC is a capability that identifies and controls critical information, indicators of friendly force actions attendant to military operations, and incorporates countermeasures to reduce the risk of an adversary exploiting vulnerabilities.[59]

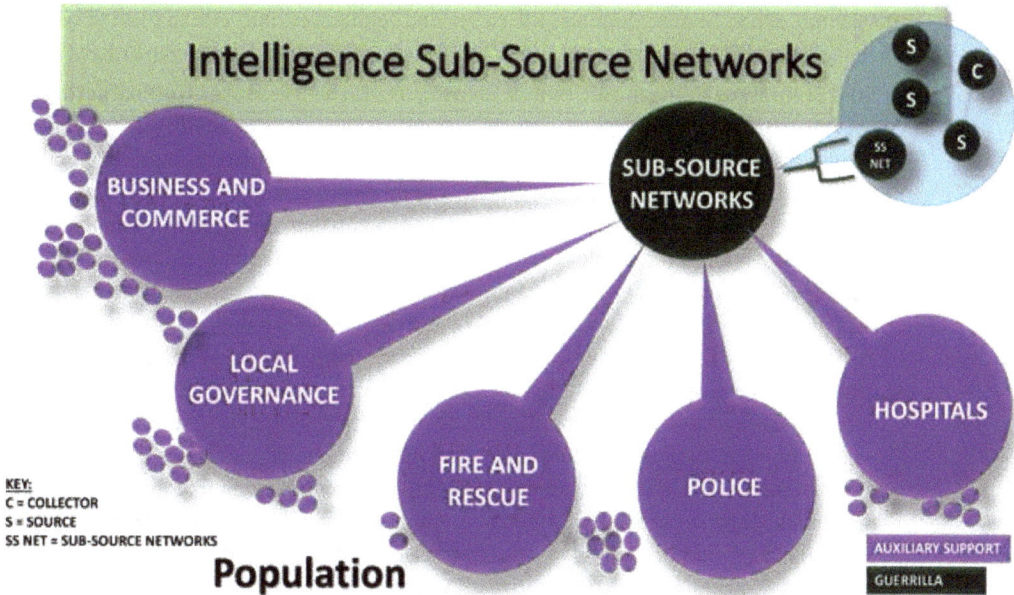

Figure 11. Intelligence Sub-Source Network Sample.

SOURCE: SOURCE: CREATED BY U.S. ARMY SERGEANT MAJOR WILLIAM DICKINSON, SENIOR OPERATIONS SERGEANT, SOJ3, SOCEUR/USED WITH PERMISSION

01. **OPSEC is a Continuous Process.** OPSEC is an iterative, five-step risk management process that enables a resistance organization to identify the most important information requiring protection and take steps to protect it. The five steps used in applying OPSEC are: 1) Identify critical information. 2) Analyze the threat. 3) Analyze vulnerabilities and identify tentative measures for protecting critical information. 4) Assess risk. 5) Implement measures, assess their effectiveness, and refine the plan as needed.

  i. **Step 1: Identify Critical Information.** Critical information is information about a resistance organization that a potential or actual adversary can use to delay or prevent attainment of the resistance organization's goals and objectives. Critical information can be divided into four broad categories—capabilities, activities, limitations (including vulnerabilities), and intentions. The identification of critical information results in the creation of a critical information list (CIL), which allows the resistance organization to focus time, effort and resources on protecting its most vital information, rather than attempting to protect all sensitive information.

  ii. **Step 2: Analyze the Threat.** Threat analysis involves the research and analysis of intelligence, CI, and open source information to identify the likely adversaries to the planned operation. A threat is a potential or actual adversary who possesses both the intent and capability to do harm to the resistance organization. The greater the combined intent and capability of the adversary, the greater the threat. Analyzing the threat can include information from a variety of sources, including the resistance organization's intelligence collection activities and open source information to identify likely adversaries and determine the extent of the threat. Consider which intelligence gathering capabilities the threat possesses

45

(Human, Open Source, Signals, Measures and Signatures, and Geospatial). Assume the capabilities the threat possesses to the greatest degree are the ones to most likely be used.

iii. **Step 3: Analyze Vulnerabilities and Identify Tentative Measures for Protecting Critical Information.** The purpose of the vulnerability analysis is to identify an operation's or activity's vulnerabilities. A vulnerability exists when the adversary is capable of collecting critical information, correctly analyzing it, and then taking timely action. In this step, leaders of a resistance organization determine where and when a potential or actual adversary is likely to obtain critical information or indicators that could lead to usage based on the intelligence gathering capabilities of the adversary. If the adversary can place the information in the correct context and take action in time to delay or prevent the attainment of a resistance organization goal or objective, the vulnerability should be reduced or eliminated through the employment of OPSEC measures. A partial list of OPSEC measures includes: communicating the CIL to organization members (so they know what to protect), and to whom they should report incidents of critical information being compromised; training on the importance of OPSEC; awareness campaigns to remind organization members to be diligent in protecting critical information; limiting critical information distribution; counter-intelligence (preventing an adversary from effectively employing intelligence gathering capabilities); and counter-analysis (using Deception in Support of OPSEC) to confuse the adversary and slow or paralyze his ability to process intelligence and take action.

iv. **Step 4: Assess Risk.** This assessment has three components. First, analyze the vulnerabilities identified in the previous action, and identify possible OPSEC countermeasures for each. Second, estimate the impact on operations such as cost in time, resources, personnel, or interference with other operations associated with implementing each possible OPSEC countermeasure. Compare this against the potential harmful effects on mission accomplishment resulting from an adversary's exploitation of a particular vulnerability. Third, select specific OPSEC countermeasures based upon a risk assessment. Calculate risk to attainment of resistance organization goals and objectives. This is done in terms of the importance of the critical information to accomplishing the goals and objectives, the threat's intent and capability, and the likelihood that the threat can exploit a vulnerability to obtain critical information or indicators in sufficient time to delay or prevent attainment of the goals and objectives. Concentrate effort on reducing or eliminating the vulnerability. If the tentative measures developed in Step 3 are not sufficient, then consider including additional measures until the vulnerability is sufficiently reduced or eliminated.

v. **Step 5: Implement Measures, Assess Their Effectiveness, and Refine the Plan as Needed.** Implement the OPSEC measures identified in Step 3 and refined in Step 4. Assess their effectiveness and continue implementation, modify implementation, discard ineffective measures or add additional measures as circumstances require.

d. **Vetting.** Vetting resistance organization members and new recruits is critical to security. This must be conducted in the pre-crisis phase for people being selected to join the organization as well as later under conditions of occupation. Individual background screenings can include biometric and forensic (I2), and biographic analysis (to determine whether a person is subject to adversary influence) and word of mouth vouching. The collected information must be secured and safeguarded from an adversary during pre-crisis and during occupation. It may be secured locally if possible, with safeguards to destroy or remove it if necessary but may also be kept at a safe foreign location such as an overseas embassy in a friendly state.

e. **OSINT.** This intelligence is based on open source information publicly available. It includes all available media (e.g., print, radio, television, and the internet—websites, blogs, communities of interest, and news sources). Other sources of information are academia and private and commercial enterprises. OSINT is susceptible to manipulation and deception (e.g., "fake news," especially as part of IO) and requires careful review and analysis. OSINT can provide context for understanding classified information, fill information gaps, gauge popular sentiment, and discern trends in friendly and adversarial governments, as well as trends within their populations.[60]

f. **Social Media.** Social network structures are composed of individuals and organizations connected by one or more types of interdependency. Some primary categories of interdependencies are friendship, kinship, common interest, group affiliation, or social relationships. Understanding these connections gives insight into a group's patterns of influence and decision-making and can help understand an organization's strengths and weaknesses. Examples of such media include Facebook, Twitter and YouTube. Such social media has been used to mobilize and coordinate protests, such as the World Trade Organization protests in Seattle, Washington, U.S., in 1999; the Arab Spring of 2010-2011; and the Ukrainian "Euromaidan" protests in 2013. Social media has also been used to share battle damage assessments such as in the ongoing Syrian civil war and to recruit fighters, such as by Islamic State in Iraq and Syria (ISIS). Al-Qaida in Iraq used social media videos depicting attacks against U.S. and coalition forces to help recruit, fundraise, and maintain the morale of its members. It also affords easy access and can quickly disseminate messages and multimedia. Video images and photographs can also be used to mitigate opponent misinformation and deception activities. This occurred in Crimea in 2014 where Russian soldiers posted images of themselves online, which allowed for identification of them as Russian soldiers in Crimea, although denied by Russia at the time. Social media infrastructure can provide inexpensive and highly effective resistance communications. This can support information environment messaging, identification and communication of opponent activity by the general population, and popular support for the resistance campaign domestically and abroad.[61]

7. **Intelligence Sharing and Training**

a. **Intelligence Sharing.** The effective sharing and use of intelligence information with friendly international partners, as well as among internal agencies and organizations, is essential to provide adequate early warning of adversary intentions and to successful resistance operations if deterrence fails. An effective intelligence sharing architecture or system requires a network for communicating important operational and threat information. Intelligence sharing and collaboration with allies and partners can reinforce their commitment in both pre-crisis and occupation situations while contributing to deterrence in peacetime. Successful intelligence sharing establishes a trusting partnership with foreign counterparts to counter a common threat and establish and maintain unity of effort. Resistance forces possessing cooperative relationships with friendly foreign intelligence services may be able to submit collection requests to them while also being given unsolicited information of advantage.[62]

b. **Intelligence Training.** During the pre-crisis stage, designated resistance intelligence personnel should receive specific training in gathering, protecting, transmitting, evaluating, and using the information they receive. Such training should continue even when under occupation to update personnel on friendly and enemy measures and countermeasures. During an occupation, a friendly foreign government can assist this process of continued training, outside the occupied area or in that foreign country.[63]

8. **Types of Intelligence: Military, Sabotage, and Political**

a. **Military Intelligence**

01. This form of intelligence focuses on the military capabilities of the adversarial or enemy state.[64] It can also include the military capabilities of proxy or surrogate forces directly or indirectly associated with the opposing foreign state. The underground can provide valuable information regarding the area of intended allied invasion or impending combat, including enemy troop numbers, unit identification and movements, as well as the nature of their arms and equipment, and even perhaps their competency in employing their arms and equipment. The underground can identify minefields and other anti-access, area denial (A2AD) emplacements, can assist in assessing their vulnerabilities, and even support actions to reduce their effectiveness or render them incapable at a critical time.[65]

02. An underground and its associated auxiliary operatives can provide valuable intelligence to allied forces preparing to liberate the country. This intelligence can also assist guerrilla and underground forces by providing valuable information to plan and guide resistance operations. This information may include enemy troop numbers, unit designations, the nature of their arms and equipment, the location of their supply depots, the pattern and routine of their patrols, and their morale. This information can also focus on various topographical factors such as swamps and ravines that inhibit access to an area or can provide concealment for attacking units or tall structures in urban areas that can be used for surveillance of movements. This information may be obtained directly by underground personnel through visual observation of the targets or through interaction with enemy troops by auxiliary members. In World War II, the French resistance reconnoitered the German coastal defenses and passed that information to the Allies in preparation for the Allied invasion of France in June 1944. During the early years of World War II, the Vietnamese resistance leader Ho Chi Minh and the Viet Minh worked with the American Office of Strategic Services against the Japanese forces. The Viet Minh provided intelligence on Japanese military dispositions and activities. Such data may also be collected by auxiliaries in the local populace, or "popular antenna," as those sources are described in a Viet Minh manual.[66]

03. Resistance organizations supported by external powers provide invaluable intelligence to those allied and partner military forces readying to defeat and oust the occupying power. Intelligence activities in this context are often conducted under the guidance of outside governments or companion military forces in the field. These sponsors often assign targets for reconnaissance (e.g., A2AD) and give technical direction to resistance personnel who lack experience in this type of work. For example, in World War II, specially trained "Jedburgh" teams (composed of an American, a Briton, and a Frenchman) were sent into France partly to advise resistance networks regarding intelligence gathering. They focused resistance intelligence collection on information necessary to Allied operations. These teams were also equipped to conduct the necessary radio communications with Britain (see appendix D). Likewise, also during World War II, the Soviets assigned Red Army personnel to partisan units to direct their activities. When such military personnel have not been available to direct and instruct, underground members have been instructed by manuals. The following is an excerpt from the Soviet *Guide Book for Partisans*, which was circulated in regions under German occupation:

> If you happen to encounter troops ... do not show that you observe the enemy ... ascertain the color of their headgear, their collar braid, and the figures on their shoulder straps. If they have questioned the inhabitants about something, try to find out what the Fascists have asked ...[67]

04. *The Guide Book for Partisans* also gave tips for ascertaining enemy intentions: "if an attack is planned, trucks will arrive loaded and depart empty; if the enemy intends to retreat, fuel and foodstuffs will be removed, roads and bridges will be demolished, telephone wires will be removed, and trains and trucks will arrive empty and depart full."[68] Similar instructions can be prepared and distributed to the general population as a form of resilience strengthening and resistance preparation during peacetime.

05. The Karen National Liberation Army in Burma used Radio Kawthulay in the 1980s and 1990s to provide casualty reports and replay information to the diaspora in Thailand. They also used captured VHF radios recovered during combat operations with the Burmese military forces to acquire tactical information about the military's operations.[69] Mass communications, such as radio and the internet can also be used for collecting and passing intelligence.

06. Resistance organizations typically operate from a position of numerical and technological inferiority as compared to the aggressor's forces. External allies and partners can help repair this gap through guidance, training, provision of equipment and information and sometimes with personnel (see Jedburghs in French Resistance, appendix D). Operations conducted by the underground and guerrillas require highly accurate intelligence in order to maximize effectiveness and maintain security of their smaller, more vulnerable forces. Undergrounds and the auxiliary members of the resistance provide this critically important military intelligence through their placement throughout their own society, facilitated by networks of clandestine operatives.

b. **Sabotage Intelligence**

01. Sabotage is an underground function that conducts destructive attacks on critical and vulnerable infrastructure, material, or human or natural resources, in an attempt to weaken occupier control and legitimacy. As with guerrilla operations, saboteurs operate in small, vulnerable groups or sometimes as a single person and require exacting intelligence in order to complete their tasks.[70]

02. Reconnaissance, transportation, and establishment of communication facilities prior to sabotage attacks occupied much of the time of the French resistance during World War II. Often working closely with Allied advisors, they surveyed targets earmarked for sabotage on D-Day. In reconnoitering a bridge for example, resistance members looked for such factors as the guard system covering the bridge and the bridge's construction. If permanent troops were evident, then their elimination had to be included in the sabotage plan. If there were only occasional patrols, the resistance would time the attack to avoid the patrol. It was also important to evaluate the bridge's construction so that the size of the explosives could be calculated. By determining the schedule of enemy train movements, saboteurs were able to destroy a stretch of railroad track while it was in use, thereby compounding the wreckage and complicating repair work.[71] Also during World War II and in direct support of Allied efforts, the Norwegian resistance worked closely with the British special operations executive (SOE) to sabotage German attempts to produce an atom bomb. This was done by sabotaging the "heavy water" production facility near Rjukan in southern Norway as well as successfully destroying the shipment of surviving amounts of that water.[72]

03. In World War II, Denmark, Danish railroad saboteurs had an elaborate system to provide this information. Throughout Jutland, underground members were stationed near major terminals to note the departures of enemy troop trains. Whenever one was seen, the observer telephoned prearranged code phrases to the sabotage cell in the next town on the railroad line. Members of the cell then proceeded to predetermined spots on the tracks to lay their mines. With this advance notice, the mines could be placed at the last moment, preventing detection by patrols.[73]

04. During the 1990s Balkan wars for independence, Kosovo resistance members used similar tactics and information to target convoys along the main Pristina–Belgrade highway and effectively controlled this main route by placing mines and using snipers to immobilize or destroy key vehicles and to limit the mobility of Serbian forces.[74]

05. In the pre-crisis environment, facilities that could be taken by an occupier, e.g., broadcast facilities, rail junctions, data-bank facilities, and other communication and logistical nodes, and used effectively to further the occupier's ends, must be assessed for potential sabotage (see appendix E, Switzerland Case Study). The actual impairment or destruction of these facilities during enemy invasion or while under occupation will then need to be weighed against the actual and information environment or strategic communication effects on the population.

c. **Political Intelligence**

01. Resistance organizations are keenly interested in political developments within the populace of their respective countries because resistance success depends on political will. They note statements and activities of persons to determine who favors the occupier or its proxy regime, so that those persons can be closely watched to determine whether their actions threaten the underground. In Belgium during World War II, the resistance kept files on collaborators and campaigned against them by using threatening phone calls and letters to dissuade these individuals from working with the enemy. If this failed, the collaborators were often assassinated.[75]

02. The late 2004 to early 2005, the Ukrainian Orange Revolution provided an excellent example of modern political intelligence and its ability to direct and fuel the actions of an insurgency. On 21 November 2004, the presidential runoff election was held in Ukraine with nonpartisan exit polling showing the challenger Viktor Yushchenko with a 52 percent lead to the incumbent Viktor Yanukovich's 43 percent vote count. When the official results were released however, Yanukovich was declared the winner with 49.5 percent to 46.6 percent for Yushchenko. More specifically, the ability to rapidly acquire and process the nonpartisan election data and distribute the results on the internet and via opposition radio and television allowed the members of the revolutionary movement to pinpoint the specific districts where voting had been rigged, thus providing the clear evidence required to mobilize the masses and garner international support.[76]

03. Resistance personnel can also note the morale of the enemy soldiers. During World War II, the Polish Home Army systematically collected data on German troops by reading their mail because there were too few Germans to handle all of the postal work. These workers would open letters and photograph contents before sending them on. These letters provided a fairly good estimate of the enemy's morale.[77]

D. **Financing**

1. Resistance organizations require money. Financing the organization while under occupation is a key underground function, with much internally gathered funding often provided by auxiliary members. The very methods of obtaining funds have a direct impact on the nature, cause, and strategy of the

organization. In the case of an extended timeframe for resistance, growing a resistance organization into something that will effect change requires time, patience, and money. Planning during peacetime for resistance under occupation requires that financial or monetary resources be predesignated with mechanisms to make those funds available at the appropriate level and to the necessary component. Even if the resistance is likely to be short-term, coordinated with allied and partner military forces, there will likely still be a requirement for access to money to fund personnel and operations.

2. Undergrounds need money to meet the following expenses: salaries of full-time workers, advances of money to persons who need money to pay human sources for information, to buy food while traversing an underground escape route, the purchase of materials for propaganda publications or internet access service, the purchase of explosive materials, and the potential purchase of communications equipment. The underground may also extend aid to families who shelter particular agents of the resistance to enable them to buy extra food. This happened in Belgium during World War II after the Nazis eliminated many resistance collaborators from the government bureaucracy. Previously, these sympathizers supplied fugitives with documents enabling them to switch identities and hold jobs. When this source of documents no longer existed, it was necessary for many evaders to go into hiding. Money to care for them was supplied by the treasury of the Armée de Belgique.[78]

3. Financial aid may be extended to the families of underground members who have been captured or forced to flee. The World War II Luxembourg resistance typified this support when it provided financial aid to the dependents of 4,200 persons who were deported and nearly 4,000 who were sent to prisons and concentration camps.[79]

4. Money may also be required to obtain the protection or silence of key officials installed by the occupier or to pay for information; it may also be needed for the bribery of corrupt officials, which is a form of subversion. Additionally, an underground may also channel funds to guerrilla units to pay salaries and buy supplies.[80]

5. Money may also be required to support services provided by the shadow government. Items such as medical care, unemployment insurance, food aid, housing subsidies, and pensions are sometimes provided by a shadow government to undermine the occupier while caring for their constituents and providing a cover, typically for auxiliary members, for activities outlawed by the new regime. These activities require a sustained and reliable income source.[81]

a. **External Sources of Funding**

01. **Foreign Governments**

    i. Resistance organizations are often aided by allied or PNs. In fact, we have stressed, these relationships are best begun prior to a crisis through joint training, information exchanges, agreements, and planning coordination. These agreements can also include means and ways for allies and partners to financially facilitate the resistance organization. In World War II, the Belgian resistance used money from Belgian reserves in London released by the British government. Similarly, many of the funds used by the French resistance were remitted from the Bank of England or from the Bank of Algiers after its liberation. The Vietcong's fight against the government of South Vietnam and its American ally was funded by China, the Soviet Union, and North Vietnam.[82]

    ii. Foreign governments extend support to resistance organizations for several reasons such as defeating a common enemy or to shaping the political outcome. Allies and partners can help plan to assist a potential resistance to defeat the common enemy. Additionally, while fighting the enemy, sometimes non-partners wish to gain intelligence information, as occurred in 1940, when the Japanese government, which was not yet allied formally with Germany and Italy, provided the Polish underground in Rome with financial aid and technical equipment in exchange for intelligence on German and Soviet occupying forces.[83] Arrangements are best planned with allies prior to the requirement.

02. **Non-State Actors**

i. Friendship societies or quasi-official aid groups can also channel funds to a resistance organization. One of the best known of the latter was the post-World War II Jewish Agency. The Agency had offices or representatives throughout the West and in the run-up to the Israeli War for Independence, Jews in Palestine, of the time, obtained critically needed financing from fellow Jews throughout the world, especially in Europe and the United States. Open appeals for money were made in newspapers, during lectures, charity balls, and other social events.[84]

ii. The PIRA also relied on funding from abroad. In 1969, the United States had five times as many Irish as Ireland. The PIRA turned to people in the United States for money and weapons as soon as it was organized enough to send agents abroad, and the Irish communities of Boston and New York proved very supportive. The Irish Northern Aid Committee was established in New York City in 1972 to provide a steady stream of money to the IRA, mostly for weapons purchases.[85]

iii. Charities and non-profit organizations are attractive sources of funding for resistance organizations. They tend to be less regulated and less scrutinized than publicly owned corporations, with many having a global presence, administering considerable sums of money.[86] These organizations can be based domestically or on foreign soil but should partake in international transactions to better conceal funding sources and distribution.

b. **Cash and Substitute Currency.** Cash can transfer internationally through informal fund transfer systems and be converted into whatever currency is best for the situation. This includes gold.[87] Hard currency, such as U.S. dollars, British pounds, or Euros, is often a good substitute for local currency because they are typically easily exchanged on the black market for local currency or goods. Hard currency is also useful when the local currency is confiscated by authorities and replaced with another currency or vouchers. More recently, the potential use of electronic credits accrued and traded via mobile phones as a means of transferring funds from one location to the next has become common in places such as East Africa. Further, the acquisition of cash value for online video gaming credits that can be acquired and traded with little visibility has also become more common. However, electronic mechanisms within an adversary-dominated operational environment can be problematic due to monitoring, jamming, or other means of electronic interference.[88]

c. **Counterfeit Money.** Counterfeit money can also fund an underground movement, especially against a new currency issued by the occupier. Planning to produce counterfeit currency requires necessary facilities and technical competency to produce such currency. Therefore, it is usually done in an allied or PN in coordination with the exiled government. If the planned resistance is likely to be short, then this method must be weighed against economic effect to the civilian population and long-term economic prospects. A resistance can also use counterfeiting to target the enemy's economy. For example, in World War II, the Nazis attempted to target the British economy with Operation Bernhard, purportedly the largest counterfeiting operation in the history of economic warfare.[89] The Polish underground chose not to do this in World War II.[90]

d. **Online Activities Targeting Aggressor.** A resistance organization can use online operations to take money or goods from the enemy state's government. This method should be carefully coordinated with the organization's information campaign and targets should be chosen for maximum political effect, while not damaging the legitimacy of the resistance cause.

e. **Value Transfer and Barter.** Resistance personnel can obtain funds directly through receipt of donated items for which they have no requirement and can resell such items on a local market. They can also exchange non-required or excess items for a needed supply item.

f. **Internal Non-Coercive Sources of Money**

01. **Gifts.** Voluntary gifts from individuals or commercial enterprises can constitute a good source of income for many undergrounds. Many industrialists and bankers provided funds for the anti-fascist underground in Italy. However, donor firms in France attempting to aid the resistance encountered difficulties in hiding these donations from the Germans which hampered the exploitation of this potential revenue source.[91]

02. **Loans.** The underground can also borrow funds. During World War II, the Service Socrates organization of the Belgian banker, Raymond Scheyven, borrowed over 200 million francs for the anti-Nazi Belgian underground. To overcome doubts from potential lenders that Service Socrates was operating on behalf of Belgian's government-in-exile, the organization invited prospective lenders to suggest a phrase to be mentioned on the British Broadcasting Corporation (BBC) on a given night. The underground passed the requests onto the London authorities who then broadcast the phrase at the designated time. The individuals then knew that they were dealing with bona fide agents of the underground.[92]

03. **Sales.** The sale of various items through door-to-door canvassing or from "front" stores can also provide money. The World War II Luxembourg resistance sold lottery tickets as well as photographs of the Grand Duchess. The Yugoslav communists raised money through sales made by party-owned clothing stores.[93]

g. **Internal Coercive Sources of Money.** Though money is critical to the sustainment of a resistance organization and its activities, a resistance organization should never rob or extort money from within its own constituency. However, if a resistance organization subsists from substantial funding from a foreign power, it can be viewed or portrayed as a puppet organization acting on behalf of that power instead of on behalf of its own people. Therefore, funding is never a matter of mere finances or economics but must always account for political perception.

01. **Forceful Appropriations from Aggressor.** Undergrounds should avoid confiscations from the general populace because such activity would greatly harm legitimacy and claims to authority (see appendix D). However, such activity directed against aggressor state targets can provide some funding while causing the aggressor state to spend more resources defending against such activity, instead of focusing those resources on defeating the resistance.

02. **Forced Contributions.** Undergrounds must not victimize their own constituency, but such a method can be used against collaborators as punishment and/or to dissuade them against continued collaboration.

03. **Taxes.** Though not recommended and likely unnecessary in a resistance of short duration awaiting allied and partner conventional force entry, taxes can be levied on a population to finance the resistance and the services provided by the shadow government. However, the amount and

method of collection must be politically and economically sustainable in order not to harm the local populace and damage the legitimacy of the resistance and its legitimate claim to power.[94]

E. **Logistics/ Sustainment.** Logistics include the critical functions of procuring, storing, and distributing supplies, maintenance, medical services, and transportation. Supplies include food, water, general supplies, fuel and oil, building materials, ammunition, explosives, major end items such as weapons and vehicles, medical supplies, and repair parts. In practice, this responsibility is shared between the underground and auxiliary. Generally, the auxiliary handles routine logistics such as food, water and fuel, while the underground often takes on the more difficult task of procuring and distributing large-caliber ammunition or other special supplies. As with most networks, the underground plans and supervises the function.[95]

1. **Procurement Methods**
   a. **Black Market.** Undergrounds can purchase some supplies on the black market. During World War II, some workers in an Italian anti-fascist underground had the specific assignment of bartering with a black market sponsored by some young fascists. This market flourished during a period when the demand for staple goods was very high and an excellent machine gun could reportedly be exchanged for 220 pounds of salt.

   b. **Legal Market.** Some items can often be purchased by an underground from legal firms through front organizations with a valid need for these items. These transactions are typically conducted in foreign countries. The items are then surreptitiously transported by resistance operatives into the occupied territory by land, sea, or air. This method is sometimes also possible within the occupied territory. In World War II Poland, the Home Army bought large quantities of artificial fertilizer from two German controlled factories at Chorzow and Moscice through agricultural cooperatives and individual farmers. The underground extracted saltpeter from the fertilizer for use in explosives.

   c. **Battlefield Recovery.** Resistance warfare often features ambushes and raids, typically by guerrilla units, against isolated CF followed by rapid withdrawal. Often, the purpose of such actions is to confiscate arms, ammunition, vehicles, communications equipment, food, medical supplies, and other supplies left behind by the CF.

   d. **Secret Confiscation.** Items can be removed secretly from plants and warehouses by workers who are sympathetic to the resistance. Italian workers during World War II were able to supply the underground with radios that were pilfered from stocks in their factories. Such confiscations do not produce a steady supply of goods. Further, the risk is high due to regular inventories. The problem of inventory checks could be avoided if sympathetic office clerks are able to account for losses by forging orders and invoices and altering bookkeeping records.

   e. **Raid.** Raids of government or commercial warehouses or other storage centers can also be used to acquire supplies and equipment. However, care should be taken to avoid theft from loyal citizens on whose behalf the resistance operates. In France during World War II, the manager of one warehouse was awakened by 12 masked resistance members who forced him to hand over his keys. There were trucks in the courtyard and 200 men stood ready to load them. A total of 38 tons of coats, sweaters, shoes, radios, and typewriters were taken. Many such raids were carried out in France after resistance workers established "understandings" with sympathetic warehouse employees.[96]

2. **Manufacturing Types and Places**
   a. Undergrounds frequently manufacture such items as mines, flamethrowers, hand grenades, incendiaries, explosives, and detonators. Rarely are they able to produce heavy equipment because of concealment problems. An exception occurred in Nazi-occupied France, when workers in a steel

mill constructed four crude tanks out of farm tractors and sheets of steel from the factory. The components were hidden separately inside the plant until they could be welded together and armed with 87 mm cannons and heavy machine guns.[97]

b. As in the previous examples, the underground can manufacture items for itself within legal facilities, manufacturing similar or similar-looking items. During World War II, the Polish Home Army used workers in legally licensed metal shops to manufacture small arms. Hand grenades, commonly referred to as "Sidelowski" because of their resemblance to round cans of Sidel polish, were produced in the same facilities as the actual cans of polish while flamethrowers were made in a factory engaged in the manufacture of fire extinguishers.[98]

c. Legal covers for manufacture of items for the underground are not always necessary or available. Small shops can be completely hidden by false walls partitioning rooms and cellars. To conceal machine noise, shops can be constructed near places where legal goods are being manufactured. Production using chemicals can be done at night to mask the appearance of colored smoke rising from chimneys.

3. **Collections from the Populace.** Goods can be systematically collected from the populace in areas with a high degree of underground influence and freedom of action. In rural areas, food is often collected for guerrilla troops. To avoid being considered "bandit" or criminal organizations by the local populace, undergrounds must either pay for requisitioned items or offer recompense with some sort of "I Owe You/IOU" process.[99]

4. **External Means**

a. A nation preparing for resistance warfare as part of its national defense planning should secure allied and partner support to that resistance effort prior to the onset of a crisis. Operational and logistical demands while under occupation, as well as the pervasive intelligence and security capabilities of the occupier, usually requires support to the resistance from a foreign source. In World War II, the British Royal Air Force parachuted arms and equipment to the French resistance. This method required coordination in advance and agreement on drop zone locations, exact drop times, and ground-to-air recognition signals. This method of supply was almost exclusively conducted at night and resistance personnel stored the items in caches near the drop zone which allowed them to leave the scene immediately and without incriminating evidence on their persons.[100]

b. Foreign governments can also support a resistance by clandestinely producing and supplying non-attributable goods in addition to items that allies want the enemy to know has been provided by them in support of the national resistance campaign. After the defeat of the Hussein regime in Iraq, the Iranian Revolutionary Guard Corps (IRGC) used this method to provide Shiite militias with material to produce improvised explosive devices. This practice gives the foreign government a degree of deniability.

c. The underground can also use businesses engaged in foreign trade to import equipment under non-contraband labels. For example, a textile firm can order textile machinery while delivering arms producing machinery or arms parts.[101]

5. **Transportation**

a. **Vehicular and Sea Transportation**

01. Arms and equipment moving about in rural areas can often be hidden among farming equipment while consignments to urban areas can be placed in compressors, gas cylinders, asphalt sprayers,

and other industrial equipment. During the Jewish fight for a state, items were sometimes hidden in trucks carrying a large amount of oranges which would then roll into any hole made while the cargo was being inspected. Contraband was also concealed by tarpaulins covered with fertilizer exuding a very disagreeable odor which often dissuaded polished and well-dressed police officers from conducting full inspections. Additionally, trucks from well-known firms such as breweries whose products were shipped to many locations and in great quantity, usually also escaped suspicion. Underground members dressed as policemen and riding motorcycles sometimes escorted heavy truckloads appearing as police escorted convoys and sometimes even succeeded in joining British military convoys, passing many roadblocks and checkpoints without inspection.

02. The New People's Army, in the Philippines, established a complex intra- and interisland network of fishing boats that appeared exactly as typical fishing boats used for legitimate island trade. This method was extremely successful because the Philippine security forces lacked a brown-water-navy and patrol boats affording the capability to secure the waterways.[102]

b. **Foot and Animal Transportation.** Guerrillas operating in remote and sometimes mountainous rural areas usually move material by foot and with animals due to the natural access difficulties. In World War II, the Greek resistance used pack animals to negotiate mountain trails and mountain dwellers to carry supplies to their resistance forces. The Viet Minh used thousands of supporters from amongst local inhabitants organized into an "auxiliary service" to move arms, equipment, food, and ammunition along trails in "human caravan" formations.[103]

6. **Storage.** Supplies can sometimes be stored in individuals' houses, but more often are stored in centralized locations to subject fewer persons to capture in the event of discovery. Caches are frequently located in remote areas. The French resistance dug and camouflaged pits near parachute drops to store equipment until it could be safely moved to more convenient hiding places. In Vietnam, local inhabitants helped "prepare the battlefield" for the guerrillas by storing food near the scene of an impending Viet Minh attack, which enabled the guerrillas to travel lightly and quickly. When supplies had to be stored for longer than a few days, the caches were ventilated and insulated against dampness with exposed ventilation pipes covered at the surface by bushes.[104]

7. **Maintenance.** Because the nature of underground logistics is often desultory, interdicted, or unreliable, resistance elements must provide for their own maintenance functions. Guerrillas usually perform unit level maintenance while the auxiliary component typically performs higher levels of maintenance (e.g., engine replacement, weapon repair). If the resistance controls a secure area, it can build dedicated maintenance facilities.[105]

8. **Medical.** Undergrounds must establish effective medical capabilities to sustain their end strengths. Wounded resistance members typically cannot go to a hospital for fear that enemy security agents or informers working for the occupier will notify the occupier's security services of the presence of the wounded resistance member. This situation can arise after a violent encounter between resistance forces and occupation forces resulting in casualties. The occupier will be watching for newly admitted people with particular types of wounds. Effective medical services also sustain morale and allow individual risk-taking in the knowledge that medical assistance is available. Prior to a crisis, resistance planning should also include support to and methods of casualty evacuation out of the country and to a friendly state. In populated areas where effective medical assistance may not exist, the resistance may maintain the loyalty of the population through the provision of medical services.[106] During resistance planning, the resistance organization should identify the requisite medical expertise and possibly recruit

appropriate doctors, nurses, and technicians for future use as auxiliary personnel. Resistance planning should not only account for evacuation to a friendly state but also the potential receipt of medical supplies and equipment from that state to supplement available supplies and equipment, possibly stored in cache sites. The practice of the Hippocratic Oath must be confirmed prior to a crisis to sustain legitimacy in the information environment. A resistance medical network can be quite extensive. It requires dedicated underground medical personnel, transportation services, supplies, equipment, and facilities. Such a dedicated network can medically sustain the resistance organization, particularly members who become injured in actions against the enemy, and can sustain resistance morale. Dedicated underground personnel, supplies, equipment, and facilities should be supplemented by auxiliary access. A sample underground medical organization is represented in figure 12.

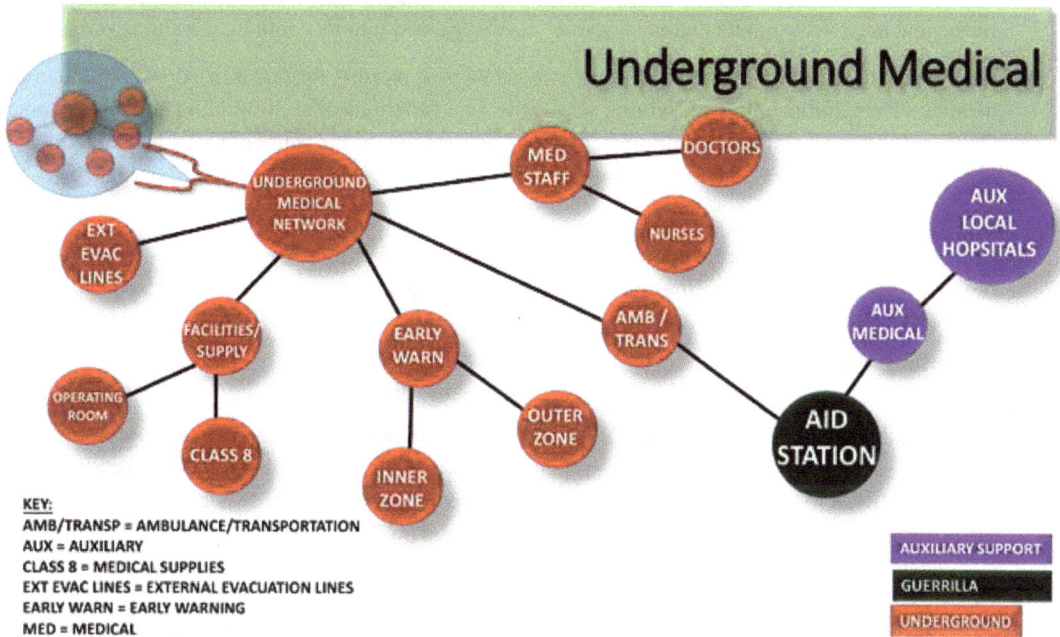

Figure 12. Underground Medical Organization Sample.

SOURCE: CREATED BY U.S. ARMY SERGEANT MAJOR WILLIAM DICKINSON, SENIOR OPERATIONS SERGEANT, SOJ3, SOCEUR/USED WITH PERMISSION

9. **Mortuary Affairs.** The evacuation of dead resistance fighters from the scene of a tactical operation such as a raid or ambush is critical for the maintenance of security. Identification of the dead could endanger the families of resistance members and could also provide the enemy with vital intelligence concerning cell membership or the larger resistance organization. In such instances, resistance personnel must remove and temporarily inter/bury the bodies of those killed in action until proper burial or disposal of the bodies is feasible. Removal and burial of the dead not only denies the enemy intelligence concerning resistance membership, numbers, and organization structure, but also denies them propaganda opportunities (see Forest Brothers, appendix D).

F. **Training**

1. Training for identified members of a resistance organization such as core cadre, should be conducted prior to a crisis as part of national defense planning and preparation to allow complete vetting, comprehensive skill building, and even cross–training. The training of resistance members necessitates the combination of hard skills and psychological preparation. While under occupation, underground members are primarily responsible for providing specific training to members of the resistance components: underground, auxiliary, and guerrillas.[107] Guerrillas, saboteurs, administrative specialists, messengers, and other operatives can accomplish their tasks only if they are competent in the use of weapons, explosives, communications equipment, etc. The training process should go beyond mere skill-building and also be used to enhance organizational security, provide sufficient indoctrination, and select and develop future leaders. Once a resistance is engaged against an occupier, it gains experience and must disseminate and incorporate lessons learned into its continuing training program.

2. During the period of initial training, leaders evaluate new recruits for reliability or risk to the organization. Thus, initial entrance training serves the dual functions of skill-building and gatekeeping. Only demonstrated ability, commitment, maturity, and fidelity to the cause of national autonomy and defense will lead to more advanced training.

3. If such training is conducted clandestinely in an occupation environment, it typically takes much longer. Training tends to occur sporadically, in intervals of short duration and focuses on individuals or very small units rather than the collective training of larger cohorts. The two most frequently used methods of training are the use of training camps and online training.[108]

   a. **Training Camps**

      01. As with the training of a nation's conventional and special forces, camps can provide an isolated environment, focused skill-building, and indoctrination to produce capable and committed individuals. In a pre-crisis environment, the government can establish the requisite facilities to provide this training to potential resistance members. While under occupation, such facilities must be inaccessible to the occupying force and possibly located in a foreign state. History demonstrates that training camps established in permissive or semi-permissive environments can evolve into highly effective institutions for producing resistance fighters. The Vietcong trained recruits at locations near the recruits' home villages. Recruits displaying the most potential received additional advanced training and were assigned to main force units. These units consisted of individuals who were extremely motivated.

      02. After the Israeli invasion of Lebanon in 1982, elements of the IRGC moved into Lebanon's Bekaa Valley and established the infrastructure for the formation of what would become Hezbollah. The IRGC conducted military training and religious indoctrination for the embryonic insurgency. By 1984, the IRGC was operating six training camps and providing salaries, medical benefits, and free education for fighters and their families.[109]

   b. **Training Online**

      01. The evolution of the internet, personal computers, portable digital devices, and wireless communications has revolutionized many aspects of society and conflict. It has also revolutionized the underground function of training. Prior to the advent of the internet, access to information sources on weapons, explosives, and tactics was fairly limited. Today, much motivational and operational information can be made available online and be only a mouse click away. However, there is significant risk to online training when considering the capabilities of the enemy

to conduct SIGINT. Measures must be taken to ensure operational security when conducting training online.

02. Hezbollah developed a videogame called Special Force in which players experienced simulated operations against Israeli soldiers based on real-life events. The game allowed players to conduct "target practice" against Israeli political leaders. Through the publication of this game, Hezbollah was able to export both its ideology and a form of skill-building.[110]

03. Governments engage in continuous online offensive and defensive measures against adversaries. During pre-crisis, the government will have access to many resources to maintain and protect its information and to securely communicate. This is the best time to train personnel engaged in these tasks. However, under occupation, these resources will likely not be securely available. Resistance forces in country, even if already trained on these methods, will need to access available online resources for continued training to assist the resistance. The most salient expression of online training is the growing community of hackers. Thousands of new websites emerge annually offering instruction and tools for hackers and are replete with operational know-how, tips, tricks, and "best practices." The visitors to the sites can learn detailed techniques for conducting denial of service attacks, stealing passwords, overloading websites, and probing networks for vulnerabilities. They can also download tools for encryption, programming, and data manipulation.[111] The use of such persons or organizations that are not part of the sovereign government must be balanced against the rules and laws bounding that government as well as the effect on legitimacy in the information environment.

c. **Conducting Training**

01. If the training of some resistance personnel is not accomplished pre-crisis, then such training becomes a core function of the underground. Through a training regime, underground leaders select, evaluate, and develop recruits to populate the underground and guerrilla forces while also perhaps conducting very specific training for potential auxiliary members. The internet can provide almost unfettered access to advanced training information on a wide spectrum of useful hard skills. Training clandestinely affects the content and conduct of training and typically aims at training individuals and very small units. The training program also has a gatekeeping role and it must rapidly assimilate lessons learned into new training techniques.

02. In case of the requirement for an extended resistance, the PIRA is an instructive example of how an organization planned, organized, and conducted training. A training department under its general headquarters was responsible for maintaining all training resources and facilities. They conducted training in three areas: new recruits training, operational skills training, and intelligence/CI/security training. New recruits training emphasized motivational information such as national history and the history of Irish resistance. The latter two training courses focused on the necessary hard skills to conduct operations and to protect the security of the organization.

    i. New recruits were required to attend training sessions about once per week during their first three months in the organization. These sessions included lectures and discussions about duties, organization history, rules of military engagement, and resisting interrogation. During this time, recruits were also evaluated on potential for service and risk to the security of the organization.

    ii. The PIRA learned to emphasize rigorous training and instruction in hard skills such as weapons, explosives, and urban and rural tactics. This resulted because previous inexperience with weapons and explosives caused numerous fratricidal deaths and premature explosions as well as ineffective attacks by impulsive and unskilled youths which merely invited arrests, interrogations, and political failure.

iii. The PIRA also sought to improve performance in bomb-making, sniping, logistics, and intelligence. Therefore, they occasionally brought specialists together to receive training and pass on lessons learned with the intent of improving safety, security, and performance.

iv. Clandestine training requirements typically do not facilitate noisy and difficult-to-conceal live fire weapons training. The PIRA therefore used remote locations throughout Ireland, including abandoned farmhouses, unused beaches, and woods. In one case, they used a beach for mortar fire using nonexplosive shells. Recruits were often kept ignorant of the exact location of these camps for security purposes. The PIRA also formed relationships with sympathetic groups abroad and conducted training in foreign areas.[112]

G. Communications

1. Communication is the exchange of information and other messaging that occurs within any organization. Underground operations, including coordination with the armed, auxiliary, and public components, cannot occur in the absence of a communications system. Underground communications are typically clandestine in nature and must balance the need for communicating with the risk of exposing personnel, plans, and facilities if the messages are intercepted.

2. Historically, the underground communication system evolves with the resistance. When resistance is included as part of national defense planning, secure and redundant communication systems can be acquired and used to train designated resistance members in the pre-crisis environment, as well as training such personnel in specific classified tradecraft methods. This avoids the pitfalls of potential lack of rigorous discipline and uniformity if this effort began after occupation and begins the resistance effort at a higher level of capability and competence. To avoid presenting a pattern that could be recognized and defeated, the underground must use various and diverse methods such as face-to-face meetings, couriers, mail, dead drops, radio, cell phones, internet, and social media. In preparation for countermeasures used by the occupying power, the underground must also be trained in security measures such as codes, frequency hopping radios, spread spectrum, and other measures.[113]

3. Effective planning for resistance also includes developing and maintaining alternative methods of communication throughout the resistance network to mitigate an adversary's disruption capabilities. If the occupier is in control of most or all of the country, it could disable, block, intercept, or alter technical communications. The occupier could also use detection methods to identify, track, arrest, or assassinate resistance leaders and members. Consequently, the national resistance and its allies and partners should not limit communication planning to technical options but should also incorporate less conventional, nontechnical communication methods.

4. **Technical Communications**

a. Resistance forces have used radios of all kinds, from ultra-high-frequency radios to ham radios to citizens' band radios. Although they can be easily jammed, intercepted, or located with direction finding devices, they are also cheap, expendable, and do not require traceable accounts as required by telephones or internet email accounts.

b. Cell phones can be useful while also carrying a high degree of risk. Connections can be monitored, locations can be found, and a plethora of information can be yielded if captured. Such disadvantages played significant roles in the targeting and killing of Abu Musab Al-Zarqawi in Iraq. In

Afghanistan, the Taliban arranged for the shutting down of cell phone towers at night to hinder American operations, while in Iraq, insurgents coerced commercial companies to maintain service because they were useful to insurgent operations.

c. Internet communications also have similar advantages of speed and accuracy with similar disadvantages of monitoring, location revelation, and information yield, if captured. However, insurgents have used them effectively for propaganda purposes by uploading captured digital video imagery of their own attacks. Resistance forces can also upload favorable images in order to help maintain the morale of the population under occupation.

d. The November 2004 to January 2005 Orange Revolution in the Ukraine featured the widespread use of internet postings and cell phone use to bolster and grow the protest against the rigged elections in Kiev. Organizers used the internet to organize demonstrations, sit-ins, and strikes while reaching an international audience to develop international support. Similar methods can be used to organize a resisting population for nonviolent protests against an occupier.

e. In 2009, the resistance in Moldova capitalized on the internet-based social messaging service Twitter to incite unrest, update protester actions, and inform on government reactions while also using Twitter to appeal to the international community for assistance. A resistance organization can use similar methods to maintain and build support among friendly foreign populations as well as its own diaspora.

f. The internet and cell phones can be used to demoralize the occupation forces by posting videos of successful guerrilla or underground attacks or large-scale demonstrations and other events. They can also be used to attract recruits to the resistance and demonstrate its resiliency, while supporting national morale against the occupier. It can also be used as a propaganda tool to influence the aggressor state's home population to reduce support for the aggressor state's actions.

g. Technical communications such as the use of the internet and cell phones can enhance recruiting, financing, training, propaganda, and even intelligence operations while attracting a large audience within cyberspace. Social media's decentralized nature is also extremely useful in reaching and mobilizing masses and is unequaled in its utility in planning and controlling day-to-day operations, unless the adversary blocks reception. Additionally, social media in particular can be used by an adversary as a deception tool against the occupied population or against the resistance organization.

h. Undergrounds must employ effective communications to survive and grow. There must be communication between the underground, the auxiliary, guerrilla forces, the internally displaced or exiled government, and, when applicable, the public component of the resistance. Furthermore, communicating with actual and potential supporters at all levels throughout the international community and attempting to demoralize the population of the aggressor state are vital to success.[114]

5. **Nontechnical Communications**

a. Nontechnical communications have the advantage of leaving no detectable electronic signature which could be intercepted, but they increase the vulnerability of the people conducting such communications. Nontechnical means include face-to-face meetings, couriers, the mail system, and dead drops. A dead drop is a secret location where messages, money, or supplies are deposited for later pickup by another agent. The advantage is that neither the person dropping the communication or items nor the person picking them up must meet or even know each other. The person making the drop at the prearranged location arranges for the display of some physical sign, at that or even another location, that only the other contact or limited organization members would recognize. The signal alerts the recipient to the fact that an item is ready for pick-up at the dead drop site. The greatest threat to this method is surveillance. Therefore, the underground must take security precautions such as varying the drop sites and signals and undertake countersurveillance.

b. In clandestine communications, when using non-face-to-face methods, acknowledgment of the receipt of a communication from receiver to sender is crucial. Without positive acknowledgment of receipt, the underground must assume that the message was intercepted, and that part of the network may be compromised. In such an event, the members most susceptible to compromise must either prepare cover activities or proceed to safe houses and await further risk evaluation.[115]

H. **Security.** In the presence of capable and competent enemy forces, security is the prime consideration for a resistance organization and permeates every aspect of its functions. Pre-crisis training for designated cadre members of the resistance is necessary to ensure its functioning and survival. Through the establishment and enforcement of security measures, the resistance enforces the integrity of the organization, the safety of its members, and the ability to conduct operations. Resistance organizational security precautions are primarily an underground responsibility, with auxiliary members usually providing early warning of threats.

1. **Compartmentalization.** Compartmentalization is the division of an organization or activity into functional segments or cells to restrict communication between them, and prevent knowledge of the identity or activities of other segments except on a need-to-know basis. This measure minimizes the danger of compromising the organization by limiting contacts between superiors and subordinates, as well as lateral communication. Using this method, any captured cell members can only divulge, at most, the identities of their few fellow cell members.

2. **Screening New Members.** New recruits, especially during occupation, present a heightened danger due to the possibility that any of them could be agents of the aggressor state. Therefore, checks must be made on their backgrounds, political activities, jobs, families, and close associates. Further, a probationary period during which they can be monitored and tested, typically with auxiliary type tasks, may also be necessary.

3. **Communications Security.** Building upon both technical and nontechnical communication methods, code words can be used to designate places, movements, and operational plans, and can also be used as a means of recognition for persons unknown to each other as resistance members. In 1998, during the onset of major conflict in Kosovo, every KLA leader assumed that the Serbian intelligence services monitored their cell phones. Therefore, cell phone discussions were kept to an absolute minimum, with strict adherence to operational security by not mentioning places, names, or specific observables which could be easily intercepted by Serbian intelligence services.

4. **Keeping Records.** Although keeping records is necessary for practically all organizations, written and electronic records within occupied territory and territory under threat of occupation must be kept to an absolute minimum. Only information that cannot be memorized or is required for future reference should be recorded. The experience of a French resistance network during World War II, known as the "Cartel," illustrates the risks of detailed records. The organization's leader made lists of its membership including names, addresses, appearances, and telephone numbers, and recorded them, un-encoded on cards for each member. In 1942, a designated organization courier was in possession of 200 of these cards in a briefcase while travelling on a train to Paris. The courier fell asleep, and while asleep a German intelligence agent took the briefcase. This ensured the downfall of the network.[116]

5. **Personal Security Measures.** During occupation, individual resistance members must continue to blend into their environment. Members must continue or establish normal routines, consistent with the routines of their neighbors, in order to avoid attracting attention. A former anti-Nazi German underground leader stated that it was difficult to avoid the surveillance of a modern police state, but that the police could be misled. The best way to mislead them was to live as conventionally and as openly as possible. "The more you resemble a normal everyday citizen in every respect the less apt you are to be suspected."[117] Today, that can mean not just physical behavior but online behavior as well. Another aspect of personal security is the establishment of random physical behavior patterns. Such adjustments to routines are effective countersurveillance methods to detect monitoring by security forces.

6. **Safe Havens.** The robust technical intelligence gathering capabilities of modern state security forces elevates the importance of secure safe havens for resistance organizations. Such safe havens must be located in places that are inaccessible or nearly inaccessible to enemy or occupation security forces. Limited safe havens may be established inside the country among reliably sympathetic populations who can also be depended upon for logistical support. Safe havens are most effective outside of the country, within friendly, preferably physically bordering, states.

7. **Actions in the Case of Capture.** The resistance membership should possess a code of conduct, to govern behavior if captured. In general, a code of conduct should prohibit divulging information such as names, cover names, addresses, or information covering past, present, or future operations. Additionally, members should be prepared for, and knowledgeable of, potential interrogation techniques employed by the enemy. Common interrogation and interview techniques include the insertion of a deception agent into the same cell to attempt to gain the confidence of the member and obtain information from him. Another technique involves demoralizing a prisoner by using information already known about him, or the organization, in order to make him believe that he was betrayed, thereby undermining his loyalty. Another technique is to promise leniency or amnesty in exchange for information.[118]

2.11. Leadership and Governance Activities and Functions, and Governments-in-Exile

A. A government-in-exile is a government displaced from its country of origin, yet which remains recognized as the legitimate sovereign authority of a nation. Exiling the national government under extreme circumstances where it cannot continue functioning within the nation should be a part of any national resistance defense planning. The purpose is to retain national representation through a sovereign, legitimate government that can speak for the nation as it fights alongside its allies and partners to restore national sovereignty. A government-in-exile will normally take up sanctuary in a nearby allied or PN. The government-in-exile should have a close relationship with, and be in command of, the shadow government in the occupied territory. Governments-in-exile can provide a ready apparatus for coordinating external support for a resistance. The functions and relationships of the government-in-exile are represented in figure 13.

B. **Resistance Leadership.** While some degree of centralized strategic direction and planning is essential for resistance organizations, the form of that leadership ranges from simple to complex, from centralized to decentralized. Through the increasing use of social media and information technology, resistance organizations can operate in a flatter, more decentralized fashion. Using the internet, if available under occupation and while considering security requirements, members can link virtually with fellow members inside and outside of the country. The relationships of resistance leadership are represented in figure 14.

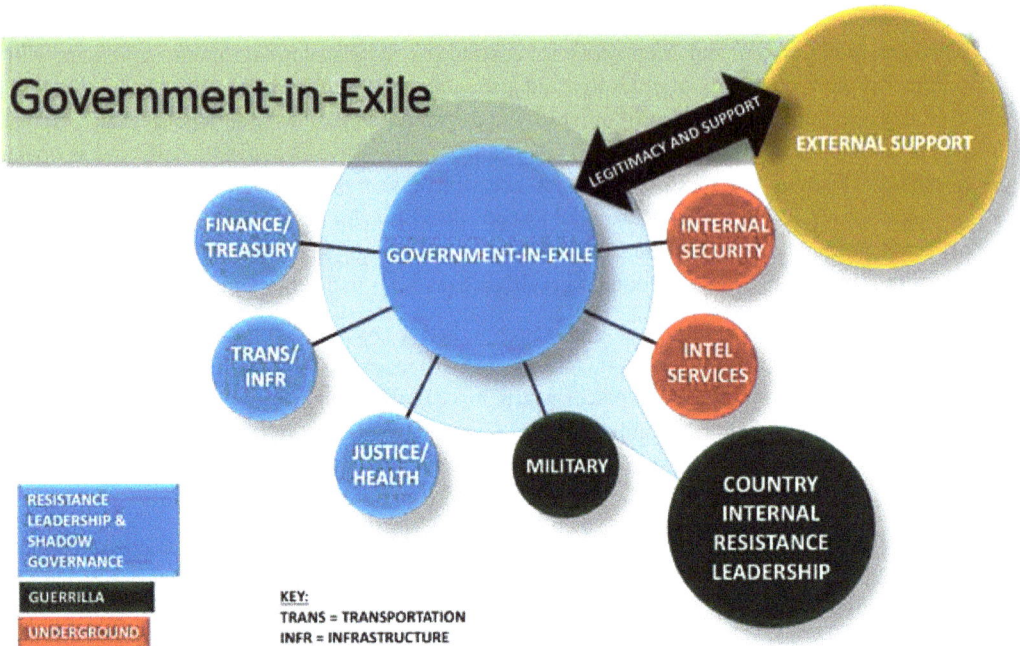

Figure 13. Generic Graphic Representation of a Government-in-Exile Sample.

SOURCE: CREATED BY U.S. ARMY SERGEANT MAJOR WILLIAM DICKINSON, SENIOR OPERATIONS SERGEANT, SOJ3, SOCEUR/USED WITH PERMISSION

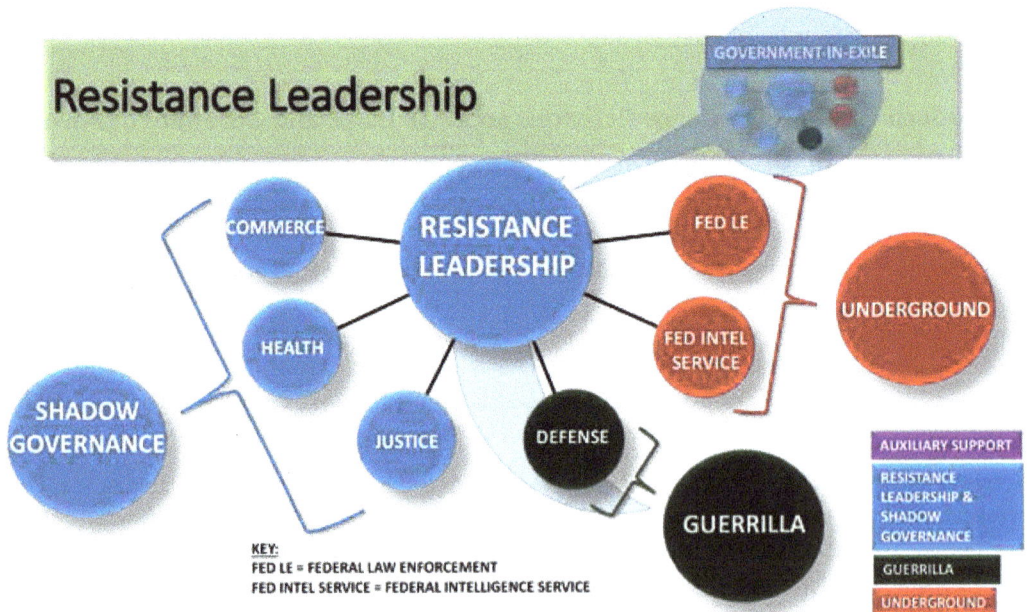

Figure 14. Resistance Leadership Relative to Other Components Sample.

SOURCE: CREATED BY U.S. ARMY SERGEANT MAJOR WILLIAM DICKINSON, SENIOR OPERATIONS SERGEANT, SOJ3, SOCEUR/USED WITH PERMISSION

C. **Shadow Government**

1. **Purpose.** The shadow government is the clandestine governance activity of the resistance and is found within the underground. It operates in competitive functional parallel with, but in opposition to, the state structure being used by the occupying regime. It acts clandestinely within the occupied area on behalf of, and in support of, the displaced or exiled government. Its primary purpose is to influence the behavior of the population. In particular, it seeks to maintain or increase popular resistance to the occupier through nonviolent and passive resistance methods, while acting to discourage collaboration (see appendix G).

2. **Structure and Role.** A shadow government, especially if it must operate for an extended term measured in years, often attempts to copy or mimic some or most of the governance functions of the imposed occupying regime (e.g., security and police, judicial processes, social and health services, and revenue generation, through taxation if required). [119] The roles of the shadow government include: adding legitimacy to the resistance organization, undermining the authority of the occupier, deterring and discouraging collaboration with the enemy by groups or individuals, and dealing with them through its judicial processes if required. The shadow government synchronizes its clandestine functions with the other resistance components. It does not act to control the other components, but rather acts to influence the behavior of the population. Shadow governance exercises a degree of supervision, control, and accountability over the population. This accountability function can extend into the restoration of sovereignty phase after removal of the occupier, where information gathered while under occupation can assist with restored judicial procedures against people charged with collaboration. The shadow government functions to discredit the occupier, support and influence the population, and lend legitimacy to the resistance. Organizationally and conceptually, it is typically an adjunct of the underground. The complex relationships of the shadow government are presented in figure 15.

3. **Maintaining Legitimacy.** The members of the shadow government will be local citizens of the same national identity as the local populace. Based on that common identity, the populace will normally grant them a large degree of legitimacy.[120] It must be perceived as acting in the best interests of the population. It should assist overt law enforcement in apprehending common criminals who harm the population or establish its own policing functions if overt functions are inadequate. [121] It can also provide or supplement social services to portions of the population in need, as the Polish Underground State (PUS) did in World War II (see appendix D). It must acquire and retain functioning empirical sovereignty and maintain strong links to the displaced or exiled government. It must also avoid the danger of doing too much, in case it becomes resource constrained and can no longer deliver functions it undertook. This would be seen as a weakening of the resistance and used by the enemy to discredit it. In the case of a nation of people resisting occupation by an enemy state, the shadow government—acting at the direction of a displaced or exiled government—should retain a large degree of legitimacy with the population as well as the international community.[122]

4. **Short-Term versus Long-Term.** National resistance is qualitatively and contextually different from insurgency. Insurgencies are internal movements which are based on grievances against a national government—in many cases an internationally recognized and legitimate one. Resistance is a form of warfare conducted by the legitimate government and population of a sovereign state to oppose occupation by a foreign power. National resistance warfare as envisioned in this concept is a short-term measure used to assist the nation and its allies and partners in the recovery of national sovereignty from the occupier. It is not part of Maoist revolutionary warfare to overthrow and replace a social and

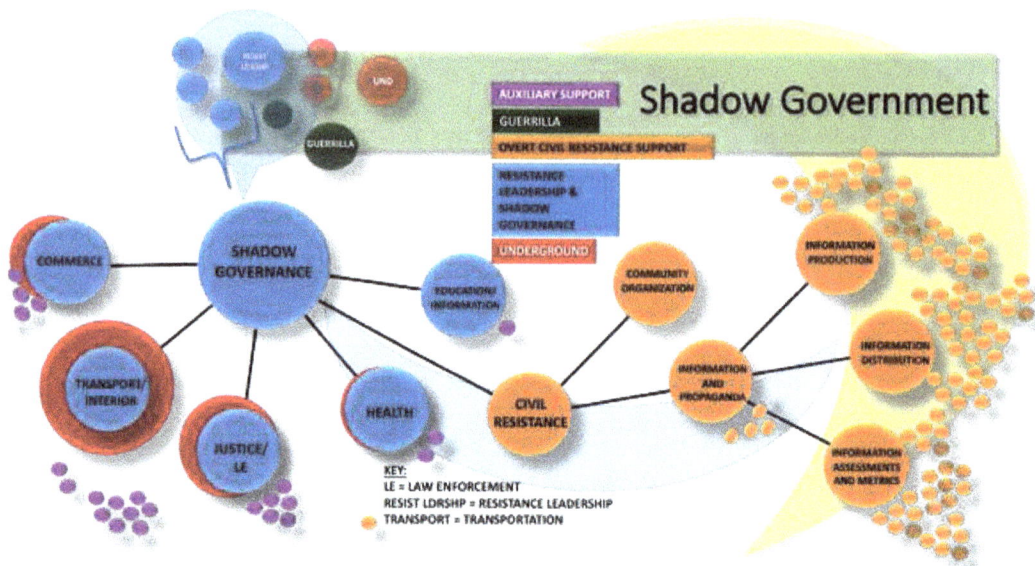

Figure 15. Shadow Government Relationship to Underground and Overt Organizations Sample.

SOURCE: CREATED BY U.S. ARMY SERGEANT MAJOR WILLIAM DICKINSON, SENIOR OPERATIONS SERGEANT, SOJ3, SOCEUR/USED WITH PERMISSION

political hierarchy. Only if the nation lacks external allies and partners committed to forcefully assist in the restoration of its national sovereignty in the short-term, will this become a long-term struggle. In such a long-term struggle, the shadow government must gradually increase its role in domestic security and service provision to grow its legitimacy domestically and internationally, grow loyalty among the population, and support the exiled government in seeking allies and partners committed to the restoration of national sovereignty.

D. **Legitimacy**

1. In the domestic context, the concept of political legitimacy is based on an understanding of the state as a political organization formed through a social contract with its citizens. In that social contract, legitimate political authority originates in the consent of the governed, while outlining a reciprocal relationship of mutual obligations and rights between the governed and the government. When a foreign state invades and occupies any or all of the national territory, it does so without the consent of the governed and thus there is no social contract. The displaced or exiled, but still legitimate, government must continuously inform its people that it exists and is working to restore national sovereignty in order to retain its legitimacy and deny legitimacy to the occupier.

2. In the international context, the legitimacy of an exiled government is formed during its governance of state territory prior to enemy incursion. The closer that government is to meeting international norms of legitimacy, through the traditional method of consensus of principal states—especially if democratically elected—then the more likely it will be to retain the support of other states as well as of its own people.

3. In the case of a nation resisting encroachment by an aggressor state, if it is in compliance with these two domestic and international conditions, then the new proxy ruling state authority emplaced by the occupier will have practically no legitimacy among the population it seeks to govern, with the exception of a potential, but small, portion of the population adhering to the occupier. The occupier will have little success in gaining any recognition of state authority for its newly emplaced regime among the international community or the domestic population. Based on domestic and international support, the legitimacy of the displaced or exiled government of the attacked state continues and does not need to be newly established. To retain this legitimacy as it fights to regain sovereignty, the resistance must avoid brutal tactics, human rights abuses, and criminal activity, while adhering to applicable laws of warfare, the Geneva Conventions and their applicable protocols, and international human rights law (IHRL).[123]

E. **Sovereignty**

As reviewed earlier in this document, sovereignty can be broken down into "juridical" and "empirical." Though it will be contested by the occupier, the original national government can hold technical juridical sovereignty over the occupied territory by preparing the necessary legal framework to transfer governing authority to an exiled governing authority, to continue representing the nation. The occupier will exercise empirical sovereignty, but the resistance can challenge this through the activities of its guerrilla, underground and auxiliary components, the displaced or exiled government, and especially the activities of a shadow government.[124]

F. **Public Component of the Resistance**

1. It is possible that in addition to the resistance components of the underground, guerrillas, and auxiliary, a public component may be able to exist within the occupied area. This public component will likely take the form of a political party and engage in nonviolent resistance to the new authorities. It should remain openly committed to nonviolent resistance in order for the new authorities to allow it to continue to exist.

2. This public component, or political wing of the resistance, can facilitate broad communication within the state to its national constituency, as well as to international audiences. It can also provide the occupation authorities with a negotiating partner that is not ostensibly connected to any violence conducted by the other components of the resistance. If there comes a time when the occupier decides to surrender its ambition to establish its governance over all, or a portion, of the state into which it encroached, then this public component or political party will be the entity with which it will most likely negotiate.

3. As an example, the PIRA originated as a radical splinter group from the official IRA in 1969, after the latter's attempts to end abstention from the political process. The PIRA opposed participation in the political process because it feared legitimizing the recognition of Northern Ireland and the Republic of Ireland as separate political entities. However, after observing the outpouring of domestic and international political support for the cause of the IRA when Bobby Sands died in his hunger strike in 1981, the PIRA realized the value of the political process. Soon afterwards, Gerry Adams became head of Sinn Fein in 1983 and the political process came to dominate the negotiations with ever decreasing violence until 2005, when the PIRA announced the end of its armed struggle and the decommissioning of its weapons.[125]

2.12. Nonviolent Resistance

A. Though the role and methods of nonviolent resistance have sometimes been obscured by religious and philosophical considerations, its tactics are fairly consistent. Typically, the resistance organizes and directs nonviolent resistance techniques and persuades ordinary citizens to carry them out. Nonviolent resistance and mass mobilization have contributed to significant resistance organizations and movements in the past 35 years. The Solidarity movement in Poland in the 1980s was a labor movement that played a decisive role in the eventual unseating of the communist regime. Ukraine's 2004-2005 Orange Revolution was a mass mobilization, with broad civil disobedience, after a fraudulent election that successfully removed a regime, replacing it with a genuinely democratic political process.

B. A resistance can use various activities to undermine the enemy's control over territory and the population. Nonviolent actions include protests, demonstrations, sit-ins, boycotts, occupation of government buildings or other locations, graffiti, symbols, media postings, and ignoring occupier orders. The resistance can also co-opt demonstrations, festivals, and institutions to align cultural identity with the cause of the resistance. These actions disrupt the smooth operation of government and civil society that is under the occupying enemy's leadership or influence. Nonviolent resistance can also provoke the enemy into responding poorly. This response could range from immediate arrests and possible violence to longer term intrusive restrictions on communities or the society at large. An excessive response or over-reaction could alienate any internal supporters, external supporters (including the occupying enemy's domestic population), and undecided individuals. In case of such a response, the resistance must take advantage of it to illuminate the enemy as an unjust and illegitimate occupier, which forms the basis of the national resistance narrative. At the same time that it undermines the occupation regime with this message supporting its narrative, the resistance shadow government can illustrate a stark contrast in efficacy and justness between the character and ethics of the occupier and those of the resistance. These actions prevent the enemy state from establishing legitimacy over the invaded nation, weaken enemy regime official morale (including security forces), occupy the enemy's time and resources, and prevent any formation of popular support for the imposed regime.[126]

C. **Nonviolent Resistance Objectives**

1. Nonviolent resistance implies an unarmed group whose activities capitalize on social norms, customs, and taboos, often provoking action by occupation security forces that will serve to alienate large segments of public opinion from the occupation regime and its agents. If the occupier does not respond to the nonviolent resisters' actions, the resisters will immobilize the processes of public order and safety, and seriously challenge the enemy's ability to govern. Nonviolent resistance rests on the basic thesis that governments and social organizations, even when they possess instruments of physical force, depend on the voluntary assistance and cooperation of a great number of individuals. This method of opposition to a power structure aims to persuade the population to refuse to cooperate with the occupier and its agents.

2. The principal tactic used to induce noncooperation is frequently described as persuasion through suffering, which is intended to withdraw popular support from the occupying regime. A misconception of this tactic is that it is designed to persuade the opponent to have a change of heart through remorse. This misconception makes the error of assuming that there are only two actors involved in this process: the suffering resister and the opponent. However, nonviolent resistance operates within a framework of three actors: the suffering nonviolent resister, the opponent (the occupying enemy and its security forces), and the larger audience (the population). In the context of national resistance to an occupying enemy, the audience is broadly expanded. It includes the populace of the enemy state

in an attempt to reduce internal support for the actions of that state, as well as extending to many populations around the world within nations supportive of the resistance, or which can be persuaded to support the resistance. Every conflict situation is dramatically affected by the extent to which the audience(s) become(s) involved.

3. This contagiousness of conflict inevitably occurs when the audience is included as the third actor. Nonviolent resistance techniques, particularly the function of suffering, provide one insight into the manipulation of the contagiousness of conflict. More than anything else, the objective of nonviolent resistance is to create situations that will involve public opinion and direct it against the occupying power or its installed regime.

4. When the nonviolent resister suffers at the hands of the occupier, it demonstrates his or her integrity, commitment, and courage while showing the injustice, cruelty, and tyranny of the occupier. If the nonviolent resister provokes a response from the occupying forces or regime that can be seen as unjust or unfair, then the charges of tyranny and persecution are confirmed. Nonviolent resistance techniques thrust the initiative and the responsibility for uninvited conflict with unarmed citizens upon an occupier. This is aikido[127] politics, where a population is mobilized by conflict between two groups in which one (here the occupying force) has aggressively or excessively responded to the other, promoting sympathy for the victim(s) and outrage at the aggressive actor. In this context, aikido politics refers to the use of the occupier's momentum against itself by the less well-resourced resistance organization.

5. There are a number of variables that affect the effectiveness of this staged suffering. One is the attitude of the occupier. The success of this method will be determined by whether the occupier cares how a population views it. If the occupier is unconcerned with popular opinion, then nonviolent resistance tactics are less likely to be effective.

6. In addition to it alienating public opinion from the occupier, underground sponsored nonviolent resistance has two other objectives. The first is to lower the morale of the occupying security forces and their government officials. Confronting unarmed and nonviolent activists can undermine the cohesive ethos of most security forces. A second objective is to tie down security forces. By organizing nonviolent resistance events, the underground can successfully divert security forces from other tasks. These coordinated operations between a nonviolent wing and guerrillas or underground can be extremely effective in disorienting security forces.[128]

D. **Nonviolent Resistance Techniques.** Nonviolent resistance can range from small, isolated challenges to specific laws, to complete disregard of occupier authority. The techniques of nonviolent resistance are classified into three general types: attention-getting devices, noncooperation, and civil disobedience. The underlying consideration in nonviolent resistance techniques is whether they serve to legitimize the position of the nonviolent resister while alienating or challenging the authority of the occupying regime.[129]

1. **Attention-Getting Devices.** Nonviolent resistance in its early stages usually takes the form of actions calculated to gain attention, provide propaganda for the cause, or be a nuisance to occupation forces. Attention-getting devices include demonstrations, mass meetings, picketing, and the creation of symbols. Demonstrations and picketing help to advertise the resistance campaign and educate the larger public about the need to maintain resistance against the occupier. Such activities provide communication to both internal and external audiences. For example, the upheaval in Tunisia that began on 17 December 2010 was begun by a 26-year-old fruit vendor who performed an act of self-immolation to protest humiliating tactics of local officials. This eventually led to large protests and clashes with Tunisian security forces, resulting in the deaths of approximately 100 people throughout the country.

As a result, then-Tunisian President Ben Ali renounced his power and departed the country. During the Nazi occupation of Denmark in World War II, King Christian became a symbol of nonviolent resistance. By maintaining his traditional morning ride through the city of Copenhagen without police or aides and keeping his royal standard flying above the palace both day and night, he indicated his presence as the rightful and legitimate sovereign representative of the people, his readiness to negotiate with the Germans, and his leadership of his people.[130]

2. **Noncooperation.** Techniques of noncooperation call for a passive nonviolent resister to perform normal activities in a slightly contrived manner to slow or impede processes, or to exaggerate difficulties, but not to the degree that the resister can be accused of breaking the law. This was also identified previously in this document as passive sabotage. Activities such as slowdowns, boycotts of all kinds, and various forms of disassociation from the occupier all exemplify acts of noncooperation. Anti-Nazi resistance in Europe during World War II included deliberate errors in adjustments of machine tools and precision instruments. Workers in shipping departments addressed shipments to wrong addresses or conveniently forgot to include items in the shipment containers. Feigned sickness was also widespread. These acts of noncooperation impeded the war effort while appearing to be honest mistakes (see appendix D). During the same period, Yugoslavian railroad workers used a particularly effective noncooperation technique: during Allied air raids they deserted their jobs, and, after the raid stayed away for 24 hours or more because of their feigned fear, seriously delaying railway traffic.[131]

3. **Civil Disobedience**

   a. Mass participation in deliberately unlawful acts, typically legal misdemeanors, constitutes civil disobedience. This is the most extreme form of nonviolent resistance because it straddles the boundary between misdemeanors and serious crimes, on the dividing line between nonviolent and violent resistance. Forms of civil disobedience include breaking specific laws, for example breaking tax laws by not paying taxes, breaking traffic laws by disrupting traffic, and breaking laws prohibiting meetings, publications, speech codes, etc., by engaging in those acts. Civil disobedience can also take the form of certain types of strikes and walkouts, resignations en masse, and minor destruction of property.

   b. In Palestine, during the Jewish uprising/resistance of 1920-1948, the population practiced an effective form of resistance though civil disobedience by preventing the capture of the Haganah raiding force by the British security forces. When the police began to search for the Haganah, people vigorously refused the security forces entry to their homes, only stopping short of using arms, often using hand-to-hand fighting, throwing bricks and stones, and injuring many security force members. At the first sign of a British cordon, a gong or siren would sound, and, at this signal, villagers from nearby settlements would rush to the area, flooding it with outsiders and effectively preventing the British from recognizing which outsiders responsible for the raid had taken refuge in the village.

   c. Civil disobedience is a powerful technique, but to be effective it must be exercised by large numbers of people. The calculated risk is that the breach of law justifies arrest by security forces and judicial punishment. However, the more massive the scale of disobedience, the more difficult and less profitable it is for the government to carry out punishments. During the Indian independence movement, Mahatma Gandhi led so many millions in acts of nonviolent civil disobedience that it proved impractical, if not impossible, for the British to jail all offenders. To the British, such widespread disrespect for a law made its enforcement ridiculous or counterproductive. Such actions can be used by a resistance organization to prevent the occupier from enforcing its will in an attempt to cause it to at least partially abdicate its authority.[132]

4. **Cyber Activism.** Cyber (or virtual) activism refers to normal, non-disruptive use of the internet in support of an agenda or cause. This is also referred to as online organizing, electronic advocacy, e-campaigning, and e-activism. Such activity includes web-based research, website design and publication, transmitting electronic publications and other materials through email, and using the web to discuss issues, form communities of interest, and plan and coordinate activities. Hacktivism—the exploitation of computer systems (hacking) for a political purpose—brings methods of civil disobedience into cyberspace. Hacktivist tactics include many evolving techniques at the leading edge of information security such as virtual sit-ins, automated email bombs, web hacks, computer break-ins, and computer viruses and worms. A virtual sit-in is the cyberspace equivalent of a blockade where the objective is to disrupt normal operations, thus calling attention to the perpetrator. Another form of hacking is website defacement, which does not necessarily seek to take information or corrupt a network but rather seeks to replace existing public content with a political message.[133] All these methods are available to a resistance organization if internet connectivity is possible within the occupied area. The drawback is the ability of the actor to be traced and discovered, especially if in occupied territory, and so this activity should only be engaged in by the expert few. This activity can also be engaged in by the displaced or exiled government against the occupier with relative impunity. An exiled government requires the assistance and ascent of a friendly nation from which this activity can be conducted.

2.13. Deterrence as an Aspect of Resistance

Government creation of deterrent factors is a part of resistance planning. Deterrence involves governments creating counter actions to potential adversary courses of action that will cause adversaries to believe that the costs outweigh the benefits of aggression. Defense options coordinated collectively among allies and partners are likely to have a greater deterrent effect than unilateral actions. Strong allied and partner-nation relations support deterrence and can solidify external support for resistance operations. For resistance to function as a deterrent, a potential aggressor must be aware of its existence. Therefore, some measure of information, particularly of the legal and policy frameworks, must be made available in the public domain for the potential aggressor to be aware of such planning. Such public knowledge also lends popular legitimacy.

2.14. Restoration of Sovereignty/ Stability Operations Considerations

A. Post Resistance Stabilization

1. Preparation and planning for post-resistance stabilization and reconstruction must begin at the outset of considering resistance as a national defense option. During its defense planning in peacetime, the government should plan this phase to the same extent that it plans for the resistance campaign. As governing sovereignty is restored to the nation, the resistance organization must cease all operations. Many of its members may desire continued anonymity. Its members, the government, and the people, must understand and demonstrate that the organized resistance remains under the closely held control of the state's governing hierarchy. It remains an integral part of, and supports, the return and restoration of the legitimate national governing authority.

2. Successful post-resistance and stabilization planning will depend on understanding the circumstances that made the nation and state vulnerable to the aggressor, and any social circumstances that created challenges during resistance operations. The government must anticipate that the post-resistance phase may not resemble its prior internal political balance of power, and many internal societal relationships

will not return to the exact same status quo ante. Relationships within the nation regarding certain portions of the population, groups, and individuals that supported the aggressor will need to be addressed. The social fabric itself may have been damaged by the intentional actions of the enemy, as well as by some resistance operations—authorized or otherwise. The government should plan toward a unity of purpose among all resistance participants, and all segments of the population. The government should be aware of and plan for the need to continue actions that maintain its legitimacy internally and with its allies and partners and the international community as a whole.

3. As during pre-crisis resiliency building, the government should focus on maintaining popular allegiance to the state, and not the particular government directing the state apparatus. This also includes assessing the effectiveness of institutions to meet the post-crisis needs of the population through equitable and inclusive delivery of services. The latter is of particular importance, since no matter how effective state institutions were, it may have been the perception of specific domestic minority groups that the government either ignored them or acted against them. This may have initially allowed the enemy to develop a foothold in the country and gain traction within those segments of the population.

4. The judicial system and law enforcement organizations require particular attention. It is during this phase that collaborators will be charged by the state. This should be a legally transparent activity based on written laws that existed prior to the conflict, and facts that emerged during the conflict. It should be processed through the state's judicial system post-conflict. This will ensure and demonstrate the protection of the rights of all citizens, according to applicable law, and retain the legitimacy of the state and the government. Additionally, this clarifies consequences for a small segment of the population that may have supported the aggressor. It will also encourage loyalty to the state by persons related to this group by ethnic, religious or other ties, by ensuring that only persons who actually collaborated with the enemy through specific violations face judicial consequences.

B. **Presence of Adversary Loyalists.** While the enemy may have withdrawn, enemy loyalists could remain in control of some territory. Such loyalists may try to continue undermining the legitimate government by using violence, threats, or other coercion to prevent cooperation with the legitimate government. In such situations, the government may need to use force against this internal adversary. Whether the state applies its military or law enforcement resources will be a decision based on the state's laws, and factors such as: the number of persons involved, degree of violence, the physical location of those persons, the degree of material support from the ousted enemy state, and the maintenance of state legitimacy.

Chapter 3. Interagency Planning And Preparation[134]

> In the long history of humankind (and animal kind too) those who learned to collaborate and improvise most effectively have prevailed. - Charles Darwin

> To be realistic, one must therefore think of a general insurrection within the framework of a war conducted by the regular army, and coordinated in one all-encompassing plan. - Carl Von Clausewitz, *On War*[135]

3.1. Introduction

Resistance planning must involve practically all government agencies or ministries, creating a whole-of-government approach. This strengthens national preparedness and thus resilience. Such preparation and planning also lends itself to strategic messaging against a potential adversary by presenting costly consequences to an adversary's attempt at sovereignty infringement. Resistance planning should also extend to organizations and other entities outside of government to increase resilience by enhancing national resolve and national integrity. First, we will cover some general principles of such planning in the form of: who, what, when, where, and why.

A. **Who.** Planning across the full range of resistance options is an inherent responsibility of every level of government. This provides layered, integrated, and mutually supporting capabilities. This effort is best led by the national government, and particularly the MOD, which is likely best suited due to the topic and intent of the plan. The lead agency also functions as the primary coordinator of the effort, but may or may not be authorized to direct other agencies. Planning by these ministries or agencies fosters unity of effort by providing shared understanding, shared references, common terminology, and shared purpose for crisis response and resistance activities. Though the national government (likely the MOD) leads this effort, commercial, private, and civil society entities outside of government should also be involved in a Total Defense effort. Portions of the plan are thus broadened to involve the whole of society in an aspect of the national defense and potential resistance.

B. **What.** A plan is a continuous, evolving instrument of anticipated actions that maximizes opportunities and guides responses. Since planning is an ongoing process, a plan is an interim product based on available information, present capabilities, and understanding. It is subject to revision based on new information and capabilities. Formation of the plan is a whole-of-government approach collaboratively integrating all facets and levels of government, vertically and horizontally. The plan may also need to include memorandums of agreement or memorandums of understanding regarding shared or overlapping areas of responsibility or necessary cooperation to effect results.

C. **When.** Plans are best described as living documents. Reviews, updates, and revisions should be regularly scheduled and involve the same entities that originated the plan. Outside of regularly scheduled reviews, plans should be updated based on an increased threat level or other significant contextual changes.

D. **Where.** Resistance planning is best hosted by the MOD. It likely has the most secure facilities for such planning and storage of such plans, and it will likely be the lead ministry during a national resistance. Certain extremely sensitive portions, such as designated resistance leadership, may be best kept outside the country, perhaps in a national embassy in a PN. The planned location of a possible government-in-exile or the embassy to one of the most secure PNs are potential options.

E. **Why.** Planning provides three principal benefits: (1) it allows agencies, ministries, departments, and jurisdictions to influence the course of events in a national crisis by determining in advance the actions, policies, and processes that will be followed; (2) it delineates and creates a mutual understanding of integrated and complementary roles and responsibilities; and (3) it contributes to unity of effort by providing a common blueprint for activity in the event of crisis through top-down integration. This collaborative planning is intended to improve coordination, assign and clarify responsibilities, reduce confusion, and increase efficiency and effectiveness. Planning is a foundational element of national preparedness and is an essential national defense activity. Planning for resistance enhances resilience and should be a national priority.

3.2. Whole-of-Government Planning Considerations

- A designated lead or primary agency to lead the planning effort (most likely MOD)
- Integration of all required instruments of national power
- Expertise of government organizations related to the plan's goal
- Participants that forge a common understanding of the operational environment and of a defined problem
- Active lines of communication appropriate to the levels of discussion to begin and foster continuous relationships throughout the planning process for ease of information sharing
- A shared goal with clearly stated objectives to achieve results through comprehensive integration and synchronization at the implementation level
- A common determination of the resources and capabilities to be aligned in order to achieve the planning objectives

3.3. Comprehensive Approach

A. **Unified Action.** Develop unified action by building a strategy to integrate all relevant instruments of national power with other government ministries and organizations, multinational partners, and nongovernmental civilian organizations as necessary. Identify the key objectives to all participants to attain the end state. Multiple organizations need a coherent plan (a plan on how the resistance planning will be accomplished) with timelines and due dates to guide the simultaneous efforts of several ministries and agencies to better understand their roles within the larger planning scheme.

B. **Establish Responsibility.** A common sense of ownership and mutual commitment can be forged among the participants when they understand what must be accomplished to achieve planning objectives and agree on how to accomplish the planning. Each ministry, organization or agency must clearly identify the resources (people and time) that they will devote to the planning effort and must agree to the responsibilities assigned to their organizations.

3.4. Government Levels

In order to create a layered, complementary, and integrated plan, each level of government, as well as the individual citizen, has certain responsibilities. Plans for resistance can differ little from emergency or disaster response plans in the initial stage of resistance. If territorial invasion occurs by an aggressor, then many of the same challenges as a natural or man-made disaster will first befall much of the population.

A. **Individuals and Local Governments.** Resilient communities begin with prepared individuals. Local leadership is provided by local government, private and commercial organizations, businesses and associations, and other nongovernmental groups. Individuals, families, and caregivers to those with special needs should develop household emergency plans to include emergency supply kits. This includes care for pets and service animals. Emergency supply kits should include food and water, first aid kits, batteries for communications devices, and possibly alternate energy sources such as generators for power. Individual citizens can also enroll in first aid courses, or even partake in government sponsored voluntary military training, as offered by some nations. Local police, fire, emergency medical services, public health and medical providers, emergency management, public works, and others in the community are often the first to respond to an incident. The local senior elected or appointed official (the mayor, city manager, or county manager) is responsible for ensuring the public safety and welfare of residents. They organize and integrate their capabilities and resources with neighboring and senior jurisdictions, NGOs, and the private sector. Businesses are vital partners within communities wherever retail locations, service sites, manufacturing facilities, or management offices are located. NGOs and not-for-profit organizations also play a key role in strengthening communities' response efforts, typically in preparation for natural or man-made disasters. Many of those organizations possess knowledge of hard-to-reach populations, conduct training and management of volunteers, and have identified shelter locations and supplies.

B. **Provincial, Mid-Level, or Intermediate-Level Governments.** This level of government has responsibility for coordinating resources and capabilities throughout its jurisdiction and typically obtaining resources and capabilities from other similar level governments. It usually has its own law enforcement personnel that can be focused on assisting specific localities. In some countries these provincial level governments are sovereign entities (e.g., each of the 50 states of the United States), and in other countries they are a subordinate level of the national level government, or they can be mixed. If these are sovereign entities, then greater coordination is required because they cannot necessarily be directed by the national government.

C. **The National or Central Government.** National governments ordinarily maintain a wide array of capabilities and resources that can supplement the above levels of government if not required for a national level crisis. The national government should also maintain working relationships with the private sector and NGOs, as well as international organizations, partner states, and allies. In the case of a foreign adversary infringing on national sovereignty, the national government is the primary defender of the nation, and primary coordinator for the other levels of government, which is why it must lead this planning effort.

3.5. Government Interagency Planning and Responsibilities

As a government-led activity, resistance planning has aspects that are shared among different agencies or ministries. Additionally, as noted above, there will be many similarities between plans covering the early stages of a threat posed by an adversarial foreign power and natural or man-made disasters. Therefore, national plans already in place to deal with such disasters should be consulted as a basis for resistance planning. Below is a sampling of suggested national government ministerial planning activities, divided between the general phases of pre-crisis or deterrence, crisis, and foreign occupation (partial or total). The following ministry titles are attempts to capture the general government divisions of most national governments. The sampling of responsibilities during the first phase pertains to domestic peacetime operations of a government. The sampling of responsibilities during crisis includes the transfer of some ministerial activities outside the nation to support the continuity of the sovereign government. The sampling of responsibilities under enemy occupation are also divided between ministries continuing to operate under occupation, and specific elements that operate externally, having been transferred as part of the exiled government to support the return of national sovereignty. Additionally, but not specified here, certain government ministries and agencies may have specific roles to fulfill upon the entry of allied forces to assist in the restoration of national sovereignty. Further, this planning may involve coordination for support from external allies and partners. These suggested responsibilities are only samples of what might be done and, accounting for administrative, traditional and cultural differences, are only intended to assist with beginning national resistance planning. A sample of responsibilities is provided in a Government Interagency Planning and Preparation chart in appendix J, and further explained in the following sections.

A. **Ministry of Defense**

1. **Pre-Crisis/Deterrence.** Potential overall lead planning ministry. Lead and provide specialized resistance-focused training for select members of SOF, CF, and specially selected persons with anticipated roles and duties in the resistance organization. Host a civilian-led national crisis planning team to organize, oversee, and lead government resistance planning and preparation. Engage in joint training and exercises with partner and allied nations for national defense and potential support to resistance. Purchase specialized communication, transportation and other equipment and supplies. Designate cache sites and store equipment and supplies as necessary, in accordance with governing laws.

2. **Crisis.** Engage enemy to defend national sovereignty. Activate select resistance underground support networks in coordination with national political authority. Stock remaining caches. Distribute specialized equipment. Disperse predesignated military leadership to external, pre-planned locations coordinated with PNs. From those locations, coordinate ongoing internal (territory under occupation) and external activities with allied and PNs and serve and advise the exiled civilian government regarding military matters.

3. **Enemy Occupation.** Resist occupier as planned. Designate small stay-behind military elements to conduct planned activities to combat occupier, maintain civilian morale, and prepare for incoming allied forces. Resistance networks conduct activities against the occupier guided by the political leadership of the exiled government. Conduct sabotage, subversion, intelligence gathering, and sharing of intelligence. Recruit, train, and equip additional underground and guerrilla members as necessary.

B. **MOI/Ministry of Justice (Internal Security and Law Enforcement)**

1. **Pre-Crisis/Deterrence.** Assist the national legislative body in writing laws as part of a national legal framework to support the establishment, development, and material support and supply of a resistance organization, as well as the potential conduct of its activities (see appendix A). Assist the national legislative body in understanding and developing laws to enforce against asymmetric adversarial penetration of societal elements and organizations, and threats to critical infrastructure. Identify, monitor, and disrupt subversive elements within the population that are assisting a hostile foreign power (CI) in gaining leverage within the domestic society.

2. **Crisis.** Conduct raids and arrests of select, known subversive elements—mostly previously monitored—that are assisting the foreign adversarial power. Increase surveillance of groups and individuals suspected of assisting the adversarial power. Employ any authorized emergency powers to counter the adversary's asymmetric warfare activities.

3. **Enemy Occupation.** Gather information on occupier activities and personnel. Members of the resistance underground pass information to be converted into intelligence. Support and mask the activities of local law enforcement agencies clandestinely cooperating with the resistance. Use access to occupier activities to monitor and collect against those activities for use against the occupier within international legal regimes upon the restoration of sovereignty. Maintain law enforcement against traditional criminal activities to protect the populace.

C. Agency for Disaster Response or Civil Emergency

1. **Pre-Crisis/Deterrence.** Educate and inform the general public regarding individual and organizational responsibilities and responses to natural or man-made disasters (e.g., United States Federal Emergency Management Agency or MSB).[136] Assist the general public with the earliest stages of preparation and response due to the early stage similarities between a natural or man-made disaster and the threat of forceful foreign occupation.

2. **Crisis.** Prepare the nation for shortages and execute most disaster protocols. The early stages of a hostile incursion will bring many of the same problems as other types of disasters. Pre-positioned disaster response supplies may need to be used. Disaster response support agreements with other nations may not be honored by those nations due to the context of war and the additional dangers to personnel and equipment.

3. **Enemy Occupation.** Continue functioning as a disaster response agency on behalf of the nation in case of disaster while under occupation. Assist and prepare the nation, as possible, for sustenance and power shortages during combat when allied and partner CF enter to restore sovereignty.

D. Ministry of Foreign Affairs

1. **Pre-Crisis/Deterrence.** Secure agreements with allies and partners to ensure legal recognition and potential placement of exiled government, if necessary, and recognition of resistance networks as operating on behalf of the sovereign exiled government.[137] Foster relationships with the diaspora for purposes of information dissemination, intelligence, funding, strategic communication and recruitment. Engage international organizations and international NGOs to develop and establish support agreements in case of crisis and for the purposes of present and future strategic communication integration.

2. **Crisis.** Coordinate execution of a plan to safely transport predetermined key members of the government to an allied or PN willing to host the government-in-exile, per agreements concluded during pre-crisis activities. Execute strategic communication plan stressing unjustified, hostile foreign invasion of sovereign nation, in coordination with office of the prime minister or president, and ministry of communication. Execute strategic communication plan, coordinated with ministry of communication, to illuminate and internationally publicize the aggressive and illegal actions of the adversary state undermining sovereignty and legitimacy.

3. **Enemy Occupation.** Continue pre-planned strategic communication to positively influence populations and governments of PNs to retain support for restoration of full sovereignty. Promote international sanctions against occupying state with governments and international organizations.

E. Ministry of Communication

1. **Pre-Crisis/Deterrence.** Develop and distribute strategic communication messages, in accordance with themes and the national narrative to domestic and international audiences as allowed by law. Prepare crisis strategic communication. Identify and disrupt hostile communication penetration of

domestic networks. Coordinate government and non-government entities to leverage existing cyberspace capabilities. Develop alternative communication distribution systems and methods in case of critical disruption.[138] Assist in development and possibly acquisition of a secure communication capability for the resistance organization. Devise alternative resistance communication means. Increase resiliency of communication networks.

2. **Crisis.** Distribute national strategic communication messages as per pre-planned and approved themes and narrative. Identify and neutralize adversarial communication capabilities within national borders as allowed by emergency laws. Attack adversary communication capability as directed by national authority. Engage partner and other international media outlets to gain international support.

3. **Enemy Occupation.** Restore and repair cyber and telecommunications infrastructure, especially emergency systems. Distribute pre-planned resistance IO products in coordination and as planned with other relevant agencies. Surreptitiously monitor occupier activities, particularly if possibly criminal. Record and transmit such information, as well as occupier communications, to exiled or partner intelligence agencies for use within strategic communication and possibly international judiciary proceedings.

F. Ministry of Education/Culture

1. **Pre-Crisis/Deterrence.** Oversee the conduct of patriotic education and events to strengthen national unity and resilience. Promote a national culture available to all citizens. Engage with neighboring nations and international organizations to conduct cultural and educational events to strengthen long-term bonds and assist strategic communication. Involve domestic civil society organizations in such activities to strengthen domestic bonds. Educate population on peaceful, passive and other resistance methods.

2. **Crisis.** Communicate message of strong national bonds and national resilience. Promote national cooperation against adversary.

3. **Enemy Occupation.** Continue general education as far as allowed by occupier. Promote surreptitious schooling in cooperation with resistance organization to subvert occupier. Engage in peaceful culturally based resistance activities against the occupier. Assist in maintaining popular morale. Disseminate clandestine information to inform population of occupier activities and motivate resistance to the occupier.

G. Ministry of Transportation

1. **Pre-Crisis/Deterrence.** Formulate plan for priority transport of military essential items in time of national crisis across air, land, and sea domains to support national defense. Coordinate and support development of alternate and clandestine transport networks in support of resistance. Identify critical nodes of transportation infrastructure for purpose of sabotage against enemy use if necessary.

2. **Crisis.** Execute national defense priority transportation plan as required by MOD. Prepare transportation networks for national resistance activities as required by plan. Conduct sabotage as required for national defense and directed by MOD.

3. **Enemy Occupation.** Restore and recover emergency transportation infrastructure. Disrupt enemy use of domestic transportation infrastructure to inhibit enemy control. Support clandestine resistance use of transportation networks domestically and for border infiltration and exfiltration. Assist sabotage of transportation systems as directed.

H. Ministry of Finance

1. **Pre-Crisis/Deterrence.** Develop plan to finance acquisition of materials, supplies, and services in support of development of a resistance organization and national resistance capability. In coordination with domestic, partner, and international banking entities, develop alternate methods of revenue generation, collection, and distribution for clandestine purposes in support of resistance.

2. **Crisis.** Activate contingency finance measures to support national defense and alternate methods of financing to support resistance. Execute diaspora financial support network for contributions to national resistance. Secure key national financial information and data to prevent manipulation by an occupier and to support a return to sovereignty.

3. **Enemy Occupation.** Continue engagement with diaspora communities and international banks and donors to sustain financial support to national resistance and the activities of the resistance organization.

3.6. Preparedness

Once a plan is formulated, it becomes part of a preparedness cycle. The cycle has several activities: plan, organize, equip, train, exercise, and evaluate, and improve. This cycle, or a similar method, is vital to understanding and improving potential resistance and thus to increasing national resilience.

A. **Plan.** Planning makes it possible to manage the life cycle of a potential crisis, determine capability requirements, and help stakeholding ministries, agencies, and other personnel learn their roles. It includes the collection and analysis of intelligence and information, as well as the development of policies, procedures, assistance agreements, and other arrangements. Planning also improves effectiveness by clearly defining required capabilities, shortening response time, and facilitating rapid information exchange. Plans should clearly define leadership roles and responsibilities, and clearly articulate the decisions that need to be made, who will make them, and when. A technique for planning collaboration is the identification and formation of boards, centers, cells, working groups, offices, elements, planning teams, and/or other cross-functional staff organizations. They manage specific processes and accomplish tasks in support of plan development. Persons assigned to these elements should be vested with the authority to make decisions, at their level, for their parent organizations to expedite overall decision-making in the planning effort.[139]

B. **Organize.** Organizing to execute resistance planning includes developing and understanding the resistance organizational structure, identifying leadership, and assembling well-qualified personnel. The resistance organization should provide command and control structures. Knowledge of the structural hierarchy allows others to identify possible counterparts for coordination and collaboration. Governments at all levels should integrate their capabilities to support the development of the resistance organization and its potential activities.

C. **Equip.** A critical component of preparedness is the acquisition of equipment, including the capability to be interoperable with equipment used by certain other government entities and PN forces. Effective preparedness requires the ability to identify, obtain, and deploy equipment, supplies, facilities, and systems in sufficient quantities necessary to perform assigned missions and tasks. The mobilization and maintenance of physical and human resources requires an effective logistics system. Government organizations responsible for providing equipment for organized resistance may store that equipment at their facilities for distribution when necessary or in prepared caches, as allowed by law and determined by plan. They must also routinely service and maintain such equipment and support the resources needed to maintain, repair, and operate the equipment in the field.

D. **Train.** Building nationwide resistance capabilities requires a systematic program to train individuals, cells or teams, and organizations. The purpose is to meet a common baseline of performance and certification standards. Professionalism and experience are the foundation of success. Rigorous, ongoing training is thus imperative. Individuals and cells or teams should meet qualification, or performance standards. Content and methods of training must produce required skills and measurable proficiency.

E. **Exercise.** Exercises provide opportunities to test plans and improve proficiency in a risk-free environment. Exercises assess and validate proficiency levels. They also clarify and familiarize personnel with roles and responsibilities. Well-designed exercises improve interagency coordination, communications, and planning, highlight capability gaps, and identify opportunities for improvement. Once a resistance plan is formed, a resistance organization structured, governmental and ministerial roles and responsibilities delineated, and the necessary equipped acquired, then portions of that plan can be exercised at various levels. Some portions of the plan involving external partner military support can be exercised with that partner in a classified environment. Domestic and partner military units can also exercise capabilities on or near potentially contested ground or at a military training center. The portion of the resistance plan involving individual and organizational disaster response can be exercised as part of a local, regional or national disaster response exercise. Other portions can be exercised among pertinent government agencies in a classified environment. The most sensitive portions of the plan may not need to be subject to a formal exercise format, but may only require certain portions be tested by a few individuals.

F. **Evaluate and Improve.** Evaluation and continual process improvement are cornerstones of effective preparedness. Upon concluding an exercise, performance is evaluated against relevant capability objectives, deficits are identified, and corrective actions instituted. Improvement planning should develop specific recommendations for changes in practice and timelines for implementation. The testing entity should institute a corrective action program to evaluate exercise participation and response, capture lessons learned, and make improvements in response capabilities. An active improvement program will provide a method and define roles and responsibilities for identification, prioritization, assignment, and monitoring of corrective actions arising from exercises and real-world events. Figure 16 illustrates a capacity building cycle to increase preparedness.

The Preparedness Cycle Builds Capabilities

Figure 16. The Preparedness Cycle.

SOURCE: United States Department of Homeland Security, National Response Framework (Washington D.C.: Department of Homeland Security, 2008)

3.7. Criteria for Successful Planning[140]

The following are criteria to measure key aspects of planning:

A. **Acceptability.** A plan is acceptable if it can meet the requirements of anticipated scenarios, can be implemented within the costs and timeframes that senior officials and the public can support, and is consistent with applicable laws.

B. **Adequacy.** A plan is adequate if it complies with applicable planning guidance, planning assumptions are valid and relevant, and the concept of operations identifies and addresses critical tasks specific to the plan's objectives.

C. **Completeness.** A plan is complete if it incorporates major actions, objectives, and tasks to be accomplished. The complete plan addresses the personnel and resources required and sound concepts for how those will be deployed, employed, sustained, and demobilized. It also addresses timelines and criteria for measuring success in achieving objectives, and the desired end state. Completeness of a plan can be greatly enhanced by including all those who could be affected in the planning process.

D. **Consistency and Standardization.** A common frame of reference with common terminology must be established. Then standardized planning processes and products must be established to foster consistency, interoperability, and collaboration.

E. **Feasibility.** A plan is considered feasible if the critical tasks can be accomplished with the resources available internally or through mutual aid. Immediate need for additional resources from other sources, such as PNs, are identified in detail and coordinated in advance. Procedures are in place to integrate and employ resources effectively.

F. **Flexibility.** Flexibility and adaptability are promoted by certain decentralized decision making and by planning for full or partial loss of sovereignty. The capability to activate regional resistance organization elements and specific networks, partial or whole, also lends flexibility.

G. **Interoperability and Collaboration.** A plan is interoperable and collaborative if it identifies other plan holders with similar and complementary plans and objectives. This is especially important when planning with allies and PNs to support a resistance and follow-on military actions for a return of sovereignty. Planning should also support regular collaboration focused on integrating with those other plans to optimize achievement of individual and collective goals and objectives. Each organization should understand the interests, policies, and values of each other's organizations. Achieving consensus on unifying goals is essential to successful planning because of the shared operational environment.

3.8. Authorities and Legal Framework

A. **Authorities.** Resistance plan development may not be among the responsibilities or missions of the agencies or ministries whose participation is required for such planning. Thus, the required staffing and/or funding may also not be immediately allocated or available. Consequently, some of these organizations may require authorization or direction, as well as additional funding allocated from their higher responsible executive or legislative bodies, in order to fulfill their responsibilities to resistance planning.

B. **Legal Framework.** Participating ministries and other organizations should have legal experts available who understand the regulatory and legal frameworks within which each ministry or agency operates. These experts can advise on the bounds of commitments and the drafting of agreements as well as relevant national laws (see appendix A).

3.9. Operations Security

OPSEC is also a critical part of the planning process. The planning process, some participating individuals, agreements, and decisions, as well as the plan itself and each of its components, must be treated as protected, or classified, information. Therefore, not all planners, especially persons without clearances and non-government civilians, will have access to all or even most planning information. Information can be compartmented within subordinate planning groups and isolated. Very few people will know the more sensitive aspects of the resistance plan, particularly the contents dealing with the crisis and occupation phases.

Conclusion

In the last analysis, luck comes only to the well prepared. - Helmuth von Moltke, the Elder

1. Resistance is a form of warfare. It is potentially conducted by the United States' allies and partners against neighboring hegemonic regimes, against whom deterrence has failed, and who possess the ability to seize some, or all, of the sovereign territory of those allies and partners. This ROC is intended to promulgate common terms, perspectives, and understanding to facilitate operational discussions in the planning, preparation, and application of resistance efforts— particularly between the United States and its allies and partners.

2. As a form of warfare conducted directly by the United States' allies and partners, the role of the United States is in support of this effort. Specifically, this form of warfare is resistance to a foreign occupier, and as such, it is distinguished from insurgencies or revolutions. Resistance warfare can be planned and prepared for by allies and partners, supported by the United States, in pre-crisis times of peace. The tactical and operational level plans and details must be closely guarded. However, the willingness of a nation to prepare to resist a foreign occupier, if necessary, through forceful and active, passive, and nonviolent methods can be part of a national deterrence strategy, as it is in the over-arching concept of Total Defense.

3. A nation's military resistance against a foreign occupier can be conducted by its active and reserve military forces (e.g., Home Guard, National Guard, Defense League). These forces can be joined, or augmented, by pre-selected networks of persons with no other relationship to, or association with, the defense or security ministries. This lack of relationship through prior military or government service can make their existence and numbers difficult to discover by an occupier with potential access to personnel records. The general population can also partake in nonviolent and passive methods of resistance.

4. To increase the deterrent value of resistance, the government must publicly promulgate legal and policy frameworks supporting its resistance planning efforts. The potential adversary must be aware, and the nation's general population must be aware and educated. Additionally, the state must have preexisting agreements for material and other support with allies and partners during the period of resistance. For example, among North Atlantic Treaty Organization (NATO) members, this would be an Article 5 circumstance, requiring an allied military response to the infringement on the sovereign territory of a member state. Nations planning resistance can also seek agreements from PNs to use their military forces to assist in reclaiming sovereign national territory.

5. While general popular resistance to an occupier, and more specific and targeted resistance by its military and designated resistance networks, is critical to eventual success, the threatened nation must engage with its allies and partners during pre-crisis to mutually plan and prepare for resistance. The period of resistance is the time necessary for allies and partners to prepare their conventional military forces to remove the enemy from occupied territory. During this time of conventional military preparation, the resistance forces, through their legitimate displaced or exiled government, can transmit intelligence from occupied territory and partake in other actions to help prepare that environment for the entry of CF.

6. The intent of planning for national resistance against an occupier is not only to ensure the return of sovereign national territory. It is also to prevent governance by any entity other than the legitimate government that was displaced by the foreign occupier. Appendix D contains several examples of internal communist competition during World War II where communists, sometimes successfully, prevented the return of the pre-war status quo. They were instead able to use the war to bring about a new government, governing

ideology, and structure. This concept advocates the continuous national focus on returning the government, state, and nation to its pre-war status. This should be the most politically palatable goal for the overwhelming majority of the nation and is integral to its resilience. Stressing and strengthening the aspects of national resilience that allow the overwhelming majority of citizens to support this singular goal are critical aspects of national deterrence.

7. Planning for resistance is at least a whole-of-government effort. For those nations that use the Comprehensive or Total Defense concept, or a variant of it, planning should also include non-government entities from the civilian sector, in addition to critical external allies and partners. This makes understanding and developing an interagency approach to, and planning process for, resistance critical (chapter 3). Such planning supports unity of effort and national cohesion, which contributes to national resilience. The initial national response to an immediate threat of foreign incursion will have many aspects of a national response to natural or man-made disaster. However, due to the circumstances in which resistance may be necessary, the nation's MOD is likely the best coordinating agency for resistance planning.

8. The appendices to this concept are included to enhance understanding, often through historical perspective (e.g., appendices D and E). They also speak about assessing population resiliency during peacetime to find societal weaknesses capable of exploitation by an aggressor. Appendix H begins the conversation regarding the development and application of resistance tactics and operations, occupier actions and responses to resistance, and their effect on the populace.

9. It is hoped that the information contained herein enlightens interested readers and policy makers. This concept strives to provide an understanding of resistance as a historical form of warfare against an occupier. Its planning and preparation strengthens the bonds between allies and partners and serves as reassurance to those potentially threatened by aggressive, hegemonic neighbors.

10. Reinforcing comprehension of resistance efforts, the enclosed selections of terminology, references, case studies, and contributing perspectives form a functional platform from which current and prospective resistance efforts can be formed and matured. This concept seeks to effectively impact the political and military conversations—strategic, operational, and tactical—with resolution and determination. It is an expressed objective that allies and partners with the requirement of situational necessity find these contents illuminative to their own purposeful implementations.↑

Appendix A. Legal Considerations[141]

Nothing distinguishes more clearly conditions in a free country from those in a country under arbitrary government than the observance in the former of the great principles known as the Rule of Law.
- Friedrich August von Hayek

A.1. Introduction

A. This section is intended to highlight the necessity of the establishment of a national legal framework to authorize and facilitate the development of a resistance organization. It is also intended to examine elements of the international legal framework that can bear upon the defending state, its resistance organization, and the individual members of that organization. Each state must consider these aspects, as well as develop its own internal, national legal framework. Such frameworks should be consistent with a state's own laws and law-making, as well as international laws, as it plans for resistance either as an additional method of warfare or as a part of its own Total Defense.

B. The legal parameters of resistance operations need to be made as clear as possible, but are also dependent on the specific context of a particular resistance operation or campaign. It is crucial that the resistance organization and its leadership pay careful attention to legal considerations and respect legal authorities in order to maintain legitimacy with the public, as well as among the international community, and in particular, its partner states and organizations. This appendix addresses legal considerations and personnel status associated with resistance operations, as well as the status of the belligerent occupier. Resistance presents numerous and complex legal issues. At the national strategic level, this means retaining national sovereignty and regaining de facto authority over occupied territory. At the operational and tactical levels, it requires supporting a resistance organization that must adhere to the law of armed conflict (LOAC), also known as international humanitarian law (IHL). Decisions involving resistance will almost always require policy determinations from the highest levels of the sovereign government conducting resistance on behalf of the nation.

A.2. Law and Legitimacy

A. Definitive laws regarding national resistance, including: funding, authority to conduct operations, legal status of personnel, and a host of other issues, must be promulgated prior to the need to employ resistance against an enemy. This is necessary in order to give maximum national legal protection, and thus legitimacy, to persons engaging in resistance and their operations. Resisters must know that they are engaging in legitimate acts, and must understand what actions are acceptable and authorized by their national government, for which they will not be prosecuted.

B. Acting in accord with international human rights norms is essential to legitimacy and success. The national resistance effort must ensure that none of its members commit violations of human rights. This assists in maintaining legitimacy among the population it seeks to reinstate to power and among its international, state and non-state supporters. Violations could jeopardize support in the form of funding, training, equipment, direct participation, planned conventional military support by allies and partners, or other assistance rendered by external partners.

A.3. National Legal Frameworks

A. Organized national resistance to a foreign occupation does not fit neatly into the various status categories originally envisioned in the international treaty law built around The Hague Conventions of 1907 and Geneva Conventions of 1949. Resistance planners and their external supporters must understand these paradigms and also provide an effective national legal framework for operations.

B. The complexity of resistance planning, preparation, and execution (involving various preparatory activities, operating environments, operational modes, and its potentially extended nature) may require several national authoritative instruments. Each will have its own unique constraints, limitations, and processes, wherein lie the legal authorities to approve a resistance campaign's actions and activities.

A.4. International Law: International Humanitarian Law and the Law of Armed Conflict

A. Formally declared wars are no longer the norm, and armed conflicts often lack obviously delimited battlefields. Resistance operations can fall into this category and present unique circumstances that can make the application of existing law difficult. The framework for analyzing conflict activities is an intricate mix of international and domestic law derived from various sources including treaties, custom, legislation, and judicial decisions.

B. Military operations involve complex questions related to international law. International law establishes certain limitations on the conduct of conflict and creates obligations for combatants. International law is generally formed through the observed custom of states, treaties (international agreements under any name or title), general principles of law accepted among the vast majority of nations, the writings of experts and scholars, and international judicial decisions. Customary international law (sometimes simply referred to as "custom") and treaties are the two primary sources of the LOAC. Judicial decisions and writings of leading authors are subsidiary means for determining the law. Customary international law constitutes those rules that derive from general practice accepted as law, as opposed to written rules. It reflects the acceptance of a practice to the degree that states no longer view compliance as discretionary but, rather, as a legal obligation. Customary international law includes the law of war (LOW) before its codification in The Hague and Geneva Conventions. In addition to covering conflict, parts of these conventions also cover belligerent occupation.

1. Among the principal IHL/LOAC international conventions or treaties covering conflict, the most important are listed below. The Hague Regulation, most of the provisions of the Geneva Conventions, and the 1977 Additional Protocols are generally recognized as customary law.
   • The 1907 Hague Conventions (especially Hague Convention IV with Hague Regulations)
   • The four Geneva Conventions of 1949 (GCs I–IV): I–Treatment of Battlefield Casualties; II–Extension of GC I Principles to War at Sea; III–Treatment of Prisoners of War; IV–Treatment of Civilians During Wartime (including the civilian population under occupation)
   • The 1977 Additional Protocols to Geneva Conventions (APs I-II): AP I–Protection of Victims of International Armed Conflicts; AP II–Protection of Victims of Non-International Armed Conflicts
   • The Weapons Conventions (e.g., Convention on Certain Conventional Weapons of 1980)

2. The principal sources of the law of occupation are:
   • The Hague Convention (IV) respecting the Laws and Customs of War on Land and its annex: Regulations concerning the Laws and Customs of War on Land (The Hague Regulations), 18 October 1907

- The Fourth Geneva Convention relative to the Protection of Civilian Persons in Time of War, 12 August 1949 (GC IV)
- The Protocol Additional to the Geneva Conventions of 12 August 1949, and relating to the Protection of Victims of International Armed Conflicts (Additional Protocol I), 8 June 1977
- The customary rules of IHL

C. One complicated aspect of customary international law is that acknowledgement by a state of certain norms as custom is not always clear and is often a matter of policy. Recognizing a norm through state practice will bind a state, if that norm is already customary international law. If the norm is not yet custom, then states can be bound by unilateral declarations showing the intent to be bound. For example, it is now generally accepted that many provisions of the Geneva Conventions are considered to have the force of customary international law, which is significant because such provisions bind states regardless of whether they have become a party to these conventions. For example, the United States has not ratified Additional Protocols I and II of the Geneva Conventions but has stated that it considers many aspects of Protocol I as custom.

D. **The International Committee of the Red Cross (ICRC).** The ICRC began as an independent Swiss organization in the mid-nineteenth century. The ICRC came into being at the initiative of a man named Henry Dunant, who lobbied political leaders to take more action to protect war victims. His two main ideas were for a treaty that would oblige armies to take care of all wounded soldiers and for the creation of national societies that would help the military medical services. In August 1864, delegates from a dozen countries adopted the first Geneva Convention, which put a legal framework around these decisions and made it compulsory for armies to care for all wounded soldiers, whatever side they were on.[142] The work of the ICRC today is primarily based on the Geneva Conventions of 1949, their Additional Protocols, and its Statutes. It is an independent, neutral organization ensuring humanitarian protection and assistance for victims of armed conflict and other situations of violence. It promotes respect for IHL and its implementation in national law. It is a unique organization because unlike NGOs, it is treaty based. That is, its activities are rooted in the Geneva Conventions, which are treaties among states, for which it is the repository, giving the ICRC unique authority among states and in conflicts.

A.5. Terminology: The Relationship between Law of War, Law of Armed Conflict, International Humanitarian Law, and International Human Rights Law

A. Law of War and Law of Armed Conflict

As discussed above, there are several specific terms used to describe the body of law applicable to participants in resistance organizations. These terms refer to separate areas of law that are related, and can be used interchangeably, but for one: the LOW, the LOAC, IHL, and IHRL. Little to no distinction exists between LOW and LOAC, with some scholars arguing that either term is acceptable, mainly because the Geneva Conventions refer to both war and armed conflict.[143] Basically, "the law of war is that part of international law that regulates the resort to armed force; the conduct of hostilities and the protection of war victims in both international and non-international armed conflict; belligerent occupation; and the relationships between belligerent, neutral, and non-belligerent States."[144] Further, "The law of war is often called the law of armed conflict. International humanitarian law is an alternative term for the law of war that may be understood to have the same substantive meaning as the law of war."[145] LOW has its origins in rules of the battlefield that have been around for centuries. LOAC is a fairly new construction and is associated with laws for the battlefield, the main difference being that laws have penalties that can be enforced within a specific jurisdiction, and rules provide a standard for conduct that do not necessarily carry consequences if violated.[146] Recognition of a state of war is no longer a prerequisite for the application of LOAC; it is the existence of armed conflict that is determinative. As noted in the commentary to the 1949 Geneva Conventions:

> The substitution of [armed conflict] for the word "war" was deliberate. It is possible to argue almost endlessly about the legal definition of "war." The expression "armed conflict" makes such arguments less easy. Any difference arising between two States and leading to the intervention of armed forces is an armed conflict. ... It makes no difference how long the conflict lasts, or how much slaughter takes place.[147]

B. International Humanitarian Law

IHL can be defined by the ICRC Commentary on Additional Protocol I (AP I) of 1987 as: "applicable in armed conflict means international rules, established by treaties or custom, which are specifically intended to solve humanitarian problems directly arising from international or non-international armed conflicts and which, for humanitarian reasons, limit the right of Parties to a conflict to use the methods and means of warfare of their choice or protect persons and property that are, or may be, affected by conflict. The expression 'international humanitarian law applicable in armed conflict' is often abbreviated to international humanitarian law or humanitarian law."[148] Its basic principles are: the distinction between civilians and combatants, the prohibition against attacking hors de combat (e.g., prisoners of war [POWs]), the prohibition against inflicting unnecessary suffering, the principle of necessity, and the principle of proportionality.

C. International Humanitarian Law and International Human Rights Law

1. IHL and the IHRL are distinct bodies of law. These terms are not interchangeable. While both are principally aimed at protecting individuals, there are important differences. IHL is applicable in time of armed conflict and occupation. IHRL is applicable to everyone within the jurisdiction of the state concerned in time of peace as well as in time of armed conflict; it applies always, to everyone.

2. IHRL is a system of international norms designed to protect and promote the human rights of all persons. It expresses the obligations of states to act in certain ways or to refrain from certain acts, in order to promote and protect the human rights and fundamental freedoms of individual persons or groups.

3. IHL is intended to limit the violence in conflicts and regulate the treatment of persons affected by armed conflict by striking a balance between humanity and military necessity.

4. Their substance is very similar and both protect individuals in similar ways. The most important substantive difference is that the protection of IHL is largely based on distinctions—in particular between civilians and combatants—unknown in IHRL.

A.6. Principles of Law of Armed Conflict

There are four key principles related to LOAC. They generally apply to what is termed "the means and methods"[149] or how the parties conduct themselves once an armed conflict is underway. The "means" refer to the weapons used to fight. The "methods" refer to the tactics of fighting. Put another way, it is the determination of who or what may be targeted, and how. The four principles central to that analysis are military necessity, distinction, proportionality, and humanity. Analyzing conduct within the context of these principles requires consulting customary law, treaties, case law, rules of engagement, and policy directives, as well as considering concerns of the coalition and HN.[150]

A. **Principle 1. Military Necessity.** Military necessity includes two elements: 1) there is a military require-ment to undertake a certain measure; 2) the measure is not forbidden by LOAC. A commander must articu-late a military requirement, select a measure to achieve it, and ensure that neither the requirement nor the measure to achieve it violates LOAC.

B. **Principle 2. Distinction.** The principle of distinction is sometimes referred to as the grandfather of all principles as it forms the foundation for much of the Geneva tradition of LOAC. The essence of the principle is that military attacks should be directed at belligerents (those who are fighting) and military targets, and not civilians (those who are not fighting or civilian property). Additional Protocol I, Article 48 sets out the rule: "parties to the conflict shall at all times distinguish between the civilian population and combatants and between civilian objects and military objectives and, accordingly, shall direct their operations only against military objectives."

C. **Principle 3. Proportionality.** The test to determine whether an attack is proportionate is found in Addi-tional Protocol I, Article 51(5) (b): if launched, "an attack which may be expected to cause incidental loss of civilian life, injury to civilians, damage to civilian objects, or a combination thereof, which would be excessive in relation to the concrete and direct military advantage anticipated" would violate the principle of propor-tionality. The same standard is echoed in Article 57 of Additional Protocol I: if the target is purely military with no known civilian personnel or property in jeopardy, no proportionality analysis need be conducted.

D. **Principle 4. Humanity.** Sometimes referred to as the principle of unnecessary suffering or the principle of humanity, this principle requires military forces to avoid inflicting gratuitous violence on the enemy. It arose originally from humanitarian concerns over the suffering of wounded soldiers and was codified as a weapons limitation: "It is especially forbidden ... to employ arms, projectiles, or material calculated to cause unnecessary suffering."[151] This principle counterbalances the principle of military necessity.

A.7. The Geneva Conventions

A. The Geneva Conventions of 1949 are four international treaties that, along with the Additional Protocols of 1977 and 2005, serve as the cornerstone of IHL and LOAC. Every state has ratified the Geneva Conven-tions of 1949, and many of these treaty provisions are now considered customary law. The purpose of these treaties is to protect victims and participants of war, including combatants and other belligerents, POWs, the sick, the wounded, the shipwrecked, and civilians.

B. Each of the four treaties that forms the Geneva Conventions of 1949 focuses on protections for a specific category of personnel and applies in any armed conflict (hereinafter referred to as GC I, GC II, GC III, and GC IV). The treaties contain an identical Article 3, which extends a minimum standard of protections of the treaties to armed conflicts not of an international character, or non-international armed conflicts (NIACs) such as insurgencies.[152] Known as "Common Article 3," it has been referred to as "the Convention in min-iature" because it requires that a range of basic humanitarian protections apply to participants and civilians during internal armed conflicts.[153] The Conventions also contain a Common Article 2, which makes their provisions applicable in "all cases of declared war or of any other armed conflict which may arise between two or more High Contracting Parties."[154]

C. The ICRC categorizes civilians as members of organized armed groups when they assume a continuing combat function involving direct participation in hostilities.[155] They are belligerents in the sense that they take up arms and fight. Yet, these groups are not armed forces and, therefore, are not combatants, mean-ing that they do not receive the same privileges that combatants are afforded. Rather, they are unprivileged belligerents. However, their unprivileged status is not unlawfulness under international law. They have no international status. If they are prosecuted for their combat activities (e.g., killing or injuring government armed forces or destroying government property) they are prosecuted under domestic law with no protection or privilege under international law.[156] Acts punishable as violations of international law are those which violate treaty-based IHL and customary IHL such as terrorism or targeting civilians.[157]

D. In 1977, the four Geneva Conventions were supplemented by Additional Protocols I and II. Additional Protocol I protects civilians during international armed conflicts (IACs) by expanding on the provisions of the Geneva Conventions, and extends these protections to national liberation movements, namely internal conflicts against colonial domination, alien occupation, and racist regimes. Additional Protocol II first sets a higher threshold than Common Article 3 for its application, and then expands the protections provided by Common Article 3, such as specific provisions for the protection of children and the protection of "objects indispensable to civilian survival."[158]

A.8. Individual Status on the Battlefield: Two Types; Belligerent or Civilian (See LOAC/IHL)

A. Civilians are a protected group unless or until they directly participate in hostilities, at which point they become belligerents and are subject to attack.

B. Within the belligerent status are categories of individuals who are considered to possess combatant's privilege, meaning they are entitled to take part in hostilities and are, therefore, afforded POW protections if captured. Belligerents, who are not entitled to fight, sometimes called unprivileged belligerents or unlawful combatants, are not recognized under LOAC as having POW protections.

C. Resistance members historically use guerrilla warfare, conspiracy, sabotage, and other methods of resistance against the occupying state. If they abide by certain conditions expressed in international treaty law, in particular The Hague Regulations of 1907 and Geneva Conventions of 1949, they can be afforded status as lawful combatants and thus POWs under international law.

1. According to the Hague Regulations annexed to the Hague Convention (IV) of 1907 respecting the Laws and Customs of War on Land, the status of combatants, and qualifications for POW treatment when captured, shall apply not only to members of state armies, but also to "militia and volunteer corps" if they fulfill the following conditions (article 1):
   - Commanded by a person responsible for his subordinates
   - Have a fixed distinctive emblem recognizable at a distance
   - Carry arms openly
   - Conduct their operations in accordance with the laws and customs of war

2. Geneva Convention (III) of 1949 relative to the Treatment of Prisoners of War (GC III, Article 4 para. A (2)), states that combatant/POW status also extends to "members of organized resistance movements belonging to a Party to the conflict" and "operating in or outside their own territory," even if this territory is occupied, provided that such organized resistance movements fulfill the following conditions:
   - Commanded by a person responsible for his subordinates
   - Have a fixed distinctive sign recognizable at a distance
   - Carry arms openly
   - Conduct operations in accordance with the laws and customs of war

3. Both Article 1 of The Hague Regulations and Article 4 of GC III are considered to be embodied within customary international law. The conditions enumerated in Article 4 para. A (2) are cumulative. In other words, all of them must co-exist for combatant status, and the status of POW, to be legitimately claimed and conferred upon a member of a resistance organization.

D. If members of an organized resistance who carried out hostile acts in occupied territory fulfill the criteria described in GC III, Article 4, Para. A (2), then they are entitled to combatant status. They must be members of an organized force, and must belong to one of the parties to the conflict, for the actions to be deemed legitimate in this context. The party to the conflict must bear responsibility for the activities of the resistance organization. Conversely, resistance members must regard themselves as subordinate to the party to the conflict, and accept its overall policy and command, especially as regards respect for the laws and customs of war. Combatant status brings three legal consequences:

- The privileged combatant is allowed to conduct hostilities and as such cannot be prosecuted for bearing arms or attacking enemy targets, unless the conduct amounts to a war crime
- The combatant is a legitimate target to the opposing forces
- In the event of capture, the combatant is afforded POW status

A.9. Members of Organized Resistance, Militias and Volunteer Corps

A. GC III, Article 4 (A), affords combatant and POW status to "members of other militias and members of other volunteer corps, including those of organized resistance movements." To achieve POW status, the members of these groups must meet the following criteria stated above:[159]

- Commanded by a person responsible for his subordinates
- Have a fixed distinctive sign recognizable at a distance
- Carry arms openly
- Conduct operations in accordance with the laws and customs of war

B. The first requirement exists to exclude from the definition individuals acting on their own. The second is required to distinguish the partisan or guerrilla from the civilian population. The third criterion means that arms must not be concealed when parties are visible to the adversary or before launching an attack by not "creating the false impression that he is a civilian ... and carry his arms openly in a reasonable way."[160] The last one exists so that combatants cannot rely on LOAC only to benefit from its protections. Conduct in accordance with LOAC is a key distinction between lawful and unlawful combatants.

1. **First Requirement: Commanded by a Person Responsible for His Subordinates**

   a. The resistance must be organized into units with a responsible command. The armed group is organized when its military engagement is of a collective nature, within the structure of the organization, under the movement's command and control, and in accordance with applicable rules. It cannot be an action undertaken by isolated individuals, without adequate preparation and training.

   b. The members must form a group having all the characteristics of a military organization, especially a system of internal discipline, hierarchy, responsibility, and honor.

      01. Internal discipline means that the rules of the organization and the orders must be strictly followed. According to AP I (article 86), the system of internal discipline must also ensure that provisions of humanitarian law are respected.

02. Hierarchy in the resistance does not necessarily mean structure as in regular military units. Resistance commanders can be both civilian and military, but their command must be recognized as military. In particular, the commander is responsible for the activities, including violations, carried out by subordinates.

03. Responsible command is military discipline. The organization must be able to prosecute and sentence members who breach disciplinary and legal rules.

2. **Second Requirement: Have a Fixed Distinctive Sign Recognizable at a Distance**
Military units of a resistance organization must have recognizable signs in the form of arm-band, hats, coat, shirt, an emblem or a colored sign worn on the chest or specific symbols displayed on items of clothing. It is not necessary to wear these signs or emblems constantly, but such signs or emblems must be visible to the targeted personnel shortly before the start of the military action and recognizable by a person at a distance not too great to permit a uniform to be recognized. The sign must be distinctive, the same for all organization members and used only by that organization.[161] If the resistance fighters are on board a vehicle, tank, plane or boat, the distinctive sign must be shown on the vehicle concerned (*Military Prosecutor v. Omar Mahmud Kassem and others*).[162]

3. **Third Requirement: Carry Arms Openly**
According to the HR of 1907 and GC III, carrying arms openly is another sine qua non condition for acquiring combatant status by members of the resistance organization. Carrying a weapon openly must be distinguished from carrying weapons in a visible or ostentatious manner. This requirement does not mean that small arms (e.g., handguns) cannot be hidden in a pocket, or in a coat, and must be held in a visible place. However, the enemy must be able to recognize an opponent preparing for an attack, regardless of the weapon he uses. In particular, it is forbidden to step into military positions as a civilian, under a false pretext and in a perfidious way, in order to gain advantage and attack the adversary (*Trial of Wilhelm List and others*).[163]

4. **Fourth Requirement: Conduct Operations in Accordance with the Laws and Customs of War**

a. The right to use force by combatants (combatant's privilege) is not unrestricted. According to article 22 of the HR, "the right of belligerents to adopt means of injuring the enemy is not unlimited." Additionally, article 35 para. 1 of AP I states that "in any armed conflict, the right of the Parties to the conflict to choose methods or means of warfare is not unlimited." Combatants must respect the laws and customs of war. If they fail to do so, combatants breach IHL and are personally responsible for their illegal acts. Furthermore, the military commander who ordered them to commit the act forbidden by IHL also bears responsibility for his order.

b. If they observe these rules, they cannot be treated by the adversary as bandits, rebels, criminals, etc. Legally, the occupying power has no excuse to refuse them the legal status of combatants. If they widely and systematically violate these rules, the occupying power may not recognize the organization as a party to the conflict, punishing them as violators of public order.

c. The occupying power cannot refuse the status of combatant to all resistance members in case of individual and sporadic violation of IHL by a single member, particularly if the violator has been punished by the commander, or by relevant organs of the organization. On the other hand, widespread disregard for the rules of IHL will affect the status of the entire resistance organization.

C. During a resistance to foreign occupation, two historically common acts by resistance members are sabotage and espionage.

1. To determine whether sabotage is legitimate as an act of warfare, it is necessary to analyze three factors: 1) the category of persons committing it (only combatants, including members of the resistance organization, may lawfully commit sabotage); 2) the target of sabotage (only military objectives may be targeted); and, 3) the means and methods used (legal unless forbidden by IHL).

2. Espionage against the occupier is a more difficult issue, and it is difficult to determine the parameters of allowable behavior. The Geneva Convention attempts to define whether the actor is protected, once espionage is determined. A member of the armed forces of a party to the conflict, resident in occupied territory, who gathers or attempts to gather information of military value within that territory, shall not be considered as engaging in espionage unless done through an act of false pretenses or in a deliberately clandestine manner. Such a resident does not lose his right to the status of POW and may not be treated as a spy, unless he is captured while engaging in espionage.[164] Further, a member of the armed forces of a party, who is not a resident of occupied territory and who has engaged in espionage in that territory, shall not lose his right to the status of POW and may not be treated as a spy, unless he is captured before he has rejoined the armed forces to which he belongs.[165]

A.10. Individual Status and Direct Participation in Hostilities

A. Significance of Individual Status

1. The terminology used in AP I (article 51), to the GC suggests that an individual person can have one of three statuses:
   - A member of the armed forces;
   - A participant in a *levée en masse* (mass uprising or mass mobilization); or
   - A civilian.

2. A member of a nation's armed forces is defined by AP I (article 43(1)), which provides: "The armed forces of a Party to a conflict consist of all organized armed forces, groups and units which are under a command responsible to that Party for the conduct of its subordinates, even if that Party is represented by a government or an authority not recognized by an adverse Party. Such armed forces shall be subject to an internal disciplinary system which, inter alia, shall enforce compliance with the rules of international law applicable in armed conflict." Significantly, especially for states using the Total Defense concept, AP I (article 43(3)) states that "Whenever a Party to a conflict incorporates a paramilitary or armed law enforcement agency into its armed forces it shall so notify the other Parties to the conflict."

3. *A levée en masse* is a special label that only regards action taken by inhabitants of an unoccupied area. That is, when facing foreign invasion, and in an attempt to prevent occupation, the civilian population (not having an opportunity to organize) is allowed to spontaneously take up arms against the invader. During this short time, only two of the four Hague conditions need be satisfied; carrying arms openly (condition 3) and respect for international law (condition 4). During this period, there is no need for armed citizens to be subordinate to a responsible commander (condition 1) or display a fixed, distinctive emblem (condition 2). A levée en masse ends, or lapses, after a relatively short time. If the territory becomes occupied, which is a question of fact and not law, then inhabitants who continue to forcibly resist the occupation can be regarded as saboteurs or as civilians directly participating in hostilities. Whereas inhabitants who participated in the levée en masse, but then laid down their arms after occupation, may return to normal civilian life.[166] [167] [168]

4. A civilian is always negatively defined as someone not falling into another category. A civilian is any person who does not belong to one of the categories of persons referred to in Article 4, paragraph A (1), (2), (3) and (6) of GC III and Article 43 of AP I. Under the law governing the conduct of hostilities (especially as contained in Article 48 et seq. of AP I), and under customary international law, civilians are entitled to general protection against the dangers arising from military operations; in particular they may not be made the object of an attack.[169] Under IHL, the civilian population and individual civilians enjoy a general protection from the effects of hostilities.

5. AP I addresses the issue of increased civilian participation in hostilities with its Article 51, which protects civilians "against dangers arising from military operations," including direct attacks.[170] This is "unless and for such time as they take a direct part in hostilities."[171] Neither the protocols nor the Conventions define direct participation in hostilities. However, whether a civilian is directly participating in hostilities is a critical determination, because doing so suspends his or her immunity, thus allowing him or her to be made the object of attack.

B. ICRC Criteria for Determining Direct Participation

1. The analysis for determining direct participation in hostilities can be broken down into two elements: 1) what is meant by "hostilities;" and, 2) what is meant by "direct participation." The first element refers to collective activities that constitute the means of injuring the enemy, and the second element refers to the individual involvement of a person. In the course of armed conflict, not all conduct constitutes hostilities. Likewise, not all actions of an individual have the quality or degree of involvement that is required to meet the threshold for direct participation.

2. The ICRC identifies three cumulative elements for a specific civilian act to constitute direct participation in hostilities:
   - Element one. "The act must be likely to adversely affect the military operations or military capacity of a party to an armed conflict or, alternatively, to inflict death, injury, or destruction on persons or objects protected against direct attack (threshold of harm);
   - Element two. There must be a direct causal link between the act and the harm likely to result either from that act or from a coordinated military operation of which that act constitutes an integral part (direct causation); and
   - Element three. The act must be specifically designed to directly cause the required threshold of harm in support of a party to the conflict and to the detriment of another (belligerent nexus)."[172]

3. For element one, threshold of harm, only an objective likelihood that the harm will occur is required. It need not actually materialize. Military harm can be any consequence adversely affecting military operations, such as sabotage, armed or unarmed activities, disturbing deployments, logistics, communications, clearing mines placed by the enemy, or wiretapping the adversary's high command. Harm is not limited to killing or wounding personnel or causing physical damage to military objects. The second clause of this element allows violent or deadly acts against nonmilitary targets to also constitute direct participation in hostilities. These include attacks against civilians, such as sniper attacks or the bombardment of residential areas.[173]

4. For element two, direct causation, the acts can go beyond the actual conduct of armed hostilities and include acts that contribute in a direct way to the defeat of an adversary. An example of this type of activity would be a civilian driving a military ammunition truck to operationally engaged fighters. However, the use of the term direct implies there can be indirect participation. Acts that may "even be indispensable to the adversary such as providing finances, food, and shelter ... and producing weapons

and ammunition" are considered conflict-sustaining activities that are not designed to directly bring about the required harm. The distinction between producing the ammunition and driving the ammunition to operationally engaged fighters is ostensibly that delivering weapons to those engaged in combat "is carried out as an integral part of a specific military operation designed to directly cause the required threshold of harm."[174] An opposing example may be the person serving as a lookout on an ambush. The contribution may not be indispensable to the causation of the harm once the ambush occurs, but he or she would almost certainly be considered to be directly participating in hostilities. Simply producing ammunition does not meet the causal link requirement, because it only maintains or increases the capacity of a party to harm an adversary. Mere participation in the war effort does not rise to direct causation.

5. The third element, belligerent nexus, refers to the objective purpose of the act carried out by the civilian. The focus is not on the subjective intent of the individual, but is "expressed in the design of the act or operation." As there is no focus on the mental capacity or willingness of the individual, even civilians forced to directly participate in hostilities, or children below the recruitment age, could be deemed combatants. Only in circumstances in which civilians are completely unaware of their roles or completely deprived of their physical freedom of action will they remain protected. Examples include an individual who is forced to act as a human shield or a driver who unknowingly transports a remote-controlled bomb.[175]

A.11. Unlawful Combatants

A. The term *unlawful combatant* does not appear in the conventions or protocols but has been frequently used by the United States as an individual status. The main characteristic of an unlawful combatant is that he or she is not entitled to POW status. The status is not universally recognized, and some argue that it is a subcategory of civilian, where the civilian takes up arms without LOAC protection. The expression unlawful combatant is thus understood as describing all persons taking part directly in hostilities without being entitled to do so and who therefore cannot be classified as prisoners of war on falling into the power of the enemy (Ex Parte Quirin et al.).[176] It would include, for instance, civilians taking direct part in hostilities or members of militias and of other volunteer groups, who are not integrated in the regular armed forces (but who belong to a party to the conflict) and who do not comply with the conditions of Article 4, paragraph A (2) of GC III.

B. The use of this status has been criticized. One scholar notes that unlawful combatants "are subject to the burdens of combatancy (they can be killed), but they have no reciprocal rights ... it is the worst of all possible worlds."[177] Unlawful combatants do not acknowledge the requirement to distinguish themselves from civilians, which could constitute an act of perfidy. Perfidy is prohibited in both IACs and NIACs and generally refers to any acts used to gain an enemy's confidence while intending to injure, kill, or capture him.[178] The act of feigning civilian status is one example. Members of the Viet Cong were considered unlawful combatants because they fought at night and purported to be civilian farmers by day. Combatants cannot wear two hats, and if they do, they lose the protections that they would be afforded under either. They are not civilians and are not lawful combatants. However, LOAC still applies to unlawful combatants in that they are entitled to the basic humanitarian protections of Common Article 3 upon capture.

A.12. Belligerent Occupation and Status under International Treaty Law

A. Article 42 of the HR states: "Territory is considered occupied when it is actually placed under the authority of the hostile army. The occupation extends only to the territory where such authority has been established and can be exercised." Article 42 lists two conditions required to apply the law of belligerent occupation:

1. Control of the territory of the occupied state by foreign armed forces;

2. The possibility of exercising authority over the population of this territory.

    a. Thus, occupation is not an act of law. It does not arise as a result of the occupant's announcement. It is a matter of fact. During the belligerent occupation, the actual authority lies in the hands of the occupant. This fact, however, does not affect the legal continuity of the occupied state.

    b. Legal sovereignty remains where it was before the territory was occupied, although the lawful government is unable to exercise its authority in the occupied territory. Even the lack of legitimate state or local authorities, in case of long-term occupation, does not result in the loss of sovereignty by the occupied state.

    c. The occupant is responsible for the occupied territory until the government of the state under occupation is able to return to power and fulfill its obligations. Actual authority over the occupied territory is, therefore, temporary, and is limited by the provisions of the law of occupation. Therefore, the occupant cannot decide unilaterally about the future of the occupied territory or change the rules of its domestic legal system.

B. Article 2 (2) of the GC IV states that: "The Convention shall also apply to all cases of partial or total occupation of the territory of a High Contracting Party, even if the said occupation meets with no armed resistance."

C. Belligerent occupation exists regardless of the question of the legality of the use of armed force by the occupier to seize the territory, and regardless of the state of war and the proclamation of the occupation.[179]

D. An effective belligerent occupation begins when organized resistance is overcome in a given area, and the occupying army is able to maintain their authority in any part of the occupied territory in a reasonable time. This does not mean that the occupying forces must be stationed in every part of the occupied territory. It is sufficient if they can react effectively, at a proper time and place, when the situation so requires. Efficacy is a constitutive element of the belligerent occupation. The occupation cannot be symbolic. It exists only where the occupant is capable of effective administrative authority, using armed forces and civilian institutions.

E. According to IHL, the occupation lasts as long as foreign troops exercise real authority over a given (occupied) territory. Occupation is a situation that is temporary, the consequence of which is the rule that the occupant should only introduce necessary changes, so as to maintain the status of the occupied territory that existed there prior to the beginning of the occupation (status quo ante).

F. An occupied territory ceases to be occupied if it is released in the course of fighting, and also if it transforms into a battlefield. The occupier cannot appoint puppet governments or proclaim a new state in the occupied territory. These actions violate of the law of belligerent occupation.

A.13. Belligerent Occupation, Hostilities and Law Enforcement

A. Article 43 of the Hague Regulations of 1907 states that: "The authority of the legitimate power having in fact passed into the hands of the occupant, the latter shall take all the measures in his power to restore, and ensure, as far as possible, public order and safety, while respecting, unless absolutely prevented, the laws in force in the country."

B. The occupied territory is still under the sovereignty of the legal state, however, it is the occupying power which exercises factual authority; thus, there is no legal basis to question the occupant's right to combat the armed resistance organization by military or military-police means—it is not a war crime, the occupant has the right to do so by virtue of the law of occupation.

C. The organized armed resistance, on the other hand, has all legal grounds for continuing the armed struggle under the conditions of invasion and occupation—it is the undeniable right of every conquered and occupied nation.

D. Neither the Hague Regulations of 1907 nor the Fourth Geneva Convention of 1949 regulate how military operations should be conducted in occupied territory. However, the general rules regarding the preparation and conduct of military operations are applicable (e.g., the principles of distinction and proportionality, the obligation to take precautionary measures, certain restrictions on means and methods of warfare, respect for persons and objects subject to special protection).

E. The responsibility for maintaining security and public order in the territory under occupation requires adequate and effective response to cases of riots, internal tensions and unrest, and prevention and prosecution of criminal offences (not related to acts hostile to the occupying power); apart from military actions, the occupant is also obliged to carry out police tasks.

F. The international legal instruments are silent on separating military and police functions in occupied territory. However, in practice, states distinguish two types of use of force by the occupying power: 1) The maintenance of security and public order, fulfilling the obligation under Art. 43 of the Hague Regulations, where the occupant, as the factual authority, is responsible for the normal functioning of territory under occupation, which allows the occupant to conduct law enforcement operations; 2) The conduct of hostilities in the occupied territory. This is based on the assumption that organized armed forces of the occupied state may still conduct an armed struggle against the occupant or there is a resistance organization which undertakes such a struggle during occupation.

A.14. Occupation and Foreign Allied Forces

A. Although both GCs and API I do not contain any precise definition of the term "international armed conflict," such a conflict exists where force is directed by one state against another, irrespective of duration or intensity.

B. During an occupation, foreign armed forces may become parties to the conflict by intervening with their own troops, by having other participants act on their behalf, or by rendering direct support to resistance military operations. If members of foreign armed forces actively support a national resistance effort, they could be considered hostile enemy forces by the occupying power. Depending on their activities, they may or may not be treated as POWs in case of capture. If foreign military personnel conduct covert operations or behave clandestinely, in case of capture, they may be accused of being spies or unlawful combatants by the occupying power.

C. Usually, anyone present in a foreign nation's territory is subject to its jurisdiction. Jurisdiction is the legal power a sovereign nation possesses to make and enforce laws without foreign direction or control. Foreign, state-sponsored military or other personnel actively supporting a resistance effort could be considered hostile enemy forces by the occupier, and may or may not be accorded POW protections, depending on their actions. Such personnel, operating covertly or clandestinely, are at risk of arrest and prosecution as spies by the occupying authority.

A.15. The Lawful (exiled) Government and Belonging to a Party to the Conflict as a Resistance

A. Resistance military units fighting against the occupier must act on behalf of a party to the conflict. There must be a link between the legal government of the occupied territory and the resistance.

B. After the adoption of the GCs of 1949, authorization may be granted by the government's tacit agreement, without written directives. This is especially the case when operations carried out by the resistance organization meet the goals of the government-in-exile and are directed against the forces of the occupier. Thus, it is possible to determine which party to the conflict is supported by the organization. The relationship between the legal government and the armed resistance would be at least a de facto relationship (*Military Prosecutor v. Omar Mahmud Kassem and others*). [180]

C. ICRC Commentary to AP I, para. 1661–"[R]esistance movements representing a pre-existing subject of international law may be 'Parties to the conflict' within the meaning of the Conventions and the Protocol. However, the authority which represents them must have certain characteristics of a government, at least in relation to its armed forces," (*Prosecutor v. Dusko Tadić*).[181]

A.16. Conclusion

A. Alliances are treaty-based organizations whereby signatories agree to take certain actions on behalf of each other. In most cases, they agree to come to each other's defense if attacked. When most of today's military alliances were formed, the methods of attack were clearer and more objective, concerning uniformed soldiers and hardware such as tanks, crossing borders, and aircraft dropping munitions. In such instances, the concept of being "attacked" was clear.

B. For example, NATO's Article 5 states that "The Parties agree that an armed attack against one or more of them in Europe or North America shall be considered an attack against them all." However, there have been recent instances of asymmetric tactics in hybrid warfare campaigns, conducted by Russia in particular within its bordering states, that may not have clearly and immediately risen to the concept of "armed attack." This is an important legal point because it means that it is legally arguable as to when or at what point an "armed attack" took place in several recent conflict areas. It also provides warning regarding potential tactics to be used against sovereign nations while attempting to remain below the threshold of incurring a major, immediate military response from external partners.

C. International customary and codified law, as well as written security treaties, may be legally insufficient to deal with situations such as described in the last paragraph, many of which were attacks using asymmetric methods as components of hybrid attacks (see appendix C). This area of legal insufficiency must be addressed by nations seeking to defend themselves against non-conventional attacks by aggressor states.

D. Additionally, as a nation prepares its plans for resistance in case of an attack, it must account for the legal status, actions and legitimacy of its resistance members within the framework of its laws. Identifying and accounting for these potential legal gaps is a necessary part of a nation's development of its national legal framework to establish, develop, and if required, authorize the conduct of resistance against an occupier.

Appendix B. Methods Of Nonviolent Resistance

> Nonviolent action is possible, and is capable of wielding great power even against ruthless rulers and military regimes, because it attacks the most vulnerable characteristic of all hierarchical institutions and governments: dependence on the governed. - Gene Sharp, "The Role of Power in Nonviolent Struggle" 1990

B.1. Introduction

A. In 2000, a Serbian student group called *Otpor!* (Resistance!) mobilized to overthrow the Balkan country's dictator, Slobodan Milosevic. The group organized around three demands: free and fair elections, academic freedom, and an unfettered news media.

B. Despite immense challenges, the youth-led group was successful, partly because it united and mobilized diverse civic associations across the country. The Serbian group appealed to large segments of the Serbian population because it maintained a strict adherence to nonviolent methods. For *Otpor!*, even if an adversary threatened, intimidated, or physically harmed its members, any form of violence, revenge, or retaliation was strictly prohibited.

C. In order to increase its power, a movement must build a cohesive coalition of organizations, groups, and institutions that are unified around common objectives. As *Otpor!* increasingly attracted support, generated sympathy, and brought in new members from among its fellow citizens, its power and leverage against Milosevic increased. The more members, resources, and external support a movement can attract, the stronger its capacity. *Otpor!* also drew support from international groups and a number of countries, including the United States and several countries in Europe, that contributed various forms of assistance to the movement.

D. Nonviolent civil action requires targeting and skill. Like so many other movements around the world, *Otpor!* could never have achieved its goals without a concerted, proactive, and determined effort. In the armed services, commanders attempt to outmaneuver and outflank their opponents, while simultaneously anticipating their responses. Nonviolent movements must also carefully plan their actions to achieve their objectives with limited resources and under conditions of uncertainty. In order to secure free and fair elections in Serbia, *Otpor!* placed monitoring teams in all voting precincts, coordinated efforts to expose electoral fraud, and mobilized mass protests in the aftermath of fraudulent elections.[182]

E. In the early 1980s, thousands of Soviet troops occupied Poland. The Poles lived under severe fear and repression by the communist government. Expressions of dissent were harshly put down by the regime. In August 1980, workers at the Lenin Shipyard in Gdansk initiated a strike in protest against a recent rise in food prices, among other grievances. Roughly 17,000 factory workers stopped work and many remained inside the factory walls in defiance of the ruling Communist Party. The group issued a list of demands that included the right to form an independent self-governing trade union, which they would name *Solidarnosc*, or Solidarity. As news of the strike spread across Poland, the movement swelled as sympathetic factory workers joined the cause. The shipyard workers, led by Lech Walesa, adhered to a strict code of nonviolent discipline. Polish strikers had learned the futility of violent confrontations against the iron-fisted communist government a decade prior when six people died in strikes. This time, within days, Solidarity successfully shut down entire industries across Poland and forced the Communist Party to the negotiating table. Government officials finally agreed to most of the workers' demands, including the establishment of a free trade union, a first in communist Europe. From a movement that started with a group of disgruntled workers in one factory, Solidarity grew to approximately ten million members in a little over a year. In the years that followed, the government reneged on parts of the labor agreement but Solidarity's power and influence paved the way for the first ever free elections in the communist bloc, in June 1989.[183]

F. Nonviolent movements have the potential to topple the most ruthless dictator. However, those in power do not typically concede their power voluntarily. Rather, they must be made to feel heightened political and economic costs to maintaining the status quo. While strategic nonviolent action has the potential to empower previously disenfranchised, disempowered, and marginalized people, there is never a guarantee that a nonviolent movement will succeed. Highly authoritarian regimes that are closed off from the global economy pose a particular challenge to would-be nonviolent resisters. In such environments, the risks and consequences for activists are high. Countless people have been arrested, tortured, and killed struggling nonviolently.

G. Burma (also known as Myanmar), under the control of a ruthless military junta, suppressed the basic freedoms of the Burmese for four decades. A highly coordinated pro-democracy movement surged both internally within Burma's borders and externally through diaspora populations and sympathizers. Burmese monks orchestrated a series of protests, demonstrations, and marches such as the 2007 Saffron Revolution. However, the junta's iron fist squashed dissident efforts and tortured and killed numerous leaders, despite international outcry. Pro-democracy leader Aung San Suu Kyi was under house arrest for over 20 years. The Burmese nonviolent struggle continued, with little success, for an extended time.[184]

B.2. A Compendium of Nonviolent Methods

A. Among the most prominent theorists of nonviolent methods of protest and revolution was Dr. Gene Sharp. Dr. Sharp compiled a list of 198 nonviolent tactics (listed in table 2). Sharp's key theme was that political power is not derived from the intrinsic qualities of those in positions of authority. Rather, the power of states is derived from the consent of the governed. The governed thus possess the moral and political authority to refuse their consent. Essentially, leaders lack power without the consent of the governed, and since most states retain a monopoly on the use of force, nonviolent resistance methods are ideal as a means for the populace to impose its will on the state.[185] Such methods can be effective when government forces may not know how to cope with nonviolent resistance. Dr. Sharp theorized that police and soldiers are trained to meet force with force but are much less psychologically prepared to fight nonviolent resistance.[186]

B. Nonviolent resistance can range from small, isolated challenges to specific laws to complete disregard for governmental authority. In *The Politics of Nonviolent Action, Part Two: The Methods of Nonviolent Action*, Gene Sharp identifies 198 methods of nonviolent action.[187] He classifies the methods into the following three categories:

1. **Protest and Persuasion:** formal statements, blogging, group presentations, distributing leaflets, wearing symbols, drama and music, joining Facebook protest groups, processions, honoring the dead, and public assemblies.

2. **Noncooperation:** social or economic boycotts, text messaging of information that is banned or censored, labor strikes, boycotting rigged elections, and refusing to recognize the legitimacy of a regime.

3. **Intervention:** hunger strikes, sit-downs in streets, live video streaming of an opponent's abuse or fraud from cell phones to internet sites, occupation of offices, seeking imprisonment, and overloading administrative facilities.

C. When conducted skillfully, a unified nonviolent movement of ordinary people can wield tremendous power, challenging even formidable military opponents or at least bringing them to the negotiating table. Yet a tremendous amount of planning, strategizing, and management is essential in order to integrate individual isolated acts of protest or defiance into a concerted, coordinated movement. Gene Sharp's 198 methods of nonviolent action are listed in table 2.

D. Very detailed and practical methods of nonviolent and passive resistance for the general population were also laid out in 1965 by Swiss Army Major H. Von Dach. He described behaviors such as; disturbing the sleep of enemy soldiers, damaging their clothing during laundering, selling them over-priced and damaged goods, and damaging fruits and vegetables during packing into grocery bags. Von Dach even broke down the population by employment and offered specific suggestions for passive resistance against occupiers. He explained how overt use of excessive drugs and medical equipment by medical personnel could conceal the extra drugs and equipment to be set aside and given to resistance members. He suggested that local policemen hide some of their uniforms and give them to active resistance members for later use as part of a deception plan during operations, to include raids and information gathering.[188]

The Methods of Nonviolent Protest and Persuasion	
Formal Statements	25. Displays of portraits
1. Public speeches	26. Paint as protest
2. Letters of opposition or support	27. New signs and names
3. Declarations by organizations and institutions	28. Symbolic sounds
4. Signed public statements	29. Symbolic reclamations
5. Declarations of indictment and intention	30. Rude gestures
6. Group or mass petitions	**Pressure on Individuals**
Communications with a Wider Audience	31. "Haunting" officials
7. Slogans, caricatures, and symbols	32. Taunting officials
8. Banners, posters, and is displayed communications	33. Fraternization
9. Leaflets, pamphlets, and books	34. Vigils
10. Newspapers and journals	
11. Records, radio, and television	**Drama and Music**
12. Skywriting and earthwriting	35. Humourous skits and pranks
Group Representations	36. Performances of plays and music
13. Deputations	37. Singing
14. Mock awards	**Processions**
15. Group lobbying	38. Marches
16. Picketing	39. Parades
17. Mock elections	40. Parades
Symbolic Public Acts	41. Pilgrimages
18. Displays of flags and symbolic colors	42. Motorcades
19. Wearing of symbols	**Honoring the Dead**
20. Prayer and worship	43. Political mourning
21. Delivering symbolic objects	44. Mock funerals
22. Protestant disrobings	45. Demonstrative funerals
23. Destruction of own property	46. Homage at burial places
24. Symbolic lights	

Table 2. Gene Sharp's 198 Nonviolent Actions (1-46).

SOURCE: Gene Sharp, Waging Nonviolent Struggle; 20th Century Practice and 21st Century Potential (Manchester: Extending Horizons Books, 2005), 51-64

The Methods of Nonviolent Protest and Persuasion (cont'd)	
Public Assemblies	**Withdrawal and Renunciation**
47. Assemblies of protest or support	51. Walk-outs
48. Protest meetings	52. Silence
49. Camouflaged meetings of protest	53. Renouncing honors
50. Teach-ins	54. Turning one's back
The Methods of Social Non-Cooperation	
55. Social boycott	63. Social disobedience
56. Selective social boycott	64. Withdrawal from social institutions
57. Lysistratic non-action	**Withdrawal from the Social System**
58. Excommunication	65. Stay-at-home
59. Interdiction	66. Total personal noncooperation
Non-Cooperation with Social Events, Customs, and Institutions	67. "Flight" of workers
60. Suspension of social and sports activities	68. Sanctuary
61. Boycott of social affairs	69. Collective disappearance
62. Student strikes	70. Protest emigration "Hijrat"
The Methods of Economic Noncooperation: (1) Economic Boycotts	
Actions by Consumers	83. Lockout
71. Consumers boycotts	84. Refusal of industrial assistance
72. Non-consumption of boycotted goods	85. Merchants' "general strike"
73. Policy of austerity	**Action by Holders of Financial Resources**
74. Rent withholding	86. Withdrawals of bank deposits
75. Refusal to rent	87. Refusal to pay fees, dues, and assessments
76. National consumer's boycott	88. Refusal to pay debts or interest
77. International consumer's boycott	89. Severance of funds or credit
Action by workers and producers	90. Revenue refusal
78. Workmen's boycotts	91. Refusal of a government's money
79. Producer's boycott	**Action by Governments**
Action by Middlemen	92. Domestic embargo
80. Suppliers' and handlers boycott	93. Blacklisting of traders
Action by Owners and Managers	94. International sellers' embargo
81. Trader's boycott	95. International buyers' embargo
82. Refusal to let or sell property	96. International trade embargo
The Methods of Economic Noncooperation: (2) The Strike	
Symbolic Strikes	**Restricted Strikes**
97. Protest Strike	108. Detailed strike
98. Quickie walkout (lightning strike)	109. Bumper strike
Agricultural Strikes	110. Slowdown strike
99. Peasant strike	111. Working-to-rule strike
100. Farm workers' strike	112. Reporting "sick" (sick-out)
Strikes by Special Groups	113. Strike by resignation
101. Refusal of impressed labor	114. Limited strike
102. Prisoner's strike	115. Selective strike
103. Craft strike	**Multi-Industry Strikes**
104. Professional strike	116. Generalized strikes
Ordinary Industrial Strikes	117. General Strike
105. Establishment strike	**Combination of Strikes and Economic Closures**
106. Industry strike	118. Hartal
107. Sympathetic strike	119. Economic shutdown

Table 2 (cont'd). 198 Nonviolent Actions (57-119).

Methods of Political Noncooperation	
Rejection of Authority	139. Noncooperation with conscription and deportation
120. Withholding or withdrawal of allegiance	140. Hiding, escape, and false identities
121. Refusal of public support	141. Civil disobedience of "illegitimate" laws
122. Literature and speeches advocating resistance	**Action by Government Personnel**
Citizens' Noncooperation with Government	142. Selective refusal of assistance
123. Boycott of legislative bodies	143. Blocking of lines of command and information
124. Boycott of elections	144. Stalling and obstruction
125. Boycott of government employment and positions	145. General administrative noncooperation
126. Boycott of government departments, agencies, and other bodies	146. Judicial noncooperation
	147. Deliberate inefficiency and selective noncooperation
127. Withdrawal from government educational institutions	by enforcement agents
128. Boycott of government-supported organizations	148. Mutiny
129. Refusal of assistance to enforcement agencies	**Domestic Government Action**
130. Removal	149. Quasi-legal evasions and delays
131. Refusal to accept appointed officials	150. Noncooperation by constituent governmental units
132. Refusal to dissolve existing institutions	**International Governmental Action**
Citizens' Alternatives to Obedience	151. Changes in diplomatic and other representations
133. Reluctant and slow compliance	152. Delay and cancellation of diplomatic events
134. Nonobedience in absence of direct supervision	153. Withdrawal of diplomatic recognition
135. Popular nonobedience	154. Severance of diplomatic relations
136. Disguised disobedience	155. Withdrawal from international organizations
137. Refusal of an assemblage or meeting to disperse	156. Refusal of membership in international bodies
138. Sitdown	157. Expulsion from international organizations
Methods of Nonviolent Intervention	
158. Self-exposure to the elements	177. Speak-in
159. The fast	178. Guerrilla theater
a) Fast of moral pressure	179. Alternative social institutions
b) Hunger strike	180. Alternative communication system
c) Satyagraphic fast	**Economic Intervention**
160. Reverse trial	181. Reverse strike
161. Nonviolent harassment	182. Stay-in strike
Physical Intervention	183. Nonviolent land seizure
162. Sit-in	184. Defiance of blockades
163. Stand-in	185. Politically motivated counterfeiting
164. Ride-in	186. Preclusive purchasing
165. Wade-in	187. Seizure of assets
166. Mill-in	188. Dumping
167. Pray-in	189. Selective patronage
168. Nonviolent raids	190. Alternative markets
169. Nonviolent air raids	191. Alternative transportation systems
170. Nonviolent invasion	192. Alternative economic institutions
171. Nonviolent interjection	**Political Intervention**
172. Nonviolent obstruction	193. Overloading of administrative systems
173. Nonviolent occupation	194. Disclosing identities of secret agents
Social Intervention	195. Seeking imprisonment
174. Establishing new social patterns	196. Civil disobedience of "neutral" laws
175. Overloading of facilities	197. Work-on without collaboration
176. Stall-in	198. Dual sovereignty and parallel government

Table 2 (cont'd). 198 Nonviolent Actions (120-198).

B.3. Organization

A. The success of most of these methods relies on the ability to secure widespread compliance. For example, boycotts require participation of large numbers of people in order to be effective. During the Orange Revolution in the Ukraine, organizers were able to mobilize hundreds of thousands of people against the government while simultaneously avoiding the use or provocation of violence. On the day after the fraudulent vote, approximately 500,000 people (many dressed in orange) gathered in Independence Square in Kiev and marched to the Ukrainian Parliament building. This globally broadcast scene sent an unambiguous message to the members of parliament, who would vote to void the election results several days later.

B. Organization is as critically important to nonviolent resistance as it is to the organization of the resistance components (auxiliary, underground, guerrilla, public). For such resistance to be successful, a significant portion of the population must be organized to participate in these methods.[189]

B.4. Factors Asserting Legitimacy

A. **Normative Factors.** Movements can cloak themselves and their techniques in the beliefs, values, and norms of society—things that the society accepts without question. For example, during World War II, the clergy of Norway led the earliest stages of the Norwegian resistance against Nazi occupation and the Quisling government. The religious leadership coalesced public opinion against the Nazis by invoking the voice of the church. Despite the establishment of a new ecclesiastical leadership, the bulk of the old established church ignored the new hierarchy. Through nonviolent action, the old establishment preserved its integrity by simply refusing to cooperate with the Nazi occupation in religious affairs. The Norwegian church and its leadership were held in high esteem, and appealed to the population within the framework of religious values. The Nazi occupation was never able to break its resistance.[190]

B. **Mystical Factors.** Factors such as charisma can also mobilize public opinion. Gandhi's leadership in India verged on mystical, and was able to mobilize large segments of the population through his fasts and religious mystique.[191]

C. **Consensual Validation**

1. This refers to when the simultaneous occurrence of events creates a sense of their validity and can coalesce public opinion. For example, if demonstrations occur simultaneously in diverse parts of the country, the cause espoused can appear valid simply because a variety of persons from different places are involved. Nonviolent resistance organizers use this psychological method of consensual validation to rally public opinion.

2. Enforcement of social compliance can be had through the method of ostracism. Ostracism can be used to apply pressure on individuals not participating in the nonviolent resistance campaign. Informal, everyday pressures can also assist in securing widespread compliance.[192]

B.5. Communication and Propaganda

A. Once it becomes necessary to launch the resistance campaign, there must be methods of continuous communication. The purpose is to inform, organize, instruct in nonviolent resistance techniques, help maintain the legitimacy of the resistance, and maintain focus on regaining sovereignty and defeating the enemy.

B. Nontraditional forms of media have effectively disseminated messages, while also serving as a source of mainstream media to a population that views its traditional media sources as corrupted by the aggressor. An advantage to the use of the internet is the ability to circumvent government censorship. However, government capabilities, though not necessarily agile, can exceed those of the resistance based on resources. The Tunisian Internet Agency allegedly hacked Facebook sites that advocated the removal of President Ben Ali and restricted access to the internet from internet protocol addresses outside of government facilities.[193] In 2011, at the height of tensions in Egypt, the government restricted access to the internet and mobile networks. Each of these protests attracted worldwide attention due to integration of and access to social media that enabled activists and spectators to communicate, coordinate, and document events as they occurred.[194]

C. During World War II in Norway, 90 percent of the teachers resigned within several months in response to Nazi pressure to teach National Socialism and join a Nazi teachers association. This rebuke was accomplished by communicating the plan to all of Norway's teachers in boxes of matches containing a statement that teachers were required to make to the Nazis. The communication was obeyed, though the leaders were never known, because each communicator was trusted and teachers were assured that others were taking similar action.[195]

B.6. Training

A. The purpose of this training is to erect a mental fence between the individuals who resist and the new authority. Instructions typically take the form of "dos and don'ts." Historically, undergrounds have found it easier to tell people what not to do than what to do.[196]

B. Gandhi placed great emphasis on nonviolent training, not only because he viewed nonviolence as a moral creed, but also because he understood that it was essential for effective resistance. He required his followers to swear an oath and he developed a code for volunteers. The American civil rights movement followed his example and applied it in its planning of the 1958 sit-in movement, by establishing special schools to train people to withstand forms of physical violence, and tolerate torment without responding violently, which could negate the strategy.[197]

C. Institutions such as the Albert Einstein Institution and the U.S. Institute for Peace are repositories of such information while also funding research, translating publications, and hosting events focusing on nonviolent political resolution. The works of Gene Sharp have been translated into over 30 languages and were employed from Ukraine to Egypt.[198]

B.7. Summary

A. Nonviolent resistance has played an increasingly prominent role in many underground activities. In the twenty-first century, the synthesis of global technological networks linking computers to the internet and social networks has resulted in innovative forms of communication and protest. Though not powerful enough alone to change a regime, the communication linkage provides a forum for a youthful demographic to engage in creative alternatives to picketing and chants, which are often social media directed. This technologically assisted form of communication and organization is likely to continue its prominence in nonviolent activity.

B. Organizing large segments of the occupied population to engage in nonviolent resistance can be an important part of the strategic communication narrative. It also allows a large segment of the population to engage in a resistance activity with minimal peril to itself, especially if engaged in by large numbers. It aids in maintaining popular morale against the occupier as well as the hope for the return of national sovereignty. Nonviolent civil action enables almost anyone, regardless of gender, age, education, or economic status to play an important part in exercising nonviolent resistance as an effective tool for resisting an aggressor state or its installed regime. At times, the challenges and risks may be great and opponents may be formidable. Through careful planning and commitment, dedicated individuals can use nonviolent civic action as a realistic method to resist aggression and reassert national sovereignty.

Appendix C. Russian Hybrid Warfare Tactics And Considerations

We, the Heads of State and Government of the North Atlantic Alliance, stand united in our resolve to maintain and further develop our individual and collective capacity to resist any form of armed attack. In this context, we are today making a commitment to continue to enhance our resilience against the full spectrum of threats, including hybrid threats, from any direction. Resilience is an essential basis for credible deterrence and defence and effective fulfilment of the Alliance's core tasks.[199] - North Atlantic Council, Warsaw, July 2016

C.1. Introduction

A. The term hybrid warfare is used here for the purpose of commonality of understanding, to express a broad range of warfare that is not limited to conventional or traditional warfare as understood for most of the last three quarters of a century. It is intended to be more inclusive of fighting styles and methods than exclusive. Hybrid warfare, as a combination of methods of warfare, includes asymmetric tactics. The term "hybrid" was effectively used by Dr. Frank Hoffman to express the combination of new technologies and fighting styles lacking traditional state structure and observance of rules, laws and customs of war.[200] Since then, it has become a widely used term to describe the combination of conventional/industrial, irregular, and unconventional methods of warfare. Combining methods of warfare, especially those we consider today as irregular, is actually typical of most wars in human history. In April 2017, the European Union established the European Centre of Excellence for Countering Hybrid Threats in Finland's capital city of Helsinki and thus "hybrid" is presently the more common term in the West to describe this type of combination warfare.

B. The topic of hybrid warfare is presented here due to states like Russia and Iran successfully using it to undermine resilience in targeted states. Hezbollah's 2006 fight against Israel attracted much attention to hybrid warfare. Hezbollah's effective use of such warfare followed the war in Chechnya, where the term originated.[201] Hybrid warfare's unique threat to a state, relevant to resilience and resistance operations, is primarily in the ability of the actor wielding the capability to initiate a multifaceted attack prior to the targeted state realizing what is occurring. This delays the attacked state's defensive efforts, resistance activation, and allied/partner response.

C. An adversary's application of hybrid warfare requires new thinking about strategic challenges, threats, and opportunities in a world of complex and interactively dynamic relationships. This requires us to develop indicators to assess, sort, form responses, and rescale security challenges much earlier in the development and risk profiles of these challenges. In particular, it requires a comprehensive understanding of the human domain (psychological and informational) and how an aggressor state accesses and leverages that domain to weaken and disable the target state by influencing perceptions and decision-making.[202]

D. During the last 25 years, non-state terrorists, insurgents, and criminal groups have increasingly used irregular methods. Hybrid warfare as a distinct challenge has arisen during the same time through an innovative mix of such irregular or asymmetric threats. Hybrid warfare combines conventional, irregular, and asymmetric means, including the persistent manipulation of political and ideological conflict. Foreshadowed by Iranian actions throughout the Middle East, and by Chinese unrestricted warfare strategists in the 1990s,[203] the hybrid warfare concept gained much attention during Russia's support for separatist insurgents in Ukraine.

C.2. General Description

A. Hybrid warfare involves a state or state-like actor's use of all available diplomatic, informational, military, and economic means to destabilize an adversary. Whole-of-government by nature, hybrid warfare places a premium on IW. Specifically, hybrid warfare incorporates the full range of modes of warfare, including conventional capabilities, irregular or asymmetric tactics and formations, terrorist acts (including indiscriminate violence and coercion), and criminal disorder. These multimodal activities are generally operationally and tactically directed and coordinated within the main battle space to achieve synergistic effects in the physical and psychological dimensions of conflict.[204] The key to hybrid warfare is convergence and coordination, allowing various actors to integrate their efforts by acting together to achieve an enhanced synergistic effect.

B. After the Russian occupation of Crimea and other aggression against Ukraine, NATO attempted a very general definition of hybrid warfare within NATO's Wales Summit Declaration of 2014. Hybrid warfare occurs when "a wide range of overt and covert military, paramilitary, and civilian measures are employed in a highly integrated design."[205]

C. Some non-kinetic aspects of hybrid warfare share methods with "influence operations" by aiming to misinform (e.g., Russia in Crimea and Syria) or to become a force multiplier (e.g., ISIS in the Middle East). These aspects can be significant enough to affect strategic calculations among belligerents, and have a long history of successful employment.[206] Sun Tzu stated "[t]o subdue the enemy without fighting is the supreme excellence."[207]

C.3. Hybrid Warfare from Antiquity to the Twentieth Century

During the Jewish Rebellion of 66 AD, a hybrid force composed of bandits, regular soldiers, and irregular or unregulated fighters employed tactics including fixed battle, roadside ambushes, and the employment of stolen siege engines against Vespasian's Roman Legions.[208] During the Peninsular War against the French forces occupying Spain from 1808-1814, a hybrid force composed of Spanish guerrillas of diverse interests (ranging from Basque independence to anti-royalist forces and bandits) combined with conventional British, Portuguese, and Spanish forces, supported by a British naval blockade, to gradually and decisively defeat the French Grand Armée.[209] During World War II, the Soviet Army integrated and synchronized an ill-equipped, irregular force with its conventional military forces to defeat the German Army and its Axis Allies on Soviet territory, and gradually push them westward.[210] The North Vietnamese Army synchronized its efforts with the irregular Viet Cong forces in the south,[211] while the government of North Vietnam conducted a highly sophisticated information campaign to defeat the technologically and conventionally superior French and then American forces by targeting the political will of both nations.[212]

C.4. Recent History

A. Since the beginning of the twenty first century, state and non-state actors have conducted hybrid warfare to coerce, disrupt, or overthrow established governments in Iraq, Syria, Afghanistan, Georgia, and other areas. Among non-state actors, Sunni Jihadist extremists formed the ISIS and claimed a boundless "Islamic State."[213] They sought to overthrow state governments through irregular methods such as the application of general terror against occupied people, assassinations of military and security officers, and guerrilla tactics against CF. ISIS raised revenue through the sale of oil, ransom from kidnapping, selling antiquities, and extortion. It recruited through the internet, and used that medium to great effect for strategic communication.

B. Hybrid threats constitute a diverse array of options with which to confront states. Among the most pressing challenges is the employment of proxies and surrogates, including terrorist and criminal networks. Iran, for example, employs an array of very capable proxy forces to extend and solidify its influence abroad, the best example of which is Hezbollah in Lebanon. It also uses surrogates such as Hamas (though Sunni) in Gaza. It has similar proxy and surrogate relationships with various groups in Afghanistan, Yemen, Iraq, and the Caucasus. Likewise, non-state actors such as Hezbollah, Hamas, and other violent groups will leverage high-end capabilities traditionally associated with state actors.

C. It is important to understand how to counter state adversaries who use modern military technologies, as well as proxies and surrogates. Difficult to detect in a timely fashion via conventional methods, countering these hybrid threats will place a premium on broad-based intelligence efforts, rapid, coordinated innovation and adaptation, and a commitment to undermining the means and will of adversaries to persist in such conduct.

C.5. Hybrid Warfare Capability Development and Application by Russia

A. Hybrid means are central to the way in which Russia operates to achieve its national objectives. In addition to using CF, or using them in a supporting role as a threat, Russian hybrid warfare combines low-level use of force, cyberattacks, economic and political coercion and subversion, and information warfare tailored specifically to the target. These actions form a complex and interconnected web of political, diplomatic, information, military, and economic activities.[214]

B. These asymmetric methods can individually be highly developed into a hybrid warfare campaign. They can be used in various combinations, depending on the strengths of each method vis-à-vis the target. These means can be conceptualized as tools in a toolbox. With multiple, highly developed tools available for different problems sets, the array of effective tools for various targets gives the holder of them many options. This range of options presents a greater problem for the defender because the defender must recognize the use of these tools and defend against a broad range of their employment. This facilitates an aggressor's cross-domain coercion.[215]

C. This conception of hybrid warfare is not new to Russia. In the 1920s, the Soviet Union developed the concept of "masked warfare" (*maskirovka*). It included various active and passive measures designed to deceive an enemy and to influence public opinion within the West.[216] The Soviet Union attempted to manipulate domestic issues within NATO states while creating ambiguity concerning the degree of its involvement. The Soviet Union employed asymmetric tactics in hybrid warfare to soften their opponents. Russia employs them today as the primary means of achieving political objectives in neighboring states.[217]

D. These different tools are tailored to the accessible points of leverage in the target state. In Western Europe (e.g., the United Kingdom, France, and Germany), Russia emphasizes business ties, particularly in energy. In countries with an Orthodox religious majority such as Romania, Bulgaria, Serbia, and Greece, Russia emphasizes the church and related organizations. In Slavic countries such as the Czech Republic, Slovakia, Poland, Bulgaria, and again Serbia, Russia emphasizes pan-Slavism, pressing the notion that "we are all Slavs." In the Baltic countries, Russia attempts to leverage the Russian speaking minority and compatriot organizations, particularly through its own government-organized NGOs, emphasizing Russian nationalism and culture. In Austria, Switzerland, Finland, and Sweden, Russia emphasizes their neutrality as leverage to influence them.[218] Russia has manipulated access to energy resources and markets, exploiting perceived economic and diplomatic leverage to disrupt freedom and stability in the Baltic countries, Central and South-ern Europe, as well as in Central Asia.

E. A recent asymmetrical tactical development in the hybrid warfare tool kit is the concept of "lawfare."[219] As the importance of international law and public perceptions of it has increased over time, states desire to be seen in compliance with international law. One working definition of lawfare is "a tactic that seeks to help states develop a legally grounded narrative that justifies their actions and couches their aggression in rational terms. This in turn, allows states to take any action they deem necessary while, at least, maintaining the ruse of compliance with the international legal system."[220] The use of lawfare in hybrid warfare serves both domestic and international ends by attempting to legitimize a state's behavior to domestic and international audiences.

F. In the case of Crimea, many Western countries, and the broader international legal community, charged Russia with violating the membership treaty of the United Nations, statutes of the Council of Europe of which Russia is a member, two regional treaties, and two bilateral treaties. Russia countered with its own legal, or lawfare, arguments such as The Kosovo Precedent, arguing that secession of Kosovo from Serbia provided international legal precedent for the secession of Crimea from Ukraine. Russia also argued that the original transfer of Crimea from Russia to Ukraine in 1954 was illegal under Soviet law. Additionally, Russia argued that an illegal seizure of power occurred in Ukraine in 2014 when power was taken from the elected Victor Yanukovych, and thus Russian actions to defend Russians in Crimea were justified. Further arguments were that Russian actions were justified by the Western Precedent, pointing to the actions of the West in Afghanistan, Iraq and Libya, providing Russia with adequate precedent to intervene.[221]

1. **Russian Actions in Ukraine**

   a. Russia's actions in eastern and southern Ukraine embrace hybrid warfare. Russia employed SOF, intelligence agents, political provocateurs, and media representatives, as well as transnational crimi-nal elements. Funded by the Kremlin and operating with differing degrees of deniability, or even acknowledgement, the Russian government used little green men (mostly Russian SOF) to achieve some physical objectives. Many Crimeans during this takeover referred to them as polite people. The objectives of these forces included instigating confusion and disrupting civil order. Addition-ally, Russian elements organized pro-Russian separatists, filling out their ranks with advisors and fighters. Russia also included funding, arming, tactical coordination, and fire support for separatist operations.[222] The Russian-aided insurgency gained local supporters while intimidating dissenters into acquiescing to a separation from the Kiev government.[223]

   b. Russian hybrid warfare thus involves CF, economic intimidation, influence operations, and dip-lomatic intervention. Sponsorship of separatist insurgency in Ukraine accords well with current Russian military doctrine and strategy, which embrace "asymmetrical actions … [including] special-operations forces and internal opposition to create a permanently operating front through the entire territory of the enemy state."[224]

   c. Russia's campaign in Ukraine is a prominent example of hybrid warfare. However, during the 2008 Russia-Georgia conflict in the breakaway regions of Abkhazia and South Ossetia, both sides used combinations of regular forces, irregular forces, and criminal elements. Prior to the war, Russian forces operating in Georgia as peacekeepers allegedly sustained a flourishing smuggling network in partnership with various Abkhaz, Ossetian, and Georgian criminal groups. This network cooper-ated with separatist militias used by Russian forces to ethnically cleanse Georgians from the two breakaway regions. Georgian military forces also cooperated with guerrilla groups operating in the area. Both sides blurred the distinction between regular government forces, criminal elements, and militias.

2. **Cyber Warfare: Estonia's Bronze Soldier of Tallinn 2007, and Georgia 2008**

   a. In January 2007, the Estonian government announced its intention to move a bronze statue (*Monument to the Fallen in the Second World War*) erected in 1947 when Estonia was held as a Soviet Socialist Republic. The move was from the center of the capital city of Tallinn to a military cemetery where some of the dead from World War II were interred on the outskirts of the city.[225] The Russian government warned against this, stating that such an action would be "disastrous for Estonians."[226]

   b. After several months of political controversy, including protest demonstrations by Estonia's ethnic Russian minority, the statue and the Red Army soldiers buried beneath were removed on 26 April 2007 and placed into the nearby military cemetery. Upon this removal, Estonia suffered riots by ethnic Russians, and denunciations from Russia, including by Russian President Vladimir Putin. Estonia, Europe's "most connected country" which pioneered e-government and was almost completely wireless, suffered cyberattacks against its internet infrastructure.[227]

   c. The attacks came in the form of distributed denial of service (DDoS) attacks, website defacements, DNS server attacks, mass email, and massive spam and phishing emails. Many attacks were conducted by hackers likely recruited by the Russian government specifically for this event. Government agencies, universities, banks, telecommunications, media and almost all societal infrastructure were attacked.[228] These attacks were the largest and most nationally disruptive cyberattacks yet launched. Though such warfare can be combined with other types of warfare, as part of a hybrid attack, this cyberattack was not combined with a physical attack.

   d. In August 2008, conventional military conflict broke out between Russia and Georgia—also another former Soviet Socialist Republic—over the Georgian breakaway provinces of Abkhazia and South Ossetia. Russian cyberattacks against Georgian political leaders, followed by attacks against Georgian government institutions, began about three weeks prior to the incursion of Russian CF. Compared to Estonia, Georgia was much less connected in cyberspace. However, unlike the attack against Estonia, the cyberattack against Georgia was combined with conventional Russian military operations.

   e. Once the incursion of Russian CF began, the cyberattacks continued, with DDoS attacks shutting down targeted websites. Other sites were defaced with Russian propaganda. As in the attack against Estonia, many hackers were likely recruited by the Russian government specifically for this attack. Additionally, Georgian hacker forums were attacked early in order to neutralize them and prevent their use in a counter-attack.[229]

   f. Russian physical target selection and cyber targeting appeared to be coordinated. This was seen by the fact that command and control centers, media outlets and other communication targets were spared from physical force if already neutralized via cyberattack.[230] However, since Georgia was much less connected than Estonia, these attacks primarily affected the ability of Georgians to connect to the outside world, and for the government to counter Russian disinformation.[231]

   g. The cyberattack on Estonia resulted in millions of dollars in damages and massive economic disruption, with short-term, temporary, and not physical effects. The cyberattack on Georgia was combined with a physical ground incursion supported by successful cyberattacks on command and control and communication centers which nullified the requirement for more destructive and expensive physical attacks. The cyberattacks on Georgia were part of an effective hybrid warfare campaign.

G. **Developments in Russia's Hybrid Warfare Capabilities.** As a way of conflict, hybrid warfare poses challenges due to its subtle nature and ability to mask the starting point of an attack. This tests a nation's resilience and can delay the defending state's ability to begin defensive measures or activate resistance capabilities. Though Russia is not the only state to develop theories on and how to engage in hybrid warfare, Russia lends itself best to the examination of this type of warfare, due to its successful use of the approach.

1. **Force Structure**

   a. Russia's cumbersome legacy divisions, designed for fighting along a frontline, were disbanded in 2009 and replaced by smaller, but more agile and better prepared brigades. Such transformation was intended to improve mobility and to conduct smaller-scale operations by better equipped units.

   b. Since 2014, Russia has worked on establishing a pool of rapid deployment forces which include special operations units, airborne forces, and other relevant brigades. In 2015, these forces were assessed to consist of approximately seventy thousand combat troops, and based on the Russian Airborne Forces.[232]

2. **National Defense Control Center**

   a. The formation of the Russian National Defense Control Center in December 2014 may be the most important innovation since the beginning of Russia's 2008 military reforms, following Russia's war with Georgia. The center's main responsibility is to maintain centralized combat control of the Russian Armed Forces, and analyze comprehensive data on military and political developments in the country and overseas. The center mainly deals with coordination of information flows among relevant state structures (MOD, General Staff, etc.) and oversees the implementation of defense and security measures.[233]

   b. The National Defense Control Center also ensures close coordination with 49 other security agencies and ministries such as the MOI, the Ministry of Emergency Situations, and Rosatom, the state nuclear energy corporation.[234] This facilitates joint operations of the armed forces, as well as interagency coordination, allowing a genuine whole-of-government approach. This center enhances Russia's capabilities for hybrid warfare.[235]

3. **National Policy/Law.** In 2009, Russia amended its law *On Defense*. Previously, the law allowed the deployment of Russian troops abroad only to fight international terrorism or to conduct peacekeeping missions. The amendments allowed greater use of Russian troops abroad to include deployment to repel an armed attack on Russian troops or citizens abroad or to assist another state by request of its leadership to repel or prevent an armed attack. Russia cited both of these changes to legitimize its operations in Crimea and the second to legitimize its operations in Syria.[236]

H. Russian Hybrid Warfare Theoretical Development[237]

1. In 2013, General Valery Gerasimov, Chief of the General Staff of the Russian Federation, published an article based on his observations of the Arab Spring. This was likely an observation of and response to NATO's Comprehensive Approach,[238] which was adopted as part of NATO's new Strategic Concept at the Lisbon Summit of November 2010. It outlined the requirement to use political, civilian and military instruments together as the approach to crisis management. He described a "new generation warfare," focusing on the combined use of diplomatic, economic, political, and other non-military methods, along with direct military force, instead of waging open war.[239] He argued that the importance of non-military means in reaching political and strategic goals has increased and that they are often more efficient than

arms alone. Gerasimov foresaw concealed, non-open use of force, such as paramilitary and civilian insurgent units, and emphasized the need to rely on asymmetric, indirect methods. He also included the information space, where real-time coordination of means and tools is possible. He placed great emphasis on targeted strikes far behind enemy lines, and destruction of enemy critical infrastructure (military and civilian elements), preferably in a short timeframe. He advocated massive use of special forces and robotized weapons such as drones. Gerasimov argued that regular forces should be used only in the late phases of the conflict, and under the guise of peacekeeper or crisis-management forces.[240]

2. In that same year, another complementary article was published by Sergei Chekinov and Sergei Bogdanov.[241] The authors agreed with Gerasimov and elaborated on his doctrine. They stressed the high importance of asymmetric actions to neutralize the enemy's military through the combined use of political, economic, technological, and information campaigns. They pointed out the need to integrate all these tools into a single, shared system of command and control to multiply their efficiency. They listed the media, religious organizations, cultural institutions, NGOs, public movements financed from abroad, and scholars engaged in research on foreign grants as possible components of a coordinated attack against the target country. In particular, they highlighted the need to gain information superiority over the target country through intensive propaganda prior to the attack, and continuous use of electronic warfare (EW) to disable enemy communications. They saw the main battleground as the information space. This type of warfare is dominated by psychological and information warfare to crush the morale of enemy troops and the opposing population, thus breaking their will to resist.

3. Chekinov and Bogdanov divided this type of warfare into an opening phase and a closing phase. The opening begins with an intensive, months-long, coordinated non-military campaign, including diplomatic, economic, ideological, psychological, and information measures. A heavy propaganda campaign is conducted to depress the enemy population, spark discontent regarding its central government, and weaken the morale of the armed forces. Government and military officers in the target country are deceived and bribed to decrease the functionality of its armed forces in advance. Concurrently, agents are deployed within the target state with funds, weaponry, and other materials to commit terrorist acts, conduct provocations, and create chaos and instability. Directly prior to the military phase (closing phase), large-scale reconnaissance and subversive missions are conducted to gather information, and locate and isolate enemy military units, key governmental facilities, and critical infrastructure. This is followed by a large-scale electronic warfare operation to disable the enemy's government and military forces. Immediately thereafter, a conventional military attack begins with precision weaponry, drones, and other automated weapons. At the end of this opening phase, the target country's main government and military control centers are destroyed, and critical infrastructure heavily damaged to render the country ungovernable, and unable to deploy its defense forces. In the next phase, the closing phase, the attackers' irregular ground forces enter the target country to isolate and destroy remaining points of resistance. This second phase is much less important than the predominantly non-military first phase. There is a striking similarity between this description and the events in Ukraine in 2014, particularly prior to, and during, the Russian operation in Crimea.[242]

4. Another way to describe this approach is the term "reflexive control" which was developed during the Soviet era. Reflexive control is designed to cause a "stronger adversary voluntarily to choose the actions most advantageous to Russian objectives by shaping the adversary's perceptions of the situation decisively."[243] This is accomplished through the application of IO/strategic communication, lawfare, political action, military operations using a tailored and timed combination of SOF and CF, and economic actions.[244]

5. This Russian approach may actually pre-date both the so-called Gerasimov Doctrine and NATO's Comprehensive Approach. Certain combinations of this style of warfare could also be termed *political warfare*. The application of political warfare can be traced back to Lenin and the early Soviet Union. In fact, an argument can be made that Russia has used this style of hybrid or political warfare when it led the Soviet Union through the twentieth century, making this style of warfare a long tradition in the Russian state.[245]

I. Russian Hybrid Warfare Application[246]

1. **The Three Main Operational Phases of Russian Hybrid Warfare.** András Rácz, of The Finnish Institute of International Affairs, analyzed Russia's operations in Crimea and in Eastern Ukraine. He described Russian hybrid warfare as being composed of three main phases: preparatory, attack, and stabilization. The following sections are based on his work: *Russia's Hybrid War in Ukraine; Breaking the Enemy's Ability to Resist.*[247]

   a. **Preparatory Phase**[248]

      01. The preparatory phase concentrates on mapping out the strategic, political, economic, social, and infrastructural vulnerabilities of the target state, and creating the necessary means for capitalizing on them. This does not differ much from traditional diplomatic and soft coercion activities. These include establishing political and cultural organizations loyal to Russia, gaining economic influence, building strong media positions, and strengthening separatist movements and other anti-government sentiments, all with the aim of pressuring the target government.

      02. This makes it practically impossible to determine whether traditional Russian influence-gaining measures may be serving as preparation for a hybrid attack before the offensive actually starts. In addition, many of the actions listed below are not explicitly or necessarily illegal in most countries, which makes it hard for the target country to defend itself against them.

      03. The preparatory phase of hybrid war, encompassing traditional aspects of foreign policy that may serve as the basis for a hybrid war, can be divided into three steps:

          i. **Step 1. Strategic preparation**
             ◘ Explore points of vulnerability in the state administration, economy, and armed forces of the target state.
             ◘ Establish networks of loyal NGOs and media channels in the territory of the target state.
             ◘ Establish diplomatic and media positions to influence the international audience.

          ii. **Step 2. Political preparation**
             ◘ Encourage dissatisfaction with the central authorities in the target state by using political, diplomatic, special operation, and media tools.
             ◘ Strengthen local separatist movements and fuel ethnic, religious, and social tensions.
             ◘ Use information measures against the target government and state.
             ◘ Bribe susceptible politicians, administrative officials, and armed forces officers.
             ◘ Establish contacts with local oligarchs and business people, making them dependent on the attacking country via profitable contracts.
             ◘ Establish contacts with local organized crime groups.

          iii. **Step 3. Operational preparation**
             ◘ Launch coordinated political pressure and disinformation actions.
             ◘ Mobilize officials, officers, and local criminal groups that have been turned.
             ◘ Mobilize the Russian armed forces under the pretext of military exercises.

04. There is no violence during the preparatory phase and the measures taken should not cross any political or legal threshold that would make the target state adopt countermeasures. However, if the targeted government detects these steps, it may feel pressure to be more receptive to Russia's demands. This could result in some political destabilization. The effects of the preparations become visible only in the next phase, following the principles that Chekinov and Bogdanov established.

b. **Attack Phase**[249]

01. An attack exploits all the weaknesses identified during the preparatory phase.

02. Open, organized, and armed violence, or its threat, begins. Crises erupted in similar ways in both Crimea and Eastern Ukraine. Unmarked units using sanitized Russian uniforms, weapons, vehicles, and equipment, appeared and started setting up barricades and checkpoints, and blocking the Ukrainian military and police barracks gates. Violence was not employed but it was clear that Ukrainian units could not leave their bases without using force against the unidentified militants.

03. Political targets are of primary importance. In Crimea, the attack started on 27 February 2014 with little green men overrunning the parliament building, and the Supreme Council of Crimea, effectively preventing local political decision-making. In Donetsk, in April 2014, the regional state administration building was among the first targets to be seized and still serves as the Donetsk National Republic headquarters. Police and local security forces failed to defend the buildings, due to the lack of clear commands, their low morale, weak leadership, and inadequate resources.

04. In Crimea, parallel to these actions, well-organized, often armed demonstrators, dressed in civilian clothes and exhibiting high tactical skills, started to take over less defended public administration buildings, media outlets, and civilian infrastructure. Capturing television and radio stations and broadcasting towers enabled the attackers to take central government media channels off the air and to replace them with Russian channels.

05. Both the attackers and the little green men consistently claimed that they were local protesters dissatisfied with the Ukrainian central government. The Russian official discourse and media consistently referred to them as the opposition or resistance. Many local inhabitants referred to them as polite people. President Putin admitted the men were Russian troops only after Crimea had been annexed, after a widely criticized public referendum.

06. These offensive operations were supported by an intensive information campaign aimed at confusing decision-makers, generating fear, and dissatisfaction vis-a-vis the central government, and weakening the resistance potential of the local Ukrainian army and police units by lowering their morale. Ukraine's command and control chains were damaged, disrupted, or jammed by sabotage attacks, corruption, and electronic warfare. As a result of these actions, almost all of the Ukrainian army, police, and navy units in Crimea ultimately surrendered. Some units sympathetic to Russia changed sides while the resistance of others was broken with low-level violence.

07. Before the attackers' affiliation with Russia became known, Moscow consistently denied its involvement, confusing Western observers and buying time. This delaying tactic allowed Russia to quickly reach a *fait accompli* in Crimea and later in Donbass. This was Moscow's policy of persistent denial in successful action.

08. This type of low-level attack, combined with persistent denial, if applied against any NATO state that is similarly vulnerable, could delay or prevent the activation of the Washington Treaty's Article 5. As most actions would remain below the threshold of NATO's collective defense guarantee, the attacker could gamble that NATO might not engage until after a significant amount of territory had been taken, causing political difficulties to NATO. The longer that territory stayed under hostile control, the more difficult it would be to politically and militarily reassert sovereignty.

09. Although most attack phase actions are conducted by nonmilitary or non-regular military means, the regular military plays a key role. In the case of Crimea and in Eastern Ukraine, Russian regular military units were placed on the border with Ukraine, posing an imminent threat of a conventional attack. This limited Ukraine's options as it did not want to risk conventional Russian incursion.

10. The legitimacy of the separatist leaders is constructed by consistent building up and strengthening by Russian diplomacy, media and public discourse. The Russian media constantly referred to the separatist authorities as if the People's Republics of Donetsk and Luhansk were properly functioning, legitimate states, and as if they represented the local population. The Soviet Union did the same when recognizing Babrak Karmal as the installed leader of Afghanistan following the 1979 Soviet invasion. However, even this constructed legitimacy may alienate locals from the true sovereign government by relying on the attacker's information monopoly once they dominate the information environment, and by establishing an alternative political power center.

11. The attack phase can be divided into three steps:

i. **Step 1. Exploding the tensions**
   - Organize massive anti-government protests and riots in the target state.
   - Infiltrate special forces, disguised as local civilians, to conduct the first sabotage attacks and capture administrative buildings (with the active or passive support of corrupt local officials and police), often in cooperation with local criminal groups.
   - Instigate provocations and sabotage attacks throughout the target country, to divert the attention and resources of the national government.
   - Employ media to launch a disinformation campaign.
   - Place Russian regular forces on the border to make it more difficult or impossible to counterattack by posing an imminent threat of an overwhelming conventional attack.

ii. **Step 2. Ousting the central power from the targeted region**
   - Disable the targeted government by capturing administrative buildings and telecommunications infrastructure in the targeted region.
   - Block the targeted nation's media, establishing communication and information monopoly.
   - Disable the local armed forces of the targeted power by blockading their barracks, bribing their commanders, breaking their morale, etc.
   - Use media to mislead and disorient the international audience, and discredit the target state.

iii. **Step 3. Establishing alternative political power**
   - Declare an alternative political center, using captured administrative buildings, by referring to real or fabricated traditions of separatism.
   - Replace administrative organs of the targeted state with newly established political bodies, thereby creating a quasi-legitimacy.

- Employ media to strengthen the alleged legitimacy of the new political bodies.
- Alienate local population from the government of the targeted state via information monopoly.
- Continuously block counterattack options of the targeted state with the threat of a conventional military attack against it.

12. Toward the end of the attack phase, the targeted state's ability to resist with its conventional military or other internal security forces is broken. Its governance, command and control, and communications capabilities are severely damaged and it loses control over some of its territory. Such success capitalizes on inherent, multiple weaknesses of the target state, explored in the preparatory phase.

c. **Stabilization Phase**[250]

01. To consolidate the results that hybrid warfare achieves, the attacking state needs to take steps to strengthen and legitimize its rule. This third phase is stabilization. In Eastern Ukraine and Crimea, initial developments followed a similar pattern, but turned out very differently.

02. Russian units stationed in Crimea exerted pressure on local elites and the population. "Independence" referendums were organized in both regions, and in both cases the results favored separatists. After less than a full day of so-called independence, Russia annexed Crimea. In Eastern Ukraine, however, separatism could not gather sufficiently strong momentum, mostly due to the low local support, and lack of support such as provided in Crimea by Russian units stationed there. In the Donbass, instead of establishing a functioning alternative power, Russia created a political, security, and social limbo.

03. Typically, hybrid warfare allows two possible options. One is the annexation of the captured territory, as occurred in Crimea, while the other option is to keep the territory inside the attacked country but deny the central government control.

04. The final political outcome, and the main goal, is achieved in either event. The strategic freedom of movement of the target country, including its freedom to choose its foreign policy, becomes severely curtailed due to the loss of territory, population, resources, and credibility. The case of Ukraine demonstrated how hybrid warfare can cripple the functioning of a state without launching a full-scale war against it.

05. The stabilization phase can be detailed in three steps:

   i. **Step 1. Political stabilization of the outcome**
   - Organize a referendum and decision about secession/independence in the target state, all with the strong diplomatic and media support from the attacking state.
   - Have the new political entity request help from the attacking state.

   ii. **Step 2. Separation of the captured territory from the target state**
   - Annex the captured territory (Crimea), or;
   - Establish an open or covert military presence, and start fighting the displaced national government in the name of the newly established political entity thereby continuing to weaken the targeted state in the political, economic, and military sense (Eastern Ukraine). A variant is an open invasion under the pretext of peacekeeping or crisis management.

   iii. **Step 3. Lasting limitation of the strategic freedom of movement of the attacked state**

117

- ◘ Loss of territory (economy, population, infrastructure, etc.) results in severe economic hardship and domestic political destabilization.
- ◘ Lacking full control over its territory, the attacked state is unable to join any political or military alliance that requires territorial integrity (e.g., NATO).

06. One of the main reasons Eastern Ukraine turned out differently from Crimea was that following the election of Petro Poroshenko as president of Ukraine on 25 May 2014, the anti-terrorist operation launched by Ukraine also gained momentum. The irregular rebel forces could not stand against the advancing Ukrainian regular military. To prevent a separatist defeat in August 2014, Russia launched an intervention by its regular forces, thus transforming the conflict into a conventional, albeit limited, interstate conflict.

C.6. Operational Factors of Success.[251] Despite Russia's partial failure in Eastern Ukraine, the employment of hybrid warfare was very effective.

A. **The Element of Surprise.** The first and probably most important reason for the effectiveness was the element of surprise. The rapid implementation of the full range of hybrid warfare (i.e., all three phases described above) took both Kiev and Western states by surprise. This form of warfare was unexpected and the means of defense against it were underdeveloped. The main tool in misleading Kiev and other governments was effective and well-coordinated information warfare, of which a major component was persistent denial.

B. **Denial of Formal Involvement**

1. Russia's denial of its involvement succeeded in confusing the Ukrainian leadership and forestalling the reactions of Western governments. Further, the attackers enjoyed a lack of accountability, as they appeared to have no connection to any foreign state.

2. Meanwhile, Ukraine was put in a situation where it formally had no one with whom to negotiate. Russia consistently refused to engage in bilateral discussions, claiming its non-involvement. Further, Moscow pushed Kiev to negotiate directly with the separatists. This would have meant the de facto recognition of the separatists as legitimate partners.

3. Additionally, the denial of formal involvement theoretically empowered Russia to stop its attack at any time in the event of difficulties. Under the guise of denial of involvement, the attack could have been stopped at almost any time, as long as no decisive political acts took place. Moscow could have exited both the Crimean and the Eastern Ukraine operations without major political loss by withdrawing its unmarked forces and ordering its agents to gradually end their activities.

C. **Attackers Indistinguishable from Civilians**

1. Another element that contributed to the efficiency of the hybrid offensive was that many of the attackers were dressed in civilian clothes. This made them practically indistinguishable from the local civilian population. This seriously limited the potential of the Ukrainian government to use force. The use of force against the demonstrators, with many local civilians among them, was a risk that most ordinary Ukrainian policemen were not willing to take.

2. Additionally, Russian propaganda was highly effective in sowing the seeds of doubt about the legitimacy of the interim government in Kiev. As a result, many police officers could not be sure which orders to follow, and whether the law was actually on their side. Most ordinary Ukrainian police officers were

not eager to risk injury or death, not to mention their career prospects, trying to defend the attacked buildings against demonstrators.

C.7. Conclusion

A. Russia effectively employed hybrid warfare against Ukraine in two instances. Foremost was the ability to identify and capitalize on the inherent weaknesses of both operational areas within Ukraine through effective exploitation. A large ethnically Russian population (though less than a majority in Eastern Ukraine) supported Russia's objectives. That population also made it easier for Russian Special Forces to disguise themselves as locals. This gave Russia plausible deniability when it claimed that local inhabitants were merely expressing dissatisfaction with the Kiev government and acting against that government. Further, a strong media presence, in the target area and abroad, gave Russia the ability to generate mistrust of the opposing central government, isolate the attacked region through information transmitted or withheld, and misinform and obfuscate Russian participation to the international community. Television and radio stations were among the first targets, where their broadcasts were immediately replaced with Russian channels, proliferating Russian messages. Russia was geographically contiguous and had military superiority, allowing for surreptitious cross-border logistical and other support. By merely placing large numbers of forces in proximity to the area of operations, on its side of the border, Russia limited its opponents' options, for fear of causing Russian forces to invade.

B. Since such hybrid methods mostly leveraged access to the population, there must be greater emphasis on engaging the public prior to crisis in order to strengthen resilience. This resilience of the population is a critical element in the Clausewitzian "trinity" of war. This could begin with instilling confidence in the population through domestic resilience measures, including popular knowledge and identification of adversary tactics such as exposing their propaganda, as well as adversarial use of methods of hybrid warfare.[252] Further, a national defense plan which includes resistance under occupation can serve to instill popular confidence against a potential adversary by demonstrating that the nation will continue to fight to regain its sovereignty and territorial integrity. Resistance planning and preparation, especially in coordination with allies and partners, can also serve as a form of deterrence. Lastly, such coordinated planning and preparation increases the likelihood of success, and of timely ally and partner action to restore lost sovereignty and territorial integrity.

Appendix D. Second World War Resistance Case Studies

Whoever wishes to foresee the future must consult the past; for human events ever resemble those of preceding times. This arises from the fact that they are produced by men who ever have been, and ever shall be, animated by the same passions, and thus they necessarily have the same results.
- Niccolo Machiavelli

D.1. Introduction

A. World War II offers excellent perspectives into resistance operations. The Axis powers easily overran most nations that they chose to invade. Resistance occurred to various degrees in different places around the world. Some resistance groups received significant assistance from the Allied powers, while others did not.

B. Appendix D focuses on four case studies, chosen for their contextual diversity. They vary in description based primarily on the availability of information. At the end of each of these studies is a summary. These sections strived to offer as much hindsight objectivity as possible in relation to resistance. The conclusions are not intended to point out deficiencies in World War II leadership thinking, nor are the conclusions intended in any way to criticize, second-guess, or dishonor any of the decision makers and brave people who fought and died. These case studies formed some of the historical basis upon which much of the twentieth century UW doctrine was based, and now corroborate recommendations covered in this resistance concept. In all, these case studies elucidate practical recommendations, and highlight problems that may occur if not taken into consideration. The cases cover French, Polish, Filipino, and Baltic resistance.

D.2. Case Studies Summaries

A. French resistance to the German invasion was ad hoc and began as grass roots efforts, even in the territories not physically occupied by the invader. Many resistance groups evolved from different ideological leanings. The socialist and communist groups wanted a vastly different post-war France than that which previously existed. Friction between the different groups reduced potential cooperation, as each group maneuvered to secure its own post-war outcomes. Added to this friction were the objectives of the British and the Americans who were very focused on defeating the Germans and less concerned with internal French conflicts affecting post-war France. General Charles de Gaulle's political skills were of premiere importance in realizing the final outcome: a non-communist France, without a military government imposed by the Allies, or civil war, and with a resumption of the sovereignty that was violated in 1940.

B. The Polish resistance to German occupation, like the French resistance, exemplifies the need to have plans in place before an invasion. Poland had a long resistance tradition that allowed a state underground organization to be established in a remarkably short time. Poland's underground organization was probably Europe's best and most integrated. Unlike France, Poland quickly prepared a government-in-exile, which was sent abroad to preserve Polish sovereignty and to continue its legitimate struggle. Poland's resistance accomplished remarkable feats, such as smuggling a V2 rocket's critical components to the British and sharing information on extermination camps. The Soviet Union was the greatest threat to the reestablishment of Poland's freedom. Poland was geographically much closer to the Soviet Union than to Western Allies, which would have decades' long consequences for Polish freedom. Poland's resistance was valiant, well organized, and highly motivated, but was unable to prevail against the steamrolling Soviet Union.

121

C. The Filipino resistance to the Japanese has several interesting aspects. Historically, resistance organization and movements must be indigenous to succeed. The Philippines was a commonwealth of the United States and thus reliant on the U.S. for its defense. Yet, despite knowledge of Japanese expansion and warnings of Japanese intentions in the Western Pacific, the U.S. did not establish a resistance capability on the archipelago. After the U.S. surrender, thousands of American and Filipino officers and soldiers escaped Japanese capture and fled to the jungles and mountains. Those military members organized remarkably effective resistance groups despite being disconnected from each other. The American leadership of Filipino guerrilla units was a direct result of the unusual cultural and political nature of the American experience in the Philippines prior to the war. The vast majority of Filipinos supported their effort, but there was domestic competition. The Hukbalahaps (Huks) were communists seeking to establish a communist post-war state. Although their numbers and capabilities never reached the necessary level to attain their post-war goals, the Huks did provide difficulties for American forces. The Moros, Filipino Muslims, were anti-Japanese but also not pro-American. Moros and Huks could be relied upon to fight the Japanese, but not in concert with American plans as an integrated force. Allied headquarters eventually saw the utility of the military led resistance groups, formed mostly by those who escaped capture by the Japanese, as part of an intelligence gathering effort to gain information on Japanese dispositions and activities, especially immediately prior to the return of American CF.

D. The Forest Brothers of the Baltics engaged in resistance against the Soviets soon after the Molotov-Ribbentrop Pact placed them under Soviet rule. When Germany attacked the Soviet Union, the Germans were initially treated as liberators by the Baltics. Yet, the Baltic people eventually came to resist Nazi domination. Soon afterwards, they were invaded again when the Soviets pushed the Germans westward and through the Baltics. However, having experienced Soviet occupation several years earlier, most of the Baltic people did not receive the Soviets as liberators. An organized resistance formed against the Soviets in the states of Estonia, Latvia and Lithuania. In each of these nations, these resisters were well organized along military lines in large units, reasonably well armed, and uniformed. They became known as the Forest Brothers. After initial clashes with Soviet forces that greatly reduced their numbers, they changed tactics for survival. They hid in the forests, eventually constructing underground bunkers. The Forest Brothers relied on the U.S. and Great Britain to stay true to the Atlantic Charter and fight for the freedom of the Baltics from Soviet domination. That assistance never arrived. Eventually, the Forest Brothers lost members who returned to normal civilian lives, were killed in fights with occupation security forces, or were betrayed by those who saw the struggle as hopeless. Their resistance survived for several years with practically no outside support.

D.3. Conclusion

These studies demonstrate both the efficacy and shortfalls of national resistance against an occupier. None of these nations had prepared resistance organizations prior to enemy invasion. Poland, due to its long history of being dominated by major powers, was able to quickly establish a basic resistance organization which became much more capable and larger in the years that followed. Also, of these nations, only Poland had an effective government-in-exile capable of representing the Polish nation. The Philippines, as a United States Commonwealth, did not have a sovereign government to exile because the United States represented its interests. Common themes emerge, such as: internal (frequently communist) competition to establish post-war governance, struggle for political legitimacy, the criticality of external material support, and the utility of the resistance conducting operations in support of a conventional invasion by friendly forces. Much can be learned from delving deeply into these few historic episodes, from which some critical points are elucidated in these case studies.

Case Study 1. France: World War II Resistance[253]

Before it could be joined, the Resistance had to be invented.[254]

The form of government of France is and remains the Republic. In law the Republic has not ceased to exist. - Algiers ordinance of 26 June 1944

CS1.1. Background

Prior to World War II the French government did not prepare for domestic resistance against occupation forces. Its focus was on the ability of its CF to deter and defend against a German invasion from the east. France constructed the defensive Maginot Line, running from Switzerland to Belgium to defend against conventional land attack. The Germans attacked on 10 May 1940 through the Ardennes Forest, avoiding the defensive line and introducing combined arms warfare and blitzkrieg. France was defeated six weeks later. After the Franco-German armistice of 22 June 1940, France was split into three distinct areas:

A. **Occupied France.** German occupation of France was expansive. Germany gained more factories, workers, minerals and the necessary air and seaports for operations against Britain by directly controlling the industrial north, the national capital of Paris, and ports on the North Sea, English Channel, and Atlantic Coast.[255]

B. **Unoccupied France, or Nominally Free Zone.** A new French government, the État Français (French State) based in Vichy, France, ruled over the southern part of France that was not occupied by German troops. Vichy France was in fact a German client state. However, after the Allied landings in the French North African territories in November 1942, followed by the surrender of French forces there, Germany assessed this surrender as a violation of the 1940 armistice allowing the existence of Vichy-controlled France. German troops then invaded Vichy France, thus occupying all of France and abrogating the distinction between Free and Occupied Zones.

C. **Annexed Territories.** The German Reich re-annexed the Alsace region and much of the Lorraine region, which had been under German control from 1871 to 1919. It contained both French and German speaking people, with dual or divided loyalties.[256]

CS1.2. Weakened Pre-War Resilience

A. The French defeat in the Franco-Prussian War of 1870 resulted in German occupation of the eastern part of France. Guerrilla warfare by *francs-tireurs* (free-shooters) sprang up in those territories. The *francs-tireurs* evolved from recreational shooting clubs and patriotic organizations that grew under the French Second Empire during the 1860s. Their sniping, ambushes, sabotage, and hit-and-run tactics against German regular forces resulted in harsh and deadly German collective reprisals against civilians for what the Germans considered to be terrorist attacks outside the laws of war. The conflict was followed by internal chaos, criminality, revolution, and the establishment of the Third Republic, which existed until 1940. In addition to the Franco-Prussian war, the French retained recent memory of the massive loss of life incurred in World War I, when approximately 1.3 million French citizens were killed. These experiences functioned to weaken French resilience and willingness to expend more lives to fight an invader.

B. In February 1934, French fascists and paramilitaries attacked the French parliament building. In reaction to this, a center-left, anti-fascist movement, dominated by socialists and communists, arose and gained some political power in 1936 as the Popular Front. Yet this group remained a minor part of French politics, and had members who were sympathetic to the Soviet Union, which raised the suspicions of many French citizens. The Front was banned in 1939 and began operating surreptitiously, which served it well as underground organizations formed during later German occupation. Though not the only organized resistance group, this movement laid the groundwork for organized resistance against German occupation, especially in and around Paris.[257]

CS1.3. Government (Internally Displaced Versus Exiled)

The French government did not have any plans for a government-in-exile in case of foreign invasion. This lack of preparation allowed the Germans to set up a client state, in the form of the French Vichy government headquartered in Vichy, France, which was militarily, economically and politically subordinated to Nazi Germany.

A. Internally Displaced Government (Free Zone)

1. In keeping with the armistice terms, the new Vichy government no longer fought Germany or assisted Germany's enemies. Indeed, by the terms of the armistice, it cooperated with Germany but was clearly subordinate to it politically and militarily. The Vichy government became the legitimate government led by the World War I French hero, Marshal Philippe Pétain, who was accepted by most French citizens. This right-leaning government was anti-communist, anti-Freemason, and, as evidence suggests, anti-Semitic. The Vichy government retained an Armistice Army of 100,000 soldiers within the Free Zone and also controlled all French military and naval forces stationed in overseas French possessions.

2. A minority of officers, including Charles de Gaulle, disagreed with the cessation of hostilities and the treaty that subordinated France. They refused to stop fighting and fled to Britain instead.

B. Nascent Government-in-Exile

1. De Gaulle was not a predesignated leader of a government-in-exile. De Gaulle was the initial voice of continuing French resistance from outside of occupied territory, and outside Vichy controlled France. Germany invaded Belgium and The Netherlands on 10 May 1940 and crossed the Franco-Belgian border on 12 May. De Gaulle was appointed as Under Secretary of War in French Prime Minister Paul Reynaud's[258] cabinet on 5 June. Germany's Axis ally Italy invaded southern France, over the Alps, on 10 June. Fearing German arrest, De Gaulle escaped to England with British assistance on 17 June. The next day, 18 June, he made his first official broadcast on BBC radio from London, urging fellow French officers and soldiers to find their way to him to continue the "flame of resistance."[259] This was probably the first significant twentieth century use of the term "resistance" in this context.[260] An armistice ending hostilities between the Axis and France was signed on 22 June.

2. De Gaulle was not the immediate and obvious center of French resistance. His support base was initially very small. Many French citizens and foreigners believed he was hasty and personally ambitious, making them wary of his government-in-exile. De Gaulle soon established an organization called Free France, followed by the Français Comité National (French National Committee), as a government-in-exile, which the British government recognized on 23 June 1940. A week later Britain recognized de Gaulle as leader of all Free French.[261] Two French military units, one Foreign Legion and the other Alpine infantry, found themselves outside of France or its colonies during the fall of France. Only 30 of 700 infantrymen, and about half of the Legionnaires, joined de Gaulle. The remainder either repatriated to France or joined the British military.[262]

3. Reliant upon Britain, de Gaulle opposed both Germany and the French Vichy government. After June 1941, the British regularly denounced the Vichy government as pro-German and traitorous to France.[263] Britain now provided limited material support as part of a 7 August 1941 agreement with de Gaulle. The Vichy government considered French citizens seeking a continuation of the war against the Germans as traitors and tried them as such. In early July 1940, a Vichy military tribunal condemned de Gaulle to death for treason.[264]

4. De Gaulle did not command much by way of military force. He attempted to demonstrate usefulness to the British, in order to retain their support, by providing military intelligence on German forces and their movements in France. In this way de Gaulle sought legitimacy by demonstrating his access to French agents and resistance networks. To this end, de Gaulle established an intelligence bureau, which subsequently became the *Bureau Central de Renseignements et d'Action* (Central Bureau of Intelligence and Action), or BCRA. The BCRA treated resistance in purely military terms, organizing networks to collect and transmit military intelligence, which gave de Gaulle credibility with the British. Later, this effort evolved to include escape lines out of France and sabotage of German installations.[265]

5. After the Allied landings in North Africa on 8 November 1942, the Americans considered replacing de Gaulle with another French military leader who could unify France. Other French officers were seen as less difficult to work with than de Gaulle.[266] Admiral Darlan was the Vichy government's seniormost officer present in Algeria, had negotiated a truce with the Americans after the North Africa landings, and was one of those potential candidates supported by the Americans, until assassinated by a fellow Frenchman.[267] Army General Giraud then assumed civil and military powers in North Africa. He was protected by the Americans, who wanted North Africa as a place to build an offensive against the Axis and did not want to be hampered by French disputes. This presented a direct challenge to the leadership of de Gaulle.[268] However, French domestic based resistance networks supported de Gaulle as leader of the Free French, increasing his legitimacy. After de Gaulle gained the support of the French Communist Party, he became the sole figure capable of uniting all resistance efforts in support of the Allied cause.[269]

CS1.4. Early Resistance Efforts

After the armistice of 1940, some people in the Alsace and Lorraine regions did not assimilate into the German Reich.[270] In occupied France, the Germans banned any type of paramilitary groups (even scouting groups such as Boy Scouts), as well as all weapons, including hunting rifles, which had to be surrendered under penalty of death. Scouting was allowed in the Free Zone.[271]

A. **Embryonic Resistance.** Resisters came from throughout the political spectrum. The left's Popular Front, formed in 1936 and banned in 1939, now expanded its anti-fascist efforts and began lethally targeting Germans. Many on the right were former World War I Soldiers who diverged from Vichy's "strategy of collaboration" and believed that military intelligence should be shared with the British.[272] Early resisters were formed and defined by existing constituency groups: combatants and veterans, friends and neighbors, pastors and priests, academics and students, businessmen and labor leaders.[273]

B. **Early Sporadic Nonviolent Resistance**

1. Early French popular resistance exhibited itself in acts of defiance such as raising the French flag over a cathedral, or laying wreaths at war memorials. In November 1940, college students gathered at the Tomb of the Unknown Soldier at the Arc de Triomphe and marched down the Champs-Élysées,

symbolically reclaiming space where German Soldiers marched daily, resulting in arrests and the closing of universities.[274]

2. Women's committees in large cities in the Free and Occupied Zones organized many protests and marches addressing food and goods shortages. Police often arrested participants, who sometimes were able to disperse. Prisoners who were moved to hospitals for treatment were often freed by resistance members or by family members. The same often occurred during the transport of prisoners. Many citizens obtained information about German movements by land, sea, and air and were able to move such information out of France to the Allies.[275]

3. A large-scale miner's protest began on 1 May 1941, the French Labor Day, in the German occupied Pas-de-Calais and Nord regions. Protesters displayed the French tricolor and red flags with hammers and sickles flying from electric wires, buildings, and conveyer belts. Slogans demanded higher wages and praised Stalin. The strike spread to other northern mining regions, and added a women's demonstration to the protest. Once the Germans saw that French security forces could not halt the activities after several weeks, they moved in and brutally repressed them, arresting about 500 miners and their wives. About half were deported to Germany as laborers, and about half of those never returned. This brutal repression resulted in miners conducting sabotage. Several saboteurs were arrested, tried, and executed. Wives who assisted were tried and sentenced to forced labor.[276]

CS1.5. Developing Resistance Networks

A. French resistance elements ranged from ideological socialists and communists to anti-communists, anti-Semitic groups, anti-Freemason, or supporters of the French military.[277] These groupings existed prior to the war, and attempted to maneuver to impose their views for a post-occupation France. However, these different groups did not usually act against each other directly.

B. These networks were basically divided between the Free and Occupied Zones. Most of the Occupied Zone resistance activities were focused on sending intelligence to the Allies or helping POWs escape. In the Free Zone, propaganda against Vichy collaboration was the primary, though not sole, effort.[278] Together, by 1944, these networks were publishing several hundred clandestine newspapers with a total circulation of about two million copies.[279]

1. **Free Zone Networks**

   a. The Free Zone had several prominent resistance networks. The *Organisation Civile et Militaire* was composed of business leaders and factory managers. It had contact with the Armistice Army's Air Force intelligence service and sent intelligence through it to London.[280] Action Ouvrière was the underground metal workers union in Lyon that organized factory strikes to undermine production for the German war economy.[281] The right wing groups, Croix de Feu and Combat, were nationalist and anti-communist.[282]

   b. Lyon became the center of resistance in the Free Zone. The city had a dense population with a church center, a university, and a large working class mostly involved in manufacturing. Lyon also had *traboules*, a succession of interior courtyards and alleys that linked streets, almost maze-like, allowing for clandestine movement or quick escape. [283]

2. **Occupied Zone Networks**

a. On 23 August 1941, three assassins from the communist Youth Battalions killed a German naval warrant officer in Paris. It was the first direct attack on an individual member of the German armed forces, violating the unspoken rule that such assassinations would not be conducted. This assassination resulted in the beginning of collective reprisals. Germans executed six communists immediately, followed by 58 within four weeks. This reprisal, in turn, inspired another assassination by a communist cell in Nantes, resulting in the execution of 48 hostages. These more spectacular gestures of violence such as blatant assassinations were predominantly practiced by communists and resulted in collective reprisals by the Germans. The communist groups did not win many adherents, alienating many other resistance groups and a majority of the French population.[284]

b. De Gaulle responded to these assassinations and reprisals on 25 October 1941. He broadcast via the BBC that "War should be undertaken by those whose business it is. I am giving the order to those in the occupied territory not to kill Germans." He called for five minutes of silence to honor the dead and later invented and awarded the *Croix de la Liberation* to Nantes.[285]

c. Several other resistance networks operated in the Occupied Zone. Among them, the *Musée de l'Homme* resistance network operated out of that museum and assisted POWs and downed Allied airmen to escape. The network transmitted intelligence to the Americans and published a paper called "Resistance." The network continued its operation until an informant's betrayal resulted in arrests, trials, deportations to Germany, and executions.[286] Two resistance groups, the *Comité d'Action Socialiste*[287] and the *Defense de la France*, were organized by Sorbonne students in Paris.[288]

d. The communist-dominated, and very independent, *Francs-Tireurs et Partisans* (FTP) committed many assassinations resulting in horrific reprisals. The FTP claimed to have killed Germans at the rate of 500 to 600 a month and accounted for about one third of resistance manpower.[289] The *Union Nationale des Combattants Coloniaux* (UNCC) was formed to assist former colonial troops, primarily black and North African, held in POW camps and hospitals. The UNCC openly provided charitable assistance while secretly trying to extract them, providing civilian clothes, safe houses, and falsified identification.[290]

3. **Réfractaires and Maquisards**

a. Germany instituted a policy known as the *Relève*[291] in June 1942 and a forced labor draft, the *Service du Travail Obligatoire*[292] in 1943. The *Relève* allowed one French POW to be released in exchange for three workers sent to Germany. The *Relève* and the *Service du Travail Obligatoire* resulted in labor strikes. Many men disappeared and were known as *réfractaires*. These *réfractaires* did not all join resistance groups; some simply hid in forests and places such as the *Vercors* plateau.

b. *Réfractaires* provided many of the recruits for resistance groups hiding in the backwoods and mountains that later became known as *maquisards* or *maquis*.[293] Initially, these groups developed a bad reputation due to their raids on farms for food, town halls for false papers, and tobacconists for cigarettes.[294] *Maquis* groups began forming in the southern mountains and forests of the Alps and the Central Massif in 1942, holding much local power by 1943.[295] Their numbers amounted to about 30,000-40,000. Many Spanish Republican veterans of the Spanish Civil War served as instructors for these young men, who were new to resistance.[296]

CS1.6. Resistance Network Dynamics

A. Internal Rivalries

1. The broadly organized internal resistance networks often competed against each other for post-war political gain. Many metropolitan networks, which tended to be socialist and communist, intended to use the conflict as an opportunity to establish an entirely new, leftist political order. The Free French under de Gaulle saw the purpose of resistance networks as facilitating Allied military activities to liberate France, and sought the reestablishment of a republic.

2. After Vichy's Armistice Army was disbanded, many of its officers established a secret organization named the Organisation de Résistance de l'Armée, intended to reveal itself when the Allies entered France. Another large group existed, the Armée Secrète. Its members went about their normal daily lives and intended to come out in units to assist the Allies during their advance.[297]

B. Franco-British Rivalries

1. The agents and the radio operators of the BCRA-controlled networks were sent from outside of France to gather military related intelligence as well as establish escape lines. Meanwhile, the British SOE established its Section F to build resistance groups in France. It interviewed French refugees for potential agents and recruited them without consulting the BCRA. SOE cared little about internal groups maneuvering for post-war political gain, and built about one hundred networks, which included communists. Resistance agents working for SOE were given sophisticated false documentation, and often had hybrid Franco-British backgrounds to ensure French cultural and linguistic identity but British loyalty. Parents were typically listed as dead and birthplaces were from locations bombed and destroyed or out of reach of German or Vichy authorities.[298]

2. After promulgation of the forced labor laws swelled the rolls of maquis, the SOE became interested in building a secret army for use during the upcoming invasion, though the British did not want to promote widespread uncontrolled guerrilla warfare. Both de Gaulle and SOE desired the resistance groups to rise up only when instructed to do so in support of the upcoming invasion. The British armed forces also organized their own SOF which operated independently, sometimes in conjunction with SOE and sometimes conflicting with it.

C. **U.S. Involvement.** Finally, in 1942, the newly created U.S. Office of Special Services (OSS) also entered this arena. Its cooperation with the SOE in France was good and OSS had a similar outlook on the French resistance, which was to focus on anti-Nazi operations and eventual assistance to invading Allied CF and not post-war political outcomes.[299]

CS1.7. Network Commonalities

A. **Cell Size.** Despite political and operational differences, the French resistance networks had multiple commonalities such as operating in small cells, with many participating only part-time.[300]

B. **Security.** Resistance members adopted *noms de guerre* for their safety.[301] Some individuals disappeared from their former lives with completely new identities. During capture or confinement, they often feigned illness to secure transfer to hospitals, which were less secure than prisons, where their escape could be organized and assisted.[302] Sometimes untrained amateurism destroyed networks. In one instance, a resistance member left a list of agents on a train, which found its way into the hands of the Germans, revealing the entire network.[303]

CS1.8. Political Unification Efforts

A. Limited unity of the various movements was finally achieved in May 1943, when de Gaulle sent his personal representative, Jean Moulin, on a clandestine mission into occupied France. Moulin succeeded in establishing a *Conseil national de la résistance* (National Council of Resistance) or CNR, joining all the major movements into one federation under the London-based leadership of de Gaulle, which added to his credibility and legitimacy.[304] The next month, de Gaulle announced the formation of the *Comité Français de la libération nationale* (French Committee of National Liberation) or CFLN, which became the provisional government of the Republic in waiting, with de Gaulle and Giraud as co-presidents.[305] These two entities demonstrated the unity of resistance in France and the colonies, and seemed to end the power struggle between de Gaulle and Giraud.

B. De Gaulle attempted to increase his credibility and legitimacy with the metropolitan communist networks with a message to them stating that he desired a "renewal" of the nation and actualization of the ideals of "liberty, equality, and fraternity."[306] In a BBC broadcast several weeks later, he stated that "national liberation cannot be separated from national insurrection." Knowing that the communists wanted to involve more of the populace in a broader social movement, and wanted a very different type of post-war government, this carefully and cleverly worded statement seemed calculated to appease them. However, this instinct to insurrection was exactly what he sought to tame later.[307]

C. The Gestapo arrested Jean Moulin in June 1943. After his subsequent death, and other arrests and deaths, the leaders of the resistance gravitated to Paris. Many of them attempted to reclaim independence from the London-based leadership. During this power struggle, the communist party gained influence within the CNR thanks to the Red Army's victory in Stalingrad in February 1943, and the earlier Vichy decision to accede to German demands to export French workers to Germany. The communists tried to channel this into support for themselves and their call for national insurrection. The BCRA and the CFLN were now challenged to contain communist resistance and to reinforce the power of the state.[308]

D. Communist resisters formed the *Comité de libération de Paris* (Paris Liberation Committee) in September 1943, whose membership soon expanded to non-communists. The communists intended to draw on popular discontent and prepare for national insurrection. The city now contained many *réfractaires*, escaping round-ups and running out of food and money. The communists sought to arm them in support of leftist insurrection.[309] Many France-based leaders saw the call to immediate action and national liberation as a way to gain independence from the leadership in London and Algiers.[310]

E. In response to the communists, the non-communist resisters in London formed the *Comité General d'Études*, or General Committee of Studies. This group selected persons who were prepared to take ministerial roles immediately upon liberation and establish legitimacy to prevent a national insurrection.[311] De Gaulle began deft internal political maneuvering by handing out the prospective portfolios. In the spring of 1944, de Gaulle ended his dual presidency with Giraud by using Giraud's association with former Vichy Vice Minister Pucheu. Pucheu was tried and executed for his role in executing communist resisters in the Free Zone. Girard was deprived of his post as commander-in-chief and departed the scene. This event drew the communists closer to de Gaulle.[312]

CS1.9. External Support

A. **Information Operations and Strategic Communication**

1. The BBC broadcasted daily into France, giving the Free French five minutes a day. The BBC was banned in Occupied France in 1940 and in the Free Zone a year later. Listening was punishable by fines or imprisonment. Yet, the number of French listeners rose from approximately 300,000 in 1941 to three million in 1942.

2. An early form of internal resistance propaganda were leaflets, called *papillons* (butterflies), that were pasted in public places such as lampposts and urinals.[313] In March 1941, the BBC urged the use of the "V" for victory sign throughout France. Its appearance gradually increased until the Germans departed in 1944.[314] In Morse code, the "V" was three dots and a dash (…-) which was also the opening beat of Beethoven's Fifth Symphony and could be hummed or tapped.[315] Busts of Marianne, symbol of the French Republic, also appeared in graffiti or physical form as a symbol of resistance.[316] The Cross of Lorrain, symbol of the territory annexed by Germany, was used by de Gaulle's Free French forces imprinted on the tricolor flag and used to decorate their equipment.[317] Its appearance in graffiti form in France also symbolized resistance.

B. Jedburgh Teams

1. During World War I, T.E. Lawrence worked with Arab guerrillas against Ottoman Turks on the Arabian Peninsula in coordination with conventional British forces. These actions tied down large numbers of Turkish military personnel guarding logistical lines, giving British CF greater freedom of maneuver. This activity inspired the World War II concept of Jedburgh teams.[318]

2. These teams parachuted into France and linked up with resistance fighters behind enemy lines. These were three-person, multinational teams consisting of a British or American officer, a French officer, and a radio operator of one of those nationalities. This multinational combination allowed for the necessary military, cultural, and political understanding, combined with necessary language capabilities and technical communication knowledge.[319]

3. The teams liaised for the purpose of guiding resistance activities in support of Allied conventional units. They coordinated air drops of arms and equipment and conducted specific skills training. They also provided leadership if required. These activities were better coordinated and executed than activities conducted by Lawrence, due to technologically improved radios and better aircraft capable of larger payload drops at night and during rougher weather.[320]

4. The Jedburgh concept was very limited in scope. The teams were conceived and designed specifically to assist Allied CF once the invasion of France began. Thus, they only deployed beginning on the night of 5 June 1944, only hours prior to the beach landings and large drops of whole conventional units.[321] Ninety-three teams were deployed, with a total of 265 members, operating until late September 1944.[322] These teams were most successful when working with well-organized *maquis*, closely linked to de Gaulle's Free French.

CS1.10. Integrating Resistance into the Conventional Force Liberation Campaign

A. The debate centering on two options of how to integrate the resistance into the liberation of France lasted well into 1944. First was a more military option: liberation by bombing and inserted agents acting against specific targets to cripple German military power. The second option was national insurrection through guerrilla warfare, allowing leftist ascendency by communist resisters.[323] During the course of over a year, the Allies gradually began to favor a provisional government with a resistance organization supporting the

Allied military advance, and were opposed to the communist model of national insurrection which could allow a communist takeover of France.[324]

B. Specialized functional resistance networks were organized to integrate with Allied efforts early in the war. These networks served a more military function in direct support of the Allies without the political goals of most other resistance networks. There were intelligence networks such as Phalanx[325] and Nimrod,[326] sabotage circuits closely linked to and directed by Britain's SOE,[327] and French organized escape lines for POWs and downed aircraft crews such as the Comet and Shelburn Networks.[328]

C. As part of D-Day planning, the Allies incorporated guerrilla warfare by resistance units into their plans. On 5 June, the day prior to the landings, the BBC broadcast coded messages to resistance forces to sabotage railways (Plan Green), roads (Plan Tortoise), and telecommunications networks (Plan Violet) to impede movement of German forces toward Normandy.

D. The Allies hoped to push German forces back into Germany with minimal political disorder in France. By D-Day, the Americans had not excluded the possibility of establishing a military government, as in Italy, but de Gaulle opposed this idea. De Gaulle sought to convince and demonstrate to the Allies that he had broad French popular support, that the French would liberate themselves with Allied help, and that the communists would not seize power—thus precluding the need for a military government. On 6 June, he broadcasted "our actions behind enemy lines must be linked as closely as possible to the front line action of the Allied and French armies." On 7 June, a delegate of the French Communist Party countered and also broadcasted a message via the BBC stating that the French should "arm themselves and fight to drive the invader from the soil of the motherland and restore liberty and independence."[329] Furthermore, the Americans sought elections for the people to choose their government, obviating the need for a provisional government led by de Gaulle.

E. The *maquis* took immediate actions that led to negative repercussions. Many donned arm bands reading "FFI," designating themselves as part of the *Forces Françaises de l'Interieur* (French Forces of the Interior), hoping to be treated as POWs if caught, and not as terrorists. Untrained and impetuous, the *maquis* made a number of mistakes, including daylight attacks and fights against French and Allied units. The Allied solution was to concentrate *maquis* elements in redoubts and supply them by air with weapons and leadership to slow German forces moving toward Allied lines.[330]

F. Britain's SOE delivered over 45,000 Sten guns, 17,000 pistols, 10,000 rifles, 300 bazookas, and 143 mortars to resistance forces. Fighters destroyed large sections of rail lines, preventing German reinforcement to Normandy by rail, and restricting them to roads, which now allowed more ambush sites, greatly slowing German movement.[331]

G. Multiple resistance groups took advantage of post D-Day turmoil, seeking different end states. The cities of Toulouse and Lyon drew close to a state of anarchy due to violent guerrilla activities conducted by disparate resistance groups, and a lack of Allied or Free French forces.[332] This included Spanish communists conducting anti-fascist violence and seeking to carry their effort into Spain, and French communists in Lyon seeking to ignite national insurrection. The Paris Liberation Committee, dominated by communists, asserted itself in Paris and took little advice from Allied and Free French delegations. This committee organized strikes and marches. They also formed neighborhood security units called *Milices Patriotiques* (Patriotic Militias), which they hoped would form the nucleus of a national insurrection.[333]

H. The first 10 weeks after the Allied invasion in Normandy resulted in untold numbers of dead *maquisards*. The reasons ranged from poor tactics to reprisals by the Germans, who regarded them as terrorists to be shot summarily. Many population centers across France, suspected of harboring *maquisards* suffered horrible reprisals including mass executions and whole towns and villages being burned.[334]

I. On 16 August 1944, the 1st Free French Division, formed with French troops in exile and attached to the American Seventh Army, landed in Provence. By early September, de Gaulle dissolved the FFIs, giving them the choice of returning to their homes or joining the regular army.[335] Meanwhile, the Americans still did not recognize de Gaulle's provisional government.

J. Resistance groups infiltrated the Paris Prefecture of Police gaining considerable freedom of movement and action. On 18 August, the Paris Liberation Committee and the CNR agreed on an insurrection order and posted it all over Paris the next day.[336] Police in civilian clothes seized the police prefecture for the resistance, the tricolor was flown around the city, and resisters patrolled in requisitioned cars. This occurred with a significant German military presence still in the city. De Gaulle feared a communist takeover or a Warsaw Uprising style bloodbath. Meanwhile, Allied and French troops moved toward Paris while in concurrent negotiations with the German commander. The 2nd Free French Armored Division entered Paris on 24 August. The Germans surrendered the next day. Insurrection and bloodbath were avoided and the city spared. On August 26, de Gaulle led a victory parade down the Champs-Élysées.

CS1.11. Post-Conflict

A. **Transition.** After the liberation of Paris, de Gaulle continued his political maneuvering to establish his provisional government in an effort to head off any communist political or military resurgence. He established the primacy of the regular military and his provisional government, headquartered in Paris. He ordered the demobilization of resistance forces including the FFI. He replaced elements of the internal resistance with members of his Free French Forces, and reestablished state power. Dashing communist hopes for a new political order, de Gaulle stated that "France is not a country that is beginning. It is a country that is continuing." The majority of de Gaulle's government contained Free French who served him in London.[337] The supremacy of the regular army and the state was further asserted through a succession of ceremonies highlighting the role of the new chivalric order, the *Compagnons de la Libération*.[338] They were proconsuls, empowered to deal with problems such as food supply, collaborators, and the restoration of order. Town level liberation committees were gradually displaced.[339]

B. **New Political Parties.** Resisters desiring post-war political power organized themselves into new political parties, with two having primacy: the *Mouvement de Libération Nationale* (non-communists) and the *Front National* (communists). Other socialist and Christian Democrat parties also formed. De Gaulle dealt with other internal forces gradually, such as Spanish Republicans who wanted to liberate Spain from Franco, German anti-fascists who wanted to establish a free Germany, and Zionist Jews who wanted to establish a Jewish state in Palestine.[340]

C. **Justice and Vengeance.** Liquidating the Vichy heritage threatened grave domestic stress. An informal and spontaneous purge of Vichy officials and supporters began in the summer of 1944, with Resistance bands conducting over 10,000 summary executions. Later, more systematic retribution followed. About 80,000 Frenchmen were killed after the liberation in the settlement of wartime scores. Special courts were established to try citizens accused of collaboration and heard approximately 125,000 cases over the next two years. Some 50,000 offenders were punished by "national degradation" (loss of civic rights for a period of years), almost 40,000 received prison terms, and between 700 and 800 were executed.[341] Marshal Pétain, the leader of Vichy France and hero of World War I, was sentenced to life imprisonment on a small coastal island, after his original death sentence was commuted by de Gaulle.

D. **Reconciliation.** Amnesty laws were passed in 1951 and 1953 allowing former Vichy officials to return to government posts.[342] After the repatriation of former prisoners and deportees, elections were held. Former resistance leaders dominated the outcomes and effected a politically leftward move: three quarters were communists, socialists, or Christian Democrats. The Fourth Republic was established but lacked long-term stability, leading to the Fifth Republic in 1958.[343]

CS1.12. Summary

A. **Resilience.** Internal political differences and the psychological effects of previous wars undermined the resilience of the French people and their government. Significant internal political divisions were compounded by the memories of Prussian reprisals for actions of irregular snipers (*Franc-Tireurs*) followed by extraordinarily high casualties in World War I.

B. **Pre-Conflict Resistance Planning**

1. **Organized Resistance.** The French government did not prepare for resistance warfare under a foreign occupation. Socialists and communists possessed the only extant networks, organized for self-protection, connected to the Communist International, and aimed at undermining the Third Republic.

2. **Government-in-Exile**

   a. There was no predesignated government-in-exile. The German partial occupation of France, annexation of some territories, and the creation of a puppet government based in Vichy led to a confusing situation. The French colonies, including the military forces located there, were placed under Vichy government control. French forces not subjecting themselves to the Vichy government were considered to be in rebellion against it. This was the situation when de Gaulle left France for Britain and sought to establish himself as a center of power opposing the Vichy government.

   b. Charles de Gaulle gathered a small military force in England and gradually won the allegiance of French people inside and outside of France. He also gained the support of the Western Allies as the center of French resistance to the Germans, and eventually established an organization that for all intents and purposes functioned as a legitimate government-in-exile.

C. **Integration of Resistance during Conflict**

1. **Integration.** The myriad of French resistance networks represented a politically disparate, and disjointed array of prospective post-war interests. Geographically, the leftist groups were centered on Paris while the rightist groups were centered on Lyon. A significant number of Spanish Republicans who fled Franco's Spain were present in southern France. They sought to return and fight, and were sympathetic to leftist French organizations. Many groups had specific purposes such as rescuing French prisoners, rescuing Jews, assisting Allied aircraft crews to escape, or gathering intelligence and sending it to England. Eventually, almost all of these groups, including reluctant communists, pledged support to de Gaulle and his Free French organization.

2. **Ideological Cleavage.** Throughout the war, de Gaulle and the Allies were always wary of the leftist groups. They understood that these groups did not seek to reestablish a French Republic but to begin a revolution for a new socialist France, aligned with the Soviet Union. Even after their pledge of support,

de Gaulle insisted that the Allies take Paris to ensure that the communist organizations around the city did not seize the capital in the wake of the German retreat.

D. External Support to Resistance

1. **Military.** The Allies lent much direct support to the resistance organizations in France in the form of material, advisors, and funding to gather intelligence, support POW and aircraft crew evasion and escape, and conduct sabotage. The multinational Jedburgh teams, formed specifically to assist Allied CF after D-Day, were the most useful to the Allied military effort. They could communicate with London for directives and advise networks on effective courses of action.

2. **Political.** De Gaulle established himself as France's political leader seeking to reestablish the pre-war French state. He gained credibility with the British SOE and American OSS by sharing information and intelligence gathered in France, and legitimacy among the French people by securing British and American support for his Free French Forces and BCRA. De Gaulle deftly maneuvered himself to the forefront of displaced French leaders by forming the Free French as an organization, using it in direct support of the Allied effort, and aligning its goals with those of the Allies. He then convinced the Americans and British to not establish a military government and gained their gradual acceptance of his ability to counter the communist threat.

E. Resistance Strategies

1. The Nazi policy of terror conditioned the actions of the resistance, though it had a lesser effect on communist organizations. The French resistance groups inflicted minimal casualties on German personnel and material. They often acted directly against collaborating countrymen instead of Germans in order to avoid reprisals.

2. De Gaulle aligned the purpose of his Free French resistance networks with those of the British and Americans. They directly supported the military effort to liberate France by gathering and sending intelligence to London to better inform invasion plans. Additionally, they built larger networks for eventual direct support of the invasion by slowing German reinforcements.

3. The communists conducted direct attacks against the Germans with the intent of bringing about harsh reprisals against the population. This was intended to radicalize the population and support the communist cause. The communists did not seek to restore France to its pre-war form but to bring about a revolution, and in some cases union with the Soviet Union. Their methods brought much suffering to the population and they united with other organizations only to expel the Germans.

4. Ultimately, the policies of de Gaulle regarding the operations of resistance forces proved victorious. He used the French resistance in direct support of allied military necessity. During the invasion of Normandy, French resistance forces conducted crucial sabotage activities, ambushes, and raids to slow the movement of German reinforcements moving toward Normandy. That gave the allies critical time to land enough men and material to build up insurmountable forces. The communists' tactics never ignited the revolution for which they hoped for, but simply caused the deaths of many French citizens through Nazi reprisals.

F. **Restoration of Sovereignty.** After the landings in North Africa, de Gaulle successfully fought to maintain his political supremacy among French officers outside of Vichy. This success carried on to the D-Day landings, after which he was able to march into Paris at the head of a small Free French contingent and prevent communist takeover. He then presented and insisted on the perspective that the French liberated themselves, though with Allied assistance, and thus did not require an Allied military government. De Gaulle's persona and political acumen allowed him to reestablish an independent and sovereign France, defeat socialist and communist groups, and avoid an Allied military government seeking elections which may have resulted in a very different government. De Gaulle may have proven himself as *"le premier résistant"* of the twentieth century.[344]

Case Study 2. Poland: World War II Resistance[345]

the largest and best organization in Europe.[346] - Harold B. Perkins, SOE

CS2.1. Background

A. Nazi Germany and the Soviet Union signed the Molotov-Ribbentrop Pact, a German-Soviet Nonaggression Pact, on 23 August 1939. On 1 September 1939, Germany invaded Poland under two pretexts: Polish persecution of ethnic Germans, and a Polish troop incursion (German staged) into Germany. This resulted in France and Great Britain declaring war on Germany on September 3. A German blitzkrieg took their forces to the edge of Warsaw by 8 September. On 17 September, the Soviet Union invaded Poland.

B. The next day, 18 September, the Polish government refused to surrender, and escaped to Romania. Military units were ordered to move south, evacuate Poland, and reconstitute in France. The Commander-in-Chief of the Polish Army appointed General Michael Tokarzewski to organize underground resistance in both parts of occupied Poland.[347] The Warsaw garrison surrendered on 28 September.

C. Soon after defeating Poland, the Germans and Soviets administratively re-organized the country. The Germans annexed Western Poland (the Polish Corridor, Lodz, and Polish Silesia), which contained large numbers of ethnic Germans. The Soviets no longer recognized Poland as an independent state and annexed eastern Poland. In the central part of the country, the German occupiers formed the General Government as an autonomous part of Greater Germany and divided it into four districts: Krakow, Warsaw, Radom, and Lublin. The German Governor-General, headquartered in Krakow, ruled directly, versus ruling through Poles, except at the lowest administrative levels.

D. During the initial invasion, as Polish security and police forces withered, robbery and looting increased. Many Poles were initially optimistic when the German Army halted the anarchy. This crackdown on the criminal element was likely the primary factor contributing to a measure of early popularity of the occupier. Then, the Nazis expropriated Jewish property and leased it to Poles, offered jobs, and began rebuilding and expanding infrastructure (to move goods, labor, and resources out of Poland and into Germany). Debts owed to Jewish creditors were canceled and initial food quotas were relatively small. Some gratitude to the occupiers accompanied accommodation to them. As the occupiers increasingly imposed regulations on traditional aspects of Polish civic life, and demanded compliance with lethal penalties against individuals and then gradually whole groups, a resistance mentality was nurtured and eventually increased.[348]

CS2.2. Polish Independence and Pre-War Resilience

A. Poland's geographic location between great powers helped form a resilient society built upon a common set of nationalist values and rules. These stemmed from a shared culture, language, religion, and a national identity cemented during a long history of resistance to invasions and occupations.[349] Up to and through World War I, much of modern Poland was occupied by Austro-Hungarian, German, and Russian forces. Several Polish organizations were formed to seek independence from foreign powers. The two primary organizations were the Polish Military Organization (*Polska Organizacja Wojskowa*), led by Józef Piłsudski and formed at the outbreak of World War I, and the National Democrats (*Narodowi Demokraci*), led by Roman Dmowski, which was formed in the late nineteenth century. In 1918, after the Treaty of Versailles, Poland regained self-rule after nearly a century and a half of foreign partitions.

B. Immediately following Poland's new independence, revolutionary Russia under the Bolsheviks sought to regain its Polish interests in the Polish-Bolshevik War of 1919-1921. Polish organizations united behind Piłsudski and defeated the Bolsheviks. The primarily Catholic nation had its national conscience reinforced by church exhortations against "godless Bolshevism." Recent Polish independence, and the defeat of Russian Bolsheviks, permeated the national consciousness and became part of education and military lore. National festivals were fused with Catholic church holidays and celebrated together. These shared experiences brought about camaraderie and national solidarity between social classes.[350]

C. The population of the new Polish state was overwhelmingly Catholic Poles, with several significant ethnic minorities such as Protestant Germans in the former East Prussia, Orthodox Christian Ukrainians (Unitarian), and primarily Orthodox Jews. Each ethnicity spoke its own language and lived in self-imposed separation, interacting mostly though trade. Social conflict among these groups existed but never rose to the national level, likely due to the overwhelming numbers of the Catholic Polish majority. Within that group, social conflict and class struggle were alleviated by a collective historical conscience, a common religion, and national solidarity fostered by the Church, voluntary organizations, schools, and the army. Class struggle was neutralized through the dedication of the elites to community welfare through direct participation in civic and social welfare organizations.[351]

D. During the struggle for independence encompassing World War I and the Polish-Bolshevik War from 1914 to 1921, the Poles established volunteer organizations to oppose the policies of the partitioning powers. Examples include the paramilitary Rifleman's Association, "Rifleman (*Zwiazek Strzelecki "Strzelec"*), its equestrian branch, the Mounted Military Preparedness Group (*Konne Przysposobienie Wojskowe "Krakus"*), and civic/aid organizations such as the Volunteer Fire Guards (*Ochotnicza Straz Ogniowa*), the Polish Red Cross (*Polski Czerwony Krzyz*), Catholic Action (*Akcja Katolicka*), and youth and women's auxiliary groups sponsored by both government and opposition groups. These organizations and their operations instilled the habit of membership in independent volunteer groups, while also aiding the formation of national identity against the centripetal forces of Poland's large neighbors. These groups also gave people a sense of self-sufficiency which translated into resistance to obligations imposed by the central government. Additionally, due to the overwhelming Christianity (Catholic, Protestant, and Unitarian) of the country, Christian references were used to evoke national unity such as referring to the new Polish independence of 1918 as the "resurrection of Poland" and the defeat of the Bolsheviks as the "Miracle on the Vistula."[352]

E. In 1926, Marshal Piłsudski led a coup and took power. He installed the Sanacja regime, which lasted until 1939. The new regime ruled through primarily one party, using governing power to marginalize the others, especially the communists. The regime controlled patronage, used police coercion, and divided its enemies. This quasi-democratic system left a legacy of mistrust in government institutions.[353]

F. As the Polish majority tried to form a new post-World War I democratic state, a substantial portion of peasants and urban socialists opposed communism but desired a new postwar political order. This resulted in socialist organizations, political and military, remaining outside the structure of what later became the Secret State and its Home Army.[354]

CS2.3. Government(s)-in-Exile

A. On 17 September 1939, during the German attack on Poland, the president of Poland authorized members of the Polish government to go abroad to continue Polish sovereignty and resist the invaders and their claims. Primary members of the Government of the Republic of Poland escaped and established a government-in-exile, first in France and then in London. This government carried significant authority and legitimacy with the Western Allies because: 1) it had substantial military units outside of Poland fighting alongside the Allies; 2) the invasion of Poland was Britain's casus belli for its declaring war on Germany, and; 3) a majority of Poles accepted the exiled officials as the true leaders of Poland, giving it *de jure* legitimacy. This government-in-exile was recognized by the Allies and was able to exert a degree of control in Poland through underground movements, excluding the communists, which consisted of a secret state that eventually became The PUS and the Home Army.[355] Allied recognition of this government was withdrawn on 5 January 1945 with Western *de facto* and later *de jure* recognition of the Soviet Union-backed government of Poland.[356] The government-in-exile continued to exist, however, until 22 December 1990 when it recognized Lech Walesa as president of Poland.[357]

B. The Union of Polish Patriots (*Związek Patriotów Polskich*), formed in Moscow, competed with the London-based government of the Republic of Poland for government-in-exile status. The Soviets dropped Polish communists into Poland in 1942 to set up a Worker's Party (*Polskiej partii robotniczej*) which included a resistance organization called the People's Guard (*Gwardią Ludową*)."[358] A Soviet proxy, the Provisional Government of the Republic of Poland (*Polski Komitet Wyzwolenia Narodowego*), established in 1944, claimed to be the legitimate government. It declared the official end of the Second Polish Republic (1918-1939) and the Polish government-in-exile. The example of Moscow's competing government-in-exile highlights the risks a legitimate government-in-exile must confront and overcome.

CS2.4. The Nazi Occupation

A. **General.** Between September 1939 and late 1944, the Nazi occupation gradually increased in brutality, moving from administrative and economic control to an attempt to totally control the populace.[359] The increasing ruthlessness stemmed from Nazi ideology, which identified Poles as Slavs, and thus sub-humans to be exploited. The two most disruptive policies imposed by the Nazis, that affected the majority of the populace, were food and labor quotas. The quotas were ever increasing, with penalties for non-compliance becoming increasingly lethal. These policies led to death through starvation for many in their homes or death by disease in labor camps. These policies engendered resistance more than any other policy of the Nazis.[360]

B. **Security Apparatus.** The Nazi security apparatus employed terror as the primary tool to force compliance. The security apparatus was split into two elements: an investigative branch, the Security Police, and a constabulary or gendarmerie, the Order Police.[361] The Security Police conducted intelligence and CI activities and investigated political activities. The Order Police enforced laws dealing with population movement, trade, health, and cultural activities and fought common and political crime. Auxiliaries consisting of ethnic Germans, Poles, Polish Ukrainians, and Soviet citizens were also used. These auxiliaries operated in small units under direct German supervision and conducted food confiscation, pacification actions, slave labor dragnets, anti-black market raids, and anti-partisan sweeps. They also conducted mass executions and deportations. The Polish Police were subordinate to the Order Police and exempt from conducting executions and round-ups. They fought common crime and coordinated night watches and fire brigades.[362] Sometimes the Polish Police were able to facilitate the escape of underground members or warn about threats from the gendarmerie or Gestapo.[363]

C. **Targeted Collective Terror.** As the Nazi occupation went on in time, it grew ruthless, increasingly engaging in deportations for slave labor and Jewish extermination. Offenses led to wholesale punishments of groups of people, due to the inherent racism of Nazi policies and their goal of destroying the Polish elite class. Large groups of probable associates of perpetrators, defined by social circles and geographic locations, were arrested and incarcerated, deported to labor camps, or killed. The Nazis targeted the Catholic church because they considered it to be the primary expression of Polish nationalism, thus facilitating resistance.[364] Sham partisan units, consisting of non-Germans, and part of the security apparatus, would often go into the suspected area to conduct reconnaissance and gather information prior to police units sweeping in.[365]

D. Use of Polish Government Bureaucrats and Institutions to Execute Policy. The occupiers used Polish bureaucrats to execute Nazi policy. The intent was to divide the Polish populace, while giving the Nazis the opportunity to claim credit for things deemed charitable. Teachers were used to issue identity documents and visit households to admonish peasants to fulfill their food quotas, and other duties, in addition to classroom instruction. Priests had to read official pronouncements from their pulpits after mass. Village quota committees made up of teachers, priests, lawyers, doctors, wealthy farmers, and prominent citizens, were tasked with ensuring fulfillment of the German food and labor service quotas. In practice, they became hostages, initially with their property at stake, and later their lives. Many Poles cooperated by accommodating for reasons such as maintaining an income, fear of being assigned a worse job, fear of deportation as forced labor, and often in the belief that they could moderate Nazi policies at their level of execution. Some required official employment cover for their underground activities, and some were able to successfully intervene when their countrymen were maltreated. Sometimes they hesitated to intercede for fear of endangering themselves and their families. Some were able to enrich themselves. Often, Poles in these official positions acted disloyally to the Nazis, thus jeopardizing themselves, but others executed Nazi policies zealously, enraging the underground and similarly jeopardizing their lives.[366] Nazi terror caused most people to accommodate as a rule, resist when necessary or able, and collaborate for self-preservation.[367]

CS2.5. Resistance Groups

A. Embryonic Resistance

1. Prior to 1939, Poland did not have an organized pre-war resistance infrastructure. In lieu of such plans, the political, fraternal, social, and veteran organizations continued to meet, often clandestinely. Organizations such as the paramilitary Rifleman's Association provided some training, while police officers, teachers, forest wardens, and bureaucrats recruited friends and relatives, forming clandestine networks. Polish bureaucrats, functioning as auxiliaries, siphoned money from government coffers, and teachers secretly taught Polish history and culture. Pre-war organizations' clandestine meetings formed the nucleus of resistance organizations that eventually conducted clandestine activities. In the meantime, political parties and their publications also went underground.[368]

2. Forbidden by occupation authorities, writers, artists, and educators continued their activities in secret. Between 1939 and 1945, over 2,000 underground papers and periodicals were published in German occupied Poland. Polish educators ran a system of secondary schools with pre-war curriculum. Warsaw had two universities and one polytechnic institution that continued secretly under occupation. Secret military officer's schools, non-commissioned officers' schools, and numerous other specialized military schools trained cadets for a clandestine army.[369]

B. Nonviolent Resistance

1. The Young Scouts (a youth organization) engaged in acts of "little sabotage," including writing anti-German slogans on walls and defacing German property. The intended effect was to make German occupiers feel unwelcome and uneasy, while bolstering Polish morale.[370] Sometimes cooperative

employees invited the organized resistance to attack shipments, warehouses, and stores to cover embezzlement and to provide supplies to the underground.[371]

2. Cheating became the predominant method of passive resistance by a majority of the population. Polish officials falsified documents to aid the resistance or alleviate hardships.[372] Food was sold on the black market for a better rate.[373] Underground shelters were built to conceal people, domestic animals, and foodstuffs like boxes of grain and salted meat in barrels, in order to survive the food quota. Millers were bribed to obtain false records of grain surrendered for the food quota. Warehouse workers mixed grain with sand to increase the weight of food delivered under quota. Stones were put in potato sacks delivered to the Germans to increase their weight. Milk surrendered under quota was diluted with water. Pigs were illegally slaughtered and affidavits were issued by officials confirming that they were stolen by bandits. Acts of sabotage were often disguised as equipment malfunction or German carelessness. The black market was the most widespread and successful form of popular resistance, serving as a release valve in a totalitarian system, and even providing the occupiers with goods and services not otherwise obtainable.[374] The fact that cheating was the most prevalent form of resistance among the majority demonstrated that accommodation was the primary mode of interaction between them and the occupiers.[375]

C. Development of the Polish Underground State

1. **Legitimacy.** Polish tradition and societal rules established during long periods of occupation, before World War I, were reinforced by underground courts and the accepted structure of an underground state. This legitimacy extended to the World War II Polish resistance organizations, united under the PUS. The vast majority of people granted legitimacy to the Resistance, and the Polish Government abroad, as the continuations of the pre-war Polish state.[376] This extension of popular legitimacy to the PUS, as an alternative source of authority, legitimized it and unified resistance under it. The PUS leadership acted in a disciplined and restrained manner to prevent reprisals. It also provided law and order in areas where occupation forces were sparse and could not check common criminality.[377] The communists remained outside of it, were loyal to the Soviet Union, and received material support from it.

2. **Structure.** Eventually, the PUS became an extraordinarily developed organization. At its head was the legitimate government-in-exile in London, with executive and consultative bodies, and representatives of the major political parties. It was able to communicate instructions to the underground inside Poland, the Home Army (guerrilla function), and a secret administration (shadow government function). The Home Army was organized by function, as was the shadow government, which had the ability to conduct investigations of collaborators and mete out justice, as well as communicate to the general population. It was likely the most developed resistance organization in Europe at the time (see fig. 17).

3. **Unification.** The process of unifying various resistance organizations began in 1940 with the formation of the Political Coordinating Committee (Polityczny Komitet Koordynacyjny), or PKP. By 1944, except for communists and a small faction of National Democrats, the resistance organizations were militarily and politically consolidated under the central authority of the PUS. By then, the PKP's delegate held the rank of deputy premier within the exiled government in London. Thus, the Polish resistance became Europe's most unified and centralized movement, with the comprehensive infrastructure of a state, including central and local civil institutions and an organized military structure.[378]

4. **Motivations.** People joined this movement typically for reasons of security and revenge, and not of ideology (excluding the communists). In areas with little law and order, people sought protection with

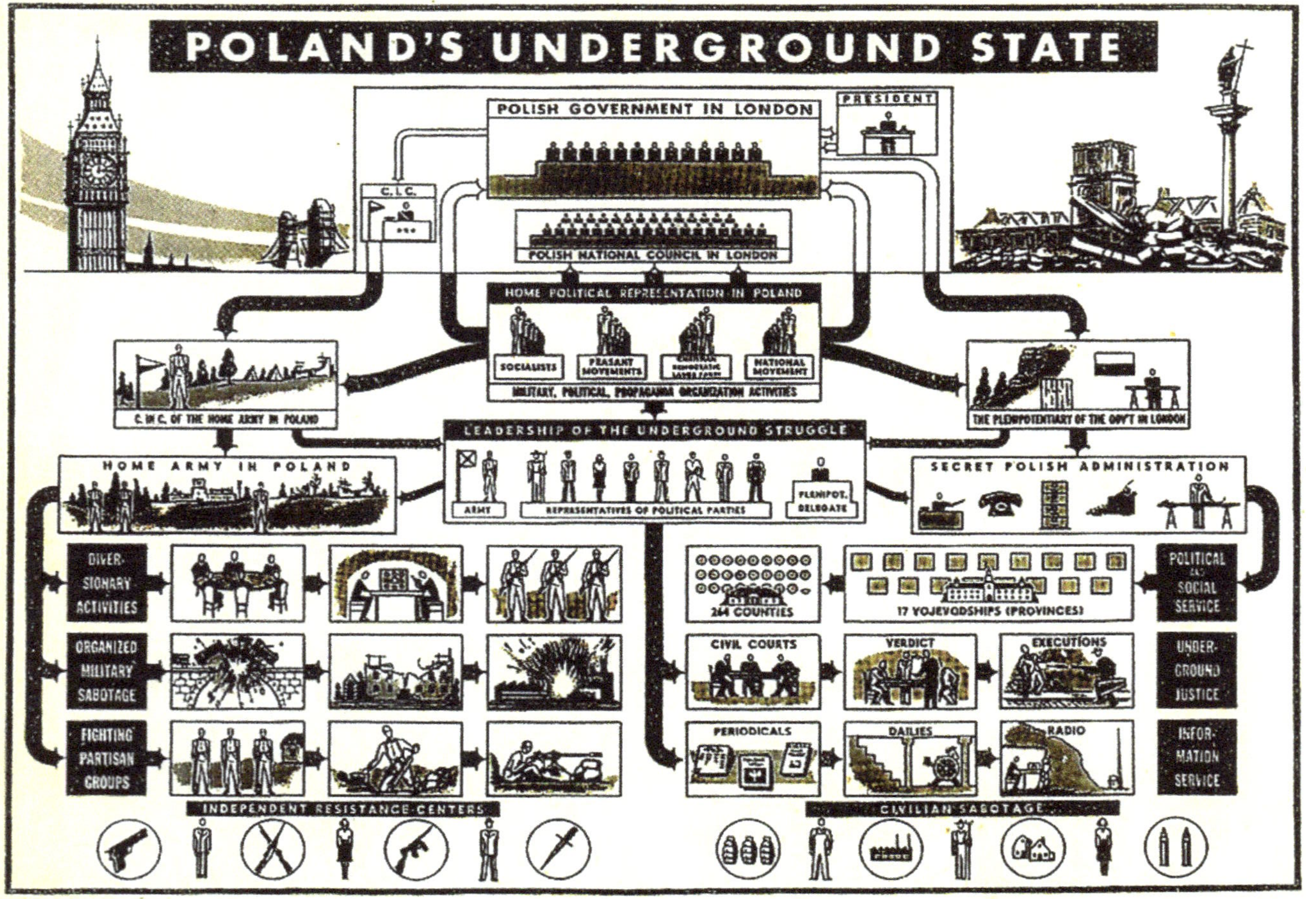

Figure 17. Poland World War II Underground State.

SOURCE: CREATED BY POLISH GOVERNMENT INFORMATION CENTER, 1941. PHOTO CREDIT: POLISH HISTORY MUSEUM/CREATIVE COMMONS CC0 the underground. The underground operated as a self-help and security organization, fighting common crime and operating as a neighborhood watch. Many who suffered under the occupation wanted to avenge themselves against the occupiers. Some joined for less noble reasons such as seeking revenge against neighbors for slights.[379]

D. **The Home Army (*Armia Krajowa*, AK)**

1. In November 1939, weeks after the Polish government's departure and his appointment to organize resistance, General Michael Tokarzewski organized the Union for Armed Struggle (*Unia walki zbrojnej*) as a vehicle to unify underground military detachments. General Tokarzewski's Chief of Staff, Colonel Stefan Rowecki, was the first commander of the Underground Army. The London Polish government-in-exile, soon renamed the Underground Army of the PUS as the Home Army (*Armia Krajowa*, AK) in 1942.[380]

2. The Americans and British never adequately armed the Home Army, which numbered about 100,000 by 1943. The Allies decided at the Teheran Conference of 1943 that Poland and Eastern Europe would fall within the Soviet sphere of influence. The Soviets supported the communist People's Army, which remained outside the PUS, and not the Home Army. Thus, the Home Army only possessed the arms they hid after the 1939 campaign along with captured weapons. The British dropped some supplies through the war, helpful for some sabotage actions, but insufficient to arm the Home Army, which totaled over 350,000 by spring 1944.[381]

E. **Communists**

1. A communist underground also existed from the beginning of Polish partition until the Soviets consolidated their power over Poland in 1947. This organization earned the enmity of the other organizations due to its support of the Nazis until 1941, when Germany broke its pact with the Soviets. From then onward, they practiced "revolutionary banditry" against the independence motivated resistance organizations and "active struggle" against the Germans. To assist the Red Army tactically and ideologically, they fostered anarchy by attacking the Germans in ways calculated to provoke harsh Nazi reprisals, hoping that such Nazi terror would eliminate the elites and radicalize the population, enticing them to join the communists.[382]

2. The communist movement consisted of the Polish Workers Party (*Polska Partia Robotnicza*) or PPR, and its People's Guard (*Gwardia Ludowa*), later renamed the People's Army (*Armia Ludowa*) or AL. Its goal was to initiate an uprising in Poland to relieve the Soviet Union, destroy the traditional Polish elite, and radicalize the people to create new recruits for its movement. Though Nazi reprisals killed many Polish elites, the communists conducted their own assassination campaigns against elites, and engaged in information warfare by replacing their goal of "class warfare" with the slogans of "active struggle" and "national liberation." In 1942, the communists began an assassination campaign of "class cleansing" to kill "reactionaries," particularly elites seeking an independent Poland. Communist attacks against the PUS and independence-minded people outnumbered their attacks against the Germans.[383]

F. **Ukrainians.** The ethnic Ukrainians living in Poland also developed a nationalist resistance group. The Poles and Ukrainians failed to coordinate activities with each other due to a hostile relationship.[384] This hostility was due to the use of Ukrainians as auxiliaries by the Germans, the excesses carried out by the Ukrainian underground, and cooperation by Ukrainian elites with the communists.

CS2.6. Lack of Unity of Effort

Initially, Polish resistance was split into populists, nationalists, Piłsudskites[385] and communists. Each group sought an independent Poland, except for the communists who sought union with the Soviets. These groups fought for Polish independence and viewed both the Third Reich and Soviet Union as their enemies. The communists received tangible Soviet support, which increased as the war continued. Meanwhile, Britain and the U.S. provided little support to the Home Army, even though they diplomatically recognized the exiled Polish government.

CS2.7. AK Operations

A. The Home Army interim military objectives were focused on self-defense, sabotage, and intelligence. These interim operations were part of a larger campaign to prepare the Polish Home Army for its primary military objective: to eventually engage in a general uprising behind German lines in support of Allied operations.[386]

B. The AK conducted a limited struggle against the Germans for two primary reasons: first, they anticipated that the Germans would lose and retreat from Poland, allowing the AK to conduct large scale fighting with Western help to reassert independence and stave off the Soviets; second, and more importantly, the AK was concerned with Nazi reprisals against the populace. Thus, their actions were well targeted and relatively moderate. During operations to capture weapons from German police, their guidance was to "unconditionally disarm and do not kill." They raided to confiscate supplies, burned food-quota documents, and crippled machinery, but they left most infrastructure undamaged, reasoning that it would soon again be in Polish hands. The food quota documentation was destroyed instead of the food to avoid suffering. Destruction to the level of loss of jobs was avoided for fear that the people who lost jobs would be shipped off as forced labor. During attacks they often pretended to be Soviet partisans, in an attempt to spare people from reprisals. They also hid the bodies of dead Germans after violent clashes to spare the people. When collaborators were too cooperative with the Nazis, they were warned. Failure to heed resulted in physical beating. Continued failure to heed resulted in review, sentencing, and death. They were more likely to target Poles than Germans for fear of reprisals.[387]

C. AK operations killed about 6,000 Germans. By June 1944, the Directorate for Sabotage (*KierownietwoDywersji*) or KeDyw, destroyed 1,167 railroad petroleum cars, 6,930 railroad engines, 19,058 railroad cars, 4,326 cars and trucks, 600 telephone and electric lines, and 443 trains. These figures exclude Operation Tempest and the Warsaw Uprising.[388]

D. The Warsaw Uprising began on 1 August 1944 and is the best known Polish resistance action. The Poles hoped to defeat the Germans and install an independent government in Warsaw, forestalling Soviet takeover of Poland. They received next to no assistance from the Western Allies. It was part of Operation Tempest which had the goal of seizing control of cities and installing Polish civil authorities ahead of the arrival of the Soviet Red Army. The Resistance intended to use the Home Army to liberate the capital within three days, ahead of the advancing Red Army. Instead, the Red Army stopped its advance, and the Resistance fought the Germans alone for 63 days. Most of Warsaw was destroyed. Hitler ordered what remained of the city razed. Polish resistance was irrevocably damaged, with over 20,000 killed and wounded and 16,000 taken prisoner. Over 250,000 civilians died. German losses were over 30,000.[389] By February 1945, the Soviets overran most of Poland, the underground and Home Army were dissolved, and many PUS leaders were arrested and imprisoned.[390] Almost 18 percent, or about 6 million of Poland's pre-war population of 35 million, perished.[391] The rising utterly failed, which also resulted in the failure of Operation Tempest.[392]

CS2.8 External Assistance

1. Poland, its secret state, the PUS, and the Home Army received very little Western Allied support. Poland was geographically difficult to access and Western Allies gradually realized that the Soviets would eventually push out the Germans. Therefore, Poland received scant Western resources, despite the large amount of intelligence Poland provided to the Allies such as V2 rocket information and components, and information concerning the extermination of people, specifically Jews at Auschwitz.

CS2.9 Summary

A. **Resilience**

1. Poland's Second Republic (1918-1939) included disparate ethnicities in distinct geographic areas. As Poland attempted to establish its new democratic state, various ethnicities' interests conflicted with those of others. Germany and Russia exploited these cleavages that weakened national resilience (e.g., ethic Ukrainians and ethnic Germans).

2. As Poland emerged from World War I and its conflict with the Bolsheviks as a newly independent state, it also developed many NGOs with civic and charitable focuses, which sometimes operated as quasi-governing institutions over the next two decades. This self-rule assisted it to quickly develop underground organizations.

B. **Pre-Conflict Resistance Planning**

1. **Organized Resistance.** The Poles did not incorporate resistance into pre-war contingency plans, and hastily established resistance as their country fell to the German and Soviet advances.

2. **Government-in-Exile**. The Poles successfully moved their government representatives out of Poland and made hasty plans for a nascent resistance directed from London. Their exiled government was able to control the activities of the PUS and Home Army to the degree that socialists and communists were their only competition for regaining independence.

C. **Integration of Resistance During Conflict**

1. **Merging into the PUS.** Disparate resistance organizations formed following the invasion of Poland. The primary, and largest, nationwide organizations were populists, nationalists, Piłsudskites, and communists. Excluding communists, these groups were independence-oriented, seeking to regain Polish sovereignty, and eventually merged into the PUS and developed a united strategy. The communists sought unification with the Soviet Union.

2. **Ideological Cleavage.** Soviet support to Polish socialist and communist resistance groups weakened Poland's overall resistance effort by preventing a cooperative, unified front. The groups supported by the Soviets were not reliant on the London-controlled underground and did not have to cooperate with them. This may have been a factor in the disastrous outcome of the Warsaw Uprising which basically destroyed non-leftist resistance. The remaining effective resistance after the Warsaw Uprising was Soviet-linked.[393]

D. **External Support to Resistance.** Western Allies gave little support to the Polish resistance, predominantly due to geography and operational reach, later exacerbated by the 1943 Teheran Conference agreement that Poland would be liberated by the Soviet Union. The Poles had to use their few hidden weapons and what they could take from the occupiers. The Soviets supported the communist resistance with advisors, weapons, and equipment, positioning them to assist the Red Army.

E. **Resistance Strategy**

1. The occupier policy of terror conditioned the actions of the resistance, though it had a lesser effect on communist organizations. The underground and Home Army inflicted minimal casualties on the Germans for fear of reprisals against the population. They often acted against collaborating Poles instead of Germans to avoid reprisals.

2. The PUS leadership based its actions and strategy on its early conviction that the Western Allies would assist them in liberating Poland. They eschewed large scale violence until the Germans retreated, which they assumed was occurring during Operation Tempest. The Warsaw Uprising destroyed the city and decimated the ranks of the resistance, resulting in fewer resistance forces to face the Soviets. Afterward, under Soviet occupation, many Poles could only await a war between the West and the Soviet Union, which never occurred, in the hope of salvation from communism.[394] Without external assistance against the might of the Soviet Union, Poland had no choice but to await a change in circumstances.

F. **Restoration of Sovereignty.** The legitimate government-in-exile in London was disenfranchised by the Western Allies, and Polish sovereignty was lost to a Soviet puppet communist administration. Many Poles fought the Soviets and their Polish, communist allies for several more years in the hope of Western assistance. The Polish state and its national sovereignty were not restored until 1989.

Case Study 3. Philippines: World War II Filipino and American Resistance

Resistance was one of the finest hours for the Philippine people. - David Joel Steinberg, *Philippine Collaboration in World War II*[395]

CS3.1. Background

A. The Japanese invaded the Philippine Islands on 8 December 1941, less than a day after the attack on Pearl Harbor, which had crippled American naval capability in the Pacific. As a United States Commonwealth, the Philippines was the responsibility of the U.S. Despite knowledge of Japanese expansion and warnings of Japanese intentions toward the Pacific islands, the U.S. pre-war effort did not establish a resistance capability for the Philippines. Plan Orange contained the U.S. Philippine defense plan, calling for U.S. forces to defend the Bataan Peninsula and Corregidor Island for up to six months while awaiting reinforcement. Poor preparation resulted in insufficient rations being stored on the peninsula to sustain a large force for any length of time against a siege, though there was adequate ammunition. Barely enough rations existed to hold out for more than a few months. Eventually, the approximately 76,000 American and Filipino forces on Bataan surrendered on 9 April 1942, and those on Corregidor surrendered on 6 May 1942.

B. After the surrender order, many U.S. Army Forces Far East personnel chose to flee to the jungle rather than surrender. Thousands of Filipino soldiers around the islands also escaped capture. However, these groups operated more as traditional guerrilla groups, based in inaccessible terrain from where they launched raids and ambushes.[396]

CS3.2. Geography

The Philippine Archipelago includes over 7,000 islands consisting of vast jungles and mountains, and considerable amounts of road-less terrain. The islands are divided into three sections: the northern islands (Luzon), the central islands collectively known as the Visayas (Panay, Negros, Cebu, Bohol, Samar and Leyte), and the southern islands (Palawan, the Sulu Archipelago and the large island of Mindanao). The southern islands are the largest but have the smallest populations. The population is concentrated along the coast.

CS3.3. Resilience

A. As a result of the Spanish-American War of 1898, the Philippine Islands, ruled by Spain prior to the war, were ceded to the United States. In 1899, resistance to American rule turned violent and became known to Americans as the Philippine Insurrection, which ended in 1902 with American victory. The Filipino *Insurrectos* and the Philippine population suffered disproportionate casualties compared to the Americans. That war ended with the capture of the insurrectionist leader, followed by an amicable peace. Afterward, American forces were stationed in the islands as part constabulary and part model for establishment of Philippine forces. The relatively mild American rule, seeking to nurture the Philippines to maturity and eventual independence, as well as deliver economic benefits, resulted in very amicable relations and a later partnership against the Japanese.

B. The racially and ethnically diverse Philippine population had multiple potential political leverage points. Racially, Filipinos were Negrito, Indonesian, and Malay but most were heavily mixed due to centuries of intermarriage. Filipinos spoke eight major languages with 60 dialects. English, although the second most spoken language, was only spoken by 28 percent of the population. There were three religions: Christians, Muslims, and Pagans. Most Christians were Roman Catholic.[397] Politically, a significant communist movement was entrenched as the Communist Party of the Philippines. Like its European counterparts, the communist party was well organized and provided a resistance structure post Japanese invasion. The Japanese failed to exploit any of these seams of potential leverage resulting from population diversity, and instead alienated all of these groups during their harsh occupation. The few collaborators were primarily in urban areas.

CS3.4. Governance of the Philippines

A. After the U.S. surrender, Japan installed a collaborationist Filipino government with a new Japanese approved constitution.[398] This nominal government had legal jurisdiction to treat guerrillas or resistance forces as either traitors or bandits. Yet, many members of the new government surreptitiously cooperated with and assisted resistance forces.[399] The Philippine Constabulary was an example of institutional level cooperation with the resistance. The U.S. had created the Constabulary decades prior to deal with guerrillas, and the Japanese reconstituted it for the same purpose. However, many Constabulary members cooperated with the resistance and provided weapons, ammunition, and intelligence. The Japanese eventually disbanded the Constabulary in October 1944.

B. The Japanese also created the *Kalabapi*, a single-party political organization, and the Japanese Army's political arm. It controlled Neighborhood Associations to ensure policing and distribution of food and commodities. The *Kalabapi* also maintained a spy network to root out resistance members and sympathizers. The loyalty of the leadership was enforced by punishing leaders and their families who collaborated by commission or omission in any way with the resistance.[400] The Japanese intended to establish a New Republic of the Philippines and have it declare war on the U.S. The Japanese effort failed at first, but later succeeded in September 1944, with the use of more cooperative collaborators.[401] Pro-Japanese Filipino organizations existed before and after the Japanese invasion. The Japanese used some to infiltrate U.S. led resistance units.

CS3.5. Resistance Groups

A. **American Led.** American and Filipino officers and soldiers who escaped Japanese capture on Bataan and Corregidor, fled into the jungles and mountains. More troops escaped later during the long marches to prison camps, after witnessing Japanese soldiers killing individuals who fell sick or collapsed. Some sought to continue to fight, while others sought escape to Australia. Some remaining U.S. civilians, operating agricultural establishments, were intermittently able to assist Americans soldiers hiding in the jungle.[402] These military members remarkably organized effective resistance groups, although disconnected from each other. The Filipino guerrilla units with U.S. leadership were organized on the pre-war military hierarchy. They were joined by small numbers of U.S. troops elsewhere on the islands.

B. **Domestic Groups.** Many local militias formed, not to combat the Japanese, but to combat banditry after the Philippine government's fall and U.S. surrender. The Japanese did not physically occupy all populated areas. Some of these local protective militias became the embryos of larger resistance groups, especially in the Visayas region. That region also had many resistance groups led by Philippine Army officers, whereas the Americans were present mostly north on Luzon and south on Mindanao.

1. The Hukbalahaps (*Hukbong Bayan Laban sa mga Hapon*), or "The People's Army Against the Japanese" (Huks) were an internal communist guerrilla movement, strongest in central Luzon. They did not directly challenge U.S.-led resistance units but were often hindrances, although they claimed to

cooperate. The Huks fought their own war against the Japanese, which they later continued against the post-war Filipino government.[403] The Huks managed to establish semi-liberated enclaves from which they harassed the Japanese and raided for weapons and supplies.[404]

2. The Moros were another domestic group that the U.S. did not control. Concentrated on Mindanao and the smaller Sulu archipelago between Mindanao and Malaysia, the Moros had been fighting for almost three and a half centuries before the Japanese invaded. They fought the Spaniards from the 1500s to 1900s and fought the U.S. from the 1900s until the 1940s. They fought all non-Moros, whether Westerners, Japanese, or the Filipinos from the north of Mindanao, with the exception of the Chinese with whom they were trading partners for centuries. U.S. Colonel Wendell W. Fertig brokered a limited alliance between the Moros and local Christians and focused them on fighting Japanese. Colonel Fertig began with a cadre of five officers and about 175 enlisted soldiers, building an organization of almost 40,000, while carving out significant areas of control on Mindanao.[405]

3. The small Chinese community, comprising less than 1 percent of the population, also had its anti-Japanese resistance organization, which received China-based communist support. Some of these Chinese communist party members traveled to China before the Japanese invaded the Philippines, to receive guerrilla warfare training. This training greatly aided the Chinese resistance effort. Chinese communists cooperated with their fellow communists, the Huks, operating primarily on Luzon, while they remained outside the cities during the occupation. Non-communist Chinese resistance groups existed but were not as militarily effective.[406]

CS3.6. Communications and Strategic Communication

U.S. and Filipino military personnel in hiding had a few short range radios, which they used to communicate with U.S. submarines surfacing offshore at night. Runners passed messages to small groups that did not have radios.[407] On Mindanao, the Church of Mindanao priests were commonly used to pass information. The priests from anti-Japanese Catholic churches required little vetting. As with European resistance organizations, some of these organizations published newspapers.[408] Every day at 1700 hours, Allies broadcasted into the Philippines. It began with General MacArthur intoning "I shall return" and ended an hour later with another promise to return.[409] The phrase "I Shall Return-MacArthur" was imprinted on match boxes, cigarette packs, and mirrors and sent into the Philippines.[410]

CS3.7. Resistance Phases

The Philippine resistance can be broken down into three phases in order to examine the events and forces that brought them about.

A. Initial/Nascent

1. On the eve of MacArthur's departure from Corregidor, he decided to "continue resistance operations by guerrilla methods," intending for resistance forces to support CF upon their return, which he assumed would happen in a matter of months. He appointed Brigadier General William Sharp as commander of the Visayan-Mindanao force to lead this resistance.[411] Colonel Claude Thorpe was sent north to form a resistance organization in central Luzon[412] and in May 1942, he established the Luzon Guerrilla Force in the newly designated Luzon Guerrilla Area.

2. Available weapons and equipment were poor. Conventional training was good, but there was no resistance training or doctrine to support it.[413] Much of the equipment, weapons, and ammunition were manufactured informally in the jungles—for example, brass curtain rods used to make bullet shell casings. Most groups were out of touch with Allied headquarters and often unaware of other resistance groups. Some groups existed more to survive than to fight. Others preyed on locals for sustenance. Communist groups fought to consolidate postwar power.[414]

3. Mindanao's resistance organization, the first well organized group, was relatively closer to Australia, and easier to reach by submarine and radio wave. Mindanao quickly benefited from its proximity to Australia, receiving supplies and agents. Other groups gradually received increased support once communications were established and submarines could deliver supplies.

4. After the surrender order, many Americans outside Bataan and Corregidor chose to flee to the jungle instead of being taken prisoner by the Japanese. Many were overcome by lack of food and disease. Others were killed or enslaved by Moros or killed by pagan Magahats.[415] Escapees who reached Australia informed the Allied command of the existence of many resistance bands with U.S. officers leading them. Radio contact was soon established, and by late 1942, submarines facilitated regular contact, bringing Allied Intelligence Bureau agents, supplies, radios and code books.[416]

B. **Mid-Term "Lay Low."** In mid-1943, MacArthur's headquarters issued the "lay low" policy to its resistance leaders. U.S. headquarters required intelligence to inform military planning for future operations. Resistance leaders were expected to collect and send information, while only engaging in sabotage of Japanese lines of communication in order to maintain Filipino morale. This limited activity was intended to avoid the reprisals that the Japanese military conducted against civilians, individually and collectively, suspected of cooperating in any way with the resistance.[417]

C. **Pre-Liberation.** Several weeks before the U.S. landings on Leyte in October 1944, instructions changed from information collection to conducting sabotage. Resistance units were directed to conduct such operations in and around Manila, to draw Japanese troops to that area and away from beach landing sites. In the fall of 1944, the U.S. headquarters required more details concerning Japanese force disposition, while also requiring an increase in activities to draw Japanese attention away from landing sites. Resistance units received orders to conduct offensives against Japanese forces, while continuing sabotage and intelligence collection. One U.S. led resistance group seized an airfield, which was then used a week prior to invasion by U.S. Marine aircraft for operations against the Japanese.[418]

CS3.8. Legal Status of the Resisting Americans

American resisters who escaped from surrendered units were regarded by the Japanese as renegades who refused to obey orders. The Japanese treated these American soldiers as illegal combatants and terrorists. Upon capture, they were tortured and executed as war criminals.[419] The U.S. military outside the Philippines continued its administrative responsibility for these inadvertent stay-behind American resisters with such things as pay and promotions, which was communicated to them from the Australian-based U.S. headquarters.

CS3.9. Networks

Unlike European World War II resistance, most Philippine-based resistance groups were concentrated in inhospitable jungles and mountains, with little activity in urban areas. However, Colonel Fertig's group on

Mindanao eventually controlled significant territory that included population centers. Political and religious differences were the greatest differentiating factors among the Filipinos of the disparate resistance groups. U.S. and Filipino army officer-led groups tried to collect specific intelligence to assist the return of regular forces, but little publicly available information has been succinctly compiled by researchers to help understand specific intelligence or escape networks.

CS3.10. Utility of Resistance

A. Approximately 260,715 guerrillas in 277 guerrilla units fought as organized, armed, and tactically employed resistance units.[420] Resistance or anti-Japanese forces controlled about 95 percent of the land, but these inhospitable interior areas were sparsely populated, and of little agricultural or other value to the Japanese war effort.[421]

B. The Australian-based U.S. headquarters viewed the Philippine resistance units as useful sources of information, and attempted to restrict the resistance unit activities to information collection. Other actions against the Japanese were intended to maintain Filipino morale. The Allied Intelligence Bureau began to send agents to establish coast-watching stations in late 1943.

C. After CF landed, many resistance personnel acted as guides. Former resistance personnel maintained contact with their intelligence sources in areas the Japanese still held. They also provided information on prison camps where Allied personnel were held in preparation for raids on those camps.[422]

D. Military led groups, and especially Filipino soldiers, who witnessed the suffering of their families, tended to want to kill Japanese enemy soldiers. However, Japanese reprisals against civilians tamed former resistance personnel's enthusiasm for retribution.

CS3.11. Summary

A. **Resilience.** Though the Filipinos had fought a war for independence against the Americans four decades earlier, the relationship between them became very amicable, resulting in partnership against the Japanese. Despite the racial, ethnic, and religious diversity within the Philippine population, the Japanese were not able to leverage the differences to their advantage. The Communist Party of the Philippines was politically well entrenched. Of the indigenous population groups opposing the Japanese, only the communists sought a complete overturn of the old political system and establishment of a new regime, but they had no nearby allies to support them. The Moros, though united by Islam, were physically separated due to being spread upon different islands, and also retained cultural and ethnic differences among them. These factors provided enough distinction among Moros that Islam was inadequate to unify them to seek physical or geographical gains.[423] Most of the other groups cooperated to fight the Japanese and relied on American promises of post-war independence.

B. **Pre-Conflict Resistance Planning.** No pre-war plans to establish resistance organizations existed in the Philippines. Additionally, a requirement for a government-in-exile was unnecessary because the U.S. held sovereignty over the islands. The U.S. had adequate knowledge of Japanese expansion plans prior to the Japanese invasion of the Philippines in 1941. This knowledge could have informed the creation of resistance or stay-behind forces. Planned resistance capability could have designated cadres, stored caches of radios, weapons, ammunition, and medical supplies. Such organization could have collected and transmitted intelligence on Japanese military and naval disposition and activities, and Japanese troop quality from the onset of war. This information may have led to speedier and more efficient Philippine-based Allied operations against the Japanese.

C. **Integration of Resistance during Conflict.** The planned resistance could have more efficiently incorporated the Huks and Moros, not by directing their activities, but by accounting for the activities that they would likely undertake within overall resistance plans, and complementing those activities. Ideally, stay-behind resistance forces could have been on an earlier trajectory to prepare for and assist the return of Allied forces.

D. **External Support of Resistance.** The U.S. provided support to resistance organizations through its own, admittedly inadvertent, stay-behind forces. Once communication was established with these Americans on the islands, supplies, equipment, weapons, and even additional personnel were delivered.

E. **Resistance Strategy.** The groups cooperating with Americans in the islands received support to maintain and to arm themselves. The American intent was to use these groups to directly assist an eventual invasion of the islands. The Americans and most of these Filipino groups became wary of direct action against the Japanese for fear of reprisals against the population. As in other parts of the world, the communists were less inhibited in this regard.

F. **Restoration of Sovereignty.** After the Japanese were ejected from the Philippines, American sovereignty over the Philippines was restored to its pre-war status. The U.S. fulfilled its promise to the Filipinos in return for their efforts against the Japanese and granted the islands independence in 1946. Concurrently, as part of the agreements granting independence, the U.S. also retained the right to maintain many military and naval bases as well as some favorable economic conditions for itself.

Case Study 4. The Forest Brothers: World War II And Post-World War II Baltic Resistance To Soviet Occupation

(Estonian: *metsavennad*, Latvian: *meža brāļi*, Lithuanian: *miško broliai*)

> The right of all peoples to choose the form of government under which they will live; and ... sovereign rights and self-government restored to those who have been forcibly deprived of them. - Atlantic Charter, August 14, 1941, Common Principle Three

CS4.1. Background

A. The Forest Brothers is the term applied to the organized, anti-communist and anti-Soviet, resistance effort in the Baltic states of Estonia, Latvia, and Lithuania. The term was first used to describe people in the Baltic region who fled to rural areas to escape the effects of the Russian Revolution of 1905. This study focuses on the post- World War II anti-Soviet effort, but will touch upon the formative resistance efforts against the Soviets in 1940-1941 and the effort against the German occupation of 1941-1944 for descriptive context.

B. Estonia and Latvia were absorbed into the Russian Empire by Czar Peter the Great in the eighteenth century. Lithuania fluctuated between a Polish duchy and an independent state. Throughout Russian dominance of the Baltics, friction accompanied the relationship, due to Baltic prosperity as compared to impoverished Russians. Estonia and Latvia became the most modern, literate, and wealthy regions of the Russian empire. In the nineteenth century, all three nations developed a strong sense of nationalism.[424]

C. When the Russian empire collapsed in 1917, the Baltics established Western style governments, and fought and defeated the Red Army, retaining their independence. Apart from a Soviet-backed attempted coup in Estonia in 1924, twenty years of independence and relative peace existed in the Baltics after World War I. However, in 1939, the Molotov-Ribbentrop Pact between the Soviet Union and Nazi Germany placed the Baltics in the Soviet sphere and allowed the Soviets to annex the region. Stalin then pressured the Baltic States to accept Soviet military bases, which was soon followed by invasion and occupation in June 1940.[425]

D. In the summer of 1940, following phony elections in each of the three Baltic States, they were annexed by, and became republics of, the Soviet Union. They soon suffered horrendous oppression and collectivization. Large numbers of priests, ministers, businessmen, intellectuals, military officers, and landowners were arrested and deported to Siberia or outright killed. Farms, industries, and businesses were collectivized. When Nazi Germany reneged on its pact with the Soviets and invaded the Baltics in June of 1941, the German troops were welcomed as liberators and seen as the lesser of two evils. Having experienced massive repression under the Soviets, many Baltic citizens cooperated and fought with the Germans against the Soviets, including the initial bands of the Forest Brothers.[426]

E. The Soviets reestablished their power over the Baltics in 1944 and began a new wave of repression, characterized by more deportations and killings. The most repressive time was from 1944 until the death of Stalin in 1953. The anti-communist Forest Brothers, in each of the Baltic States, fought the Soviets until they were effectively destroyed as organized entities by the Red Army and People's Commissariat for Internal Affairs (NKVD).

F. Each of the three Baltic States attempted to form governments-in-exile, to varying degrees, but with little success. Lithuanian President Antanas Smetona, escaped abroad on 15 June 1940,[427] and attempted to form a government-in-exile. He was not successful because he lacked legitimacy with his people, and because there was no such allowance in Lithuanian law.[428]

G. The last independent Latvian government, prior to the German and Soviet invasions, did not arrange for, or empower a government-in-exile. Though, on 17 May 1940, the Latvian government authorized its Envoy to Britain to represent Latvian national interests under a special powers act "in case of war" when diplomats abroad were unable to contact the Latvian government under war-time conditions.[429] However, the envoy refused to form a government-in-exile, insisting that it would not achieve recognition and would place him, as a diplomat of Latvia, in an untenable position.[430]

H. An Estonian government-in-exile was not established by the government, but one was established by the Estonian diaspora in 1953, lasting until 1991.[431] Former members of the government of Estonia found refuge in Sweden and established what they claimed was a government-in-exile, identifying ministers and a Governor-in-exile of the Estonian Central Bank. This government was never recognized by the Allies, internationally, or by the Estonian Foreign Service. However, it continued to exist, and symbolically reunited with the Government of Estonia on 8 October 1992.[432]

I. The resistance in Lithuania was numerically the largest and best organized. All three Baltic States experienced very similar situations regarding each Soviet and Nazi occupation. During the first Soviet occupation, the Soviets assimilated all three states as Soviet Socialist Republics (SSRs), denying them their individual sovereignty, unlike their treatment of other eastern European states such as Poland. This situation reoccurred in the second Soviet occupation. Even under the German Reich, the three countries were absorbed into *Ostland* and not recognized by Germany as separate sovereign states.

CS4.2. The First Soviet Occupation

A. In June 1940, as a result of the Molotov-Ribbentrop Pact, the Soviets began their occupation of Estonia, Latvia, and Lithuania. In each country, the Soviets established new electoral rules, thus rigging the elections, and ensuring new legislatures full of communists. In each country, those new legislatures requested that their countries become absorbed into Soviet Union directly as Soviet Socialist Republics.[433] Each request was immediately granted.[434]

B. People engaged in various forms of resistance. Only 15 percent of eligible Lithuanian voters cast ballots in the Diet elections. Many voted for cartoon characters. Political rallies were sparsely attended. Portraits of Lenin and Stalin were stolen. A Red Army concert was interrupted by attendees singing patriotic songs and national flags often made appearances.[435]

C. Seeking to eliminate competing sources of authority throughout the Baltics, the Soviets sent their secret police, the NKVD, to target certain institutions. Property belonging to the Roman Catholic church (the Lithuanian population was 90 percent Catholic) and synagogues of the Jewish religion were confiscated and monasteries closed. The clergy had to pay special taxes and some were executed. Members of former government cabinets and senior members of the major political parties were deported. Large farms were confiscated and collectivized or converted into state farms. Large bank deposits were impounded and local currency was banned.[436] The armed forces of each state were disbanded, purged and incorporated into the Red Army, staffed with Russian commissars. Tens of thousands of people were deported to Siberia.[437]

D. On 9 October 1940, the Lithuanian Activist Front (LAP) resistance group was organized with hundreds of three-man cells across the country. The LAP stockpiled weapons with the goal of rising up and inciting revolt when the time was right. It absorbed other groups such as the Iron Wolf and the Lithuanian Freedom Army, and grew to a strength of 36,000 members.

E. In Latvia, the Soviets arrested 3,353 people based on Article 58 of the Russian Criminal Code (enforced throughout the Baltics), which defined punishments for so-called "counterrevolutionary crimes," including "treason." This repression later reached its peak on the night of 13-14 June 1941, when 15,443 men, women and children were taken to the Union of Soviet Socialist Republics (USSR). Among them were many soldiers and officers of the former Latvian Army, who were shot or arrested and deported to the USSR.[438]

F. In Estonia, active resisters became known as Forest Brothers, and in the summer of 1941, their resistance became known as the "Summer War."[439] Gradually, they engaged in military actions against Red Army units as well as against collaborators. These irregular tactics caused some confusion and panic among many Soviet soldiers who feared being surrounded by the Forest Brothers and would flee.[440] The main strengths of the Forest Brothers were their loose organization, ability to blend into their environment, and a generally supportive population.[441]

CS4.3. The Nazi Occupation

After the German invasion of June 1941, the Lithuanian LAP formed a provisional government, which was also attempted by Estonia. Those governments were soon disbanded by the Nazis, and all three Baltic States became part of *Ostland*, which also comprised Belorussia. Riga hosted the institutions of civil administration and military command for *Ostland*, as well as the main SS and police establishments.[442] As in other occupied territories, the Nazis demanded laborers for work throughout the Reich. Resistance to this led to abductions, imprisonment and executions.

CS4.4. The Second Soviet Occupation

A. By early 1944, Soviet guerrillas were operating in eastern Lithuania, still under German occupation. The Lithuanians feared Soviet return and the destruction of Lithuania as a state. Thus, the political arm of the Lithuanian resistance, the Supreme Committee for the Liberation of Lithuania, with German permission, formed the Lithuanian Territorial Defense Force, which became the Home Guard.[443] The Germans sensed a threat to their rule and attempted to exert direct control over the Home Guard, causing most of its members to flee to rural areas to prepare to conduct guerrilla warfare against the oncoming Soviets.[444] During June to August 1944, the Soviets ostensibly liberated Lithuania from German occupation once again. The massive Red Army prevented the formation of organized resistance, allowing the Soviets to quickly transition to power consolidation.[445] During the next five months, the Soviets executed or deported tens of thousands of people as they exacted a toll for collaboration with the Germans.[446]

B. As the Soviets retook Estonia, they formed People's Defense Units, consisting of Estonians who were intimidated into joining, or who volunteered. These units were designed to crush the resistance. They guarded local government buildings, executed arrests and deportations, and requisitioned property.[447] Many of the members were thieves and opportunists who joined for social leadership. They were able to intimidate their fellow citizens through threats of prosecution and deportation.[448]

C. On 17 July 1944 the Soviet Army reached the border of Latvia, and soon the Nazi occupation was replaced by the second Soviet occupation. The Latvian SSR government returned from Russia as an instrument of

Soviet occupation policy.[449] In the autumn of 1944, an armed, organized resistance movement started to develop in Latvia.[450]

CS4.5. The Three Stages of Resistance

A. The anti-Soviet resistance in the Baltics can be broken down into three stages. The first stage (July 1944 to May 1946) saw the largest battles as well as caused the greatest number of casualties. Partisans in the forests gathered in the hundreds in well-fortified camps. They attacked occupation authorities in small towns, disarmed the military units of local collaborators in rural districts, released arrested people, and destroyed draft and quota records.[451]

B. During the second stage (May 1946 to November 1948), large-scale battles against Soviet security forces were avoided. Only about 4,000 partisans remained in the forest, divided into smaller groups, and moved into purpose-built camouflaged bunkers, built under barns, dwelling houses, on the slopes of streams, under potato pits, and under the roots of trees. Entrances were in and under stoves, and in both wet and dry wells. When discovered within these bunkers, the occupants were trapped because these structures were not suited for defensive purposes and often had no means of escape.[452]

C. The third and last stage (November 1948 to May 1953) saw the resistance movement losing strength. During this time, greater attention was devoted to propaganda work through print. This work was intended to develop national and social awareness of the population, to maintain their belief in the successful end of the struggle resulting in regaining independence.[453]

D. During the second and third stages, civil disobedience increased as partisan military actions decreased. These political actions interfered with Soviet-sponsored elections and collectivization programs, and encouraged people not to recognize the Soviet regime or institutions representing it. The resistance even issued instructions for citizen behavior while under this occupation. However, despite thousands of copies of dozens of periodicals, songs, prayer books, and proclamations printed in cramped underground bunkers, the resistance could not overcome Soviet information dominance among the population.[454]

E. Food was typically the most critical concern of the partisans in the forests. Local farmers usually supported them, but as deportations increased, this support structure dwindled. This caused the Forest Brothers to requisition food and material from the better-off inhabitants and collectivized farms, often issuing certificates in return, acknowledging the requisitions.[455]

F. The numbers of guerrillas declined precipitously after the early large-scale battles, with the loss of life, deportation of supporters, and some members returning to legal lives. The NKVD constantly improved its ability to locate and penetrate resistance groups. The violence and terror applied by the Soviets, through torture and execution of captured fighters and display of mangled corpses in public squares, also took their toll on the morale of the Brothers and the population. Finally, the realization that the Western powers would not uphold the Atlantic Charter caused active resistance to become increasingly futile, ceasing by 1953.

CS4.6. The Forest Brothers

A. Lithuania

1. The Lithuanian Forest Brothers group was the most organized and widespread. Approximately 4 percent of the Lithuanian population was directly or indirectly engaged in the movement. Most of

the men who went into the forests were of military age. Women who joined performed clerical, communications, and medical work. Farmers constituted the largest portion of the Brothers. Only a few were professional Servicemen, mostly junior military officers.[456] The resistance was comprised of individuals across the social spectrum.[457] Approximately 100,000 people were part of the resistance movement between 1944 and 1956.[458]

2. The main tier of the Forest Brothers, the "active front line soldiers" were composed of fighters who wore military uniforms to project legitimacy as combatants engaged in warfare and not bandits behaving criminally. They lived in the forests in underground bunkers and tunnels and were armed with captured German and Soviet weapons.[459]

3. The second tier consisted of "inactive fighters" whose participation was more passive. They were armed but led open legal lives, fighting only when an opportunity presented itself.

4. The last tier were the "supporters," who comprised a substantial portion of the population. They provided supplies, shelter, and intelligence, and supported an underground press.[460] Similarly corresponding tiers existed in both Estonia and Latvia.[461] In terms of this ROC, the first tier were guerrillas, the second corresponded to the underground, and the third were the auxiliary.

5. Motivations for joining the Forest Brothers included avoiding conscription into the Red Army, fear of persecution for previous political associations, families fleeing deportation, farmers resisting collectivization, and enemies of the Soviets based on knowledge and experience of the previous Soviet occupation.[462]

B. Estonia

In the early part of the resistance in Estonia, the Armed Resistance Union was the dominant organization seeking restoration of independence. However, it was very hierarchical and conventionally minded, and registered its members, posing a danger to them. People who wanted to resist but were wary of such an organization joined other small groups, which became the Forest Brothers.[463] The largest Estonian Forest Brothers' organization was the Armed Resistance League, which operated from 1946-1949, operating in groups of 5-10.[464]

C. Latvia

In Latvia, anti-Soviet resistance organization also began under Nazi occupation in anticipation of Soviet return.[465] Former soldiers, who served in Latvian or German units, constituted the core.[466] They were later joined by peasants, workers, and intellectuals, totaling about ten to fifteen thousand active combatants.[467] The Latvian Forest Brothers were most active in the heavily wooded and swampy areas of Latvia's borders. Local peasants supplied them with sustenance and intelligence.[468] As with the other Baltic Forest Brothers, the NKVD gradually infiltrated them and destroyed them as groups.[469]

CS4.7. Forest Brothers' Strategy, Operations, and Tactics

A. Strategy

1. Initially, a common underground authority to create a strategy, represent the movement, and seek support from abroad did not exist in any Baltic state. As resistance developed throughout the Baltics beginning in 1944, the movements considered it vital to establish credibility and legitimacy to attain Western assistance. They expected to receive weapons, ammunition, medicine, communications

equipment, and political support in fulfillment of the Atlantic Charter, which affirmed that all nations had a right to regain their lost independence.[470]

2. In 1946 the Supreme Committee on the Reestablishment of Lithuania issued a proclamation defining its objectives, strategy, and tactics.[471] On 16 February 1949, it established the Movement of the Struggle for Freedom of Lithuania, declaring it as the highest legal authority in Lithuania, with its goal being to reestablish an independent and democratic Republic.[472] In effect, it established a shadow government. Lithuanian partisans adopted the military structure of the Lithuanian Army, taking oaths, maintaining military discipline, and subjecting themselves to courts martial. Most of them wore Lithuanian military uniforms, adorned with the symbols of Lithuania, such as the mounted knight, tricolored flag, and Cross of Vytis.[473]

3. In 1945, groups of Estonian Forest Brothers began attacking smaller NKVD units, village councils and executive committees of the parishes, and collaborators. Food, clothes, and cash were requisitioned from cooperatives and other state enterprises in the name of the Republic of Estonia.[474] Initially, the Forest Brothers' connections inside the security agencies allowed them to avoid many anti-resistance operations. The Brothers also disseminated political information through leaflets and newspapers.[475]

4. Seventy five percent of Forest Brother targets were civilian collaborators, collective farms, and other public entities. The motivation behind selecting those targets was most commonly revenge against informants and collaborators. Many Forest Brothers who surrendered or were arrested by the Soviets were punished by death or deported to Siberia.[476]

5. Despite the fact that organized resistance in Latvia and Lithuania was substantially more widespread, active and organized, Estonia still had an estimated 15,000 to 20,000 people hiding in the forests, which was over one percent of the population at the time.[477] Many of these Estonians believed that the Western powers would soon assist them (the so-called "waiting for the white boat").[478] They were also joined by those wishing to avoid conscription to the Red Army and by former members of the Defence League and the Home Guard. They formed the core of the Forest Brothers movement. There were about 30,000 Forest Brothers in post-war Estonia.[479]

6. In response, the Soviet authorities started massive military operations to capture the Forest Brothers and their supporters. The network of NKVD agents rapidly expanded, and raids in forests and farmhouses became increasingly frequent.[480]

7. The strategy of the leadership of the Estonian Forest Brothers was formed by a small core of men named Jerlet, Oras, Raadik, Saaliste and his brother Saalists. One writer used their names to form the acronym JORSS and referred to their strategy as the JORSS strategy. Like the partisans in Lithuania and Latvia, their ultimate success depended on Western intervention. Thus, they awaited a war between the Soviets and the west to trigger an uprising in Estonia.[481]

8. The Estonian JORSS organization was created in 1947 and was eradicated in 1949. Its primary objective was the survival of the Forest Brothers as an armed force prepared and ready to join the West in a war against the Soviets. The second objective was to convince the population to overthrow Soviet rule. The third objective was to gather information useful to the West and relay it to assist their war preparations. These tasks were performed mostly from 1947 onwards.[482] Their strategy can be broken down into further detail (see table 3).

The JORSS Strategy: Ends, Ways, and Means			
Ends	**Ways**	**Means**	**Risks**
he Forest rothers are repared or combat pon the utbreak of ar between he Western ountries nd the JSSR.	The survival of the Forest Brothers and their weaponry by stealth. Contact between groups for the purpose of mobilizing the Forest Brothers. Provisioning with the help of supporters and through confiscation. A loosely structure organization for the sake of security. Long-term instructions to coordinate the activity of different groups.	Existing Forest Brothers. Terrain outside of settlements. Mediation between different groups. Supporters in the form of local inhabitants. State institutions. Discipline and morale of the Forest Brothers.	*Acceptance*: The Soviet authorities do not accede and respond with violence. *Acceptance/feasibility*: The support of the inhabitants is crucial (they accept as long as they can see the merit or do not suffer). *Acceptance/feasibility*: Controlling different groups i complicated. *Compliance*: Avoidance of war by the Western countries changes the strategy as a whole.
he Estonian opulation an be nobilized or an armed prising gainst oviet rule at he outbreak f war, while nobilization y the JSSR is bstructed.	Information operations to explain the objectives of the Forest Brothers' activities, the criminal nature of the Soviet rule, and the international political situation. Taking into account the interests of the populations. Clarifying the objectives by means of actions (armed attacks on Soviet representatives and sites). Identifying members of the general population suited for mobilization (physical capability and attitude). Arming the population for mobilization with outside help.	Leaflets, brochures. Facilities for compiling media. Disseminators of oral information. Data for information operations (crimes by the Soviet authorities, peoples's attitudes, anti-Soviet activities of foreign countries). Information to identify targets (physical parameters, justification for the selection of target). Resources for carrying out armed activities. Contacts with Western countries.	*Feasibility*: The dissemination of written information requires printing facilities. *Feasibility*: The availability of resources attacking the Soviets and arming the population. *Feasibility*: Lack of contact with the Western countries.
he Western ountries nd Forest rothers ossess trategic nformation eeded to repare for nd conduct he war.	Gathering information (the political and economic situation in the Estonian SSR, people's attitudes, location/nature of the Soviet Security agencies and armed units). Relaying the information to foreign countries.	Local inhabitants. Collaborators with the Soviet authorities (by making them double agents). Forest Brothers capable of gathering intelligence (skills, discipline, connections, ability to move around). Communication channels (radio stations, postal service, courier, state media).	*Feasibility*: Persuading collaborators to cooperate (with compromising circumstances, material resources, a strong ideological stance). *Feasibility*: Establishing a system of communication (channels, methods, and willingness of both parties).

Table 3. The JORSS Strategy: Ends, Ways, and Means.

)URCE: Martin Herem, "The Strategy and Activity of the Forest Brothers: 1947-1950," *Combatting Terrorism :change* vol. 3. no. 3 (August 2013), https://globalecco.org/226

B. **Operations and Tactics**

1. Resistance operations were intended to support the strategic goal of national liberation by the Western powers, based on the declaration of self-determination in the Atlantic Charter.[483] Thus, group members believed that they only had to continue resistance until help arrived. Only then would they focus on overthrowing Soviet rule. This caused a less aggressive strategy of minimal violence in order to avoid large scale Soviet attacks.[484]

2. Lithuanian activities sought to: 1) prevent Sovietization of the country by annihilating communist activists and NKVD forces in the countryside; 2) safeguard the public order, protect the population from robberies by civilians or Red Army soldiers; 3) free political prisoners; 4) enforce the boycott of the elections to the Supreme Soviet of the USSR, or the leadership of the puppet state; 5) disrupt the drafting of Lithuanian youth into the Red Army; 6) obstruct the nationalization and collectivization of land, and; 7) prevent the settling of Russian colonists on the land, and in the homesteads, of deported Lithuanians.[485]

3. In application, this resulted initially in large partisan units of up to 800 men or more who fought pitched battles against the Red Army in the northeast of Lithuania. The Soviet forces could easily replace their losses, while the partisans could not, and large partisan formations were easy to locate and destroy. This method of fighting lasted only until mid-1945.[486] The Brothers continued terrorizing Soviet officials, attacking government buildings, ambushing NKVD troops, assassinating communist leaders, and attacking prisons to free their captured comrades.[487]

4. The JORSS strategy (table 3) called for an organizational structure. The developed organization covered all of Estonia (see fig. 18) and divided the country into three sections.[488] Each section had a staff that oversaw approximately one-third of Estonian territory. Communication was through proxy, mail, and coded radio messages. Each of these one-third staffs of 3 to 5 persons managed branches of 3 to 5 persons beneath them, as well as coordinating activities between them. The primary activities supported survival and de-conflicting activities.[489]

5. Outside of the formal organization were supporters and sympathizers. Supporters sustained the Forest Brothers with material goods, accommodations, medical aid, and transportation. Many farmers had arrangements with the Forest Brothers, exchanging farm labor for food. Each band was self-supporting, with no centralized funds held at a higher level and distributed downwards.[490]

6. This decentralized organization required strong leadership and discipline at the local level. This was sustained by the social status that accompanied being a Forest Brother, and the developed discipline of hiding in the forest and undertaking surreptitious actions over the course of years, while surrounded by the enemy. The organizational structure (see fig. 18) was cellular.[491]

7. When the Forest Brothers engaged in robbery and theft in order to sustain themselves, they gave the Soviets the opportunity to use effective propaganda against them, which cost them some support among the population, and enabled the Soviet authorities to develop their intelligence network. This was what the Brothers tried to avoid by exchanging labor for material support.[492]

CS4.8. Soviet Tactics and Operations

A. The Soviets acted against the partisans initially with pitched battles against large formations where they could afford to expend many lives. Resistance numbers decreased from those battles as well as from the

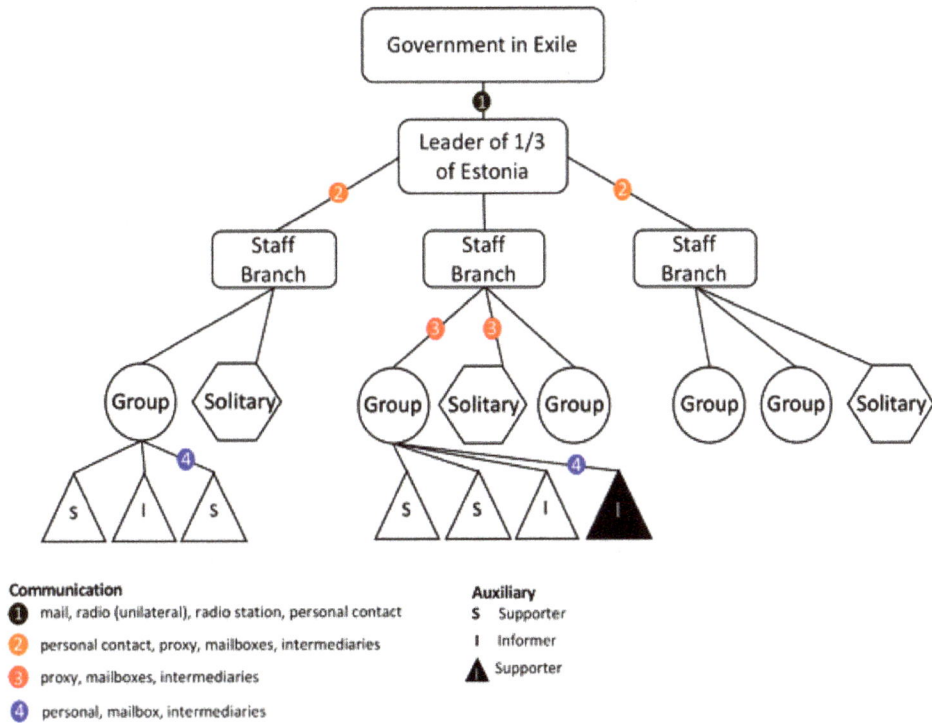

Figure 18. Outline of the Strategy of the Estonian Forest Brothers by Martin Herem (The Comfort Zone Model).

SOURCE: Martin Herem, "The Strategy and Activity of the Forest Brothers: 1947-1950," *Combatting Terrorism Exchange* vol. 3. no. 3 (August 2013), https://globalecco.org/226

Forest Brothers returning to civilian lives. They realized that the survival of the resistance and the Forest Brothers in particular relied on rural community support. Thus, the Soviets also acted on the population to reduce popular support.

B. First, Soviet land reforms took land from the middle class and wealthy, and divided it among the lower classes and landless. Middle class farmers were the most likely to assist the partisans. This created an internal class struggle and allowed the Soviets to portray themselves as champions of the poor. This resulted in at least some accommodation and neutrality toward the Soviets, while greatly reducing the resources available to the resistance. In Lithuania, the Soviets awarded the city of Vilnius, a part of pre-war Poland, to Lithuania, gaining some political support and attenuating some opposition.

C. Second, in the late 1940s, the Soviets engaged in massive deportations in the Baltics. Tens of thousands of people were deported to Siberia. This had two effects: first, it immediately increased the number of people fleeing deportation and joining the resistance, second, the deportations concurrently destroyed the support networks providing food, supplies, and information, from within the general population. This forced the partisans in the forests to forage and requisition food and supplies from the local populace. Requisitioning, in turn, alerted the authorities to their presence and increased the burden on the population.

D. Third, land collectivization was pushed more rapidly in the Baltics than elsewhere, reducing the number of people sympathetic to the resistance. In the beginning of 1949, throughout the Baltics, collectivization of land ranged from 3.9 to 8 percent. By the end of that year, collectivized land ranged from 62 percent in Lithuania, 80 percent in Estonia, and to 93 percent in Latvia. Farmers who resisted were deported. Moral and material support to the resistance declined.[493]

E. The NKVD also recruited local, ethnic militias to give the political impression of a civil war instead of a war by partisans against Soviet occupation. They also continuously described the Forest Brothers as bandits. This was easy to refute immediately after the occupation when the Brothers wore uniforms, were organized in large units, and fought conventional battles.

F. A critical tool of intelligence gathering was torture. Its effective use yielded much useful information however, it also resulted in many resistance members preferring to kill themselves rather than surrender. This accounts for the many stories of surrounded partisan units heroically attempting to shoot their way out of hopeless situations.[494]

G. Other methods of intimidation employed by the Soviets were public hangings of local partisans, and the public display of the dead bodies of resistance fighters in market squares, sometimes in obscene poses. Local residents were often forced to closely view the bodies so that NKVD agents could observe them for emotional responses, and thus pick out compatriots and family members who were otherwise not known. Those persons were then arrested and interrogated for further information and often deported to Siberia. To avoid this occurrence to friends and family, in situations where capture was imminent, many Forest Brothers killed themselves by holding grenades to their heads to avoid identification.[495] Burning of homes and farms suspected of belonging to resistance members was also widespread.[496]

H. In 1946, the Soviets added the tactic of pseudo-partisans. These groups would claim to be resisters from another region, arrange a meeting with actual resisters and then kill them all. They also robbed and murdered in the name of the Forest Brothers in order to discredit them.[497]

I. Publicly, the Soviets granted immunity from prosecution if resisters simply agreed to renounce their opposition and turn in at least one weapon. In practice, when these resisters surrendered, they were threatened with deportation unless they provided information on their former comrades or even agreed to re-join them as NKVD agents. The majority of resisters presented that choice agreed to cooperate, but then did nothing to assist the NKVD, and avoided deportation for themselves and their families without betraying comrades. However, some did betray their comrades.[498] The Soviets were also successful in inserting agents into actual resistance cells. Some of these agents were Baltic ex-patriot communists living in the Soviet Union, who could easily blend among their countrymen.

J. By the early 1950s, the Soviets had established specialized groups to find and monitor resistance cells. These groups built detailed pictures of the cells they monitored, such as actual names, code names, behaviors, communication methods, and contacts. Once they were satisfied that they knew practically all the members, they acted to eliminate them at once. As resistance numbers dwindled, so did their support among the population.[499]

CS4.9. Resistance Unravels

A. Throughout the Baltics, a rapidly changing social order, deteriorating economic conditions, and lack of personal security turned people from focusing on ideology and national independence toward personal survival. Former members of paramilitaries, or those who served in the German army, were excellent targets for Soviet recruitment as informants based on their compromising past and fear of persecution. Heavy blows to the movement were dealt by the mass deportations and collectivization in March 1949. Many supportive farmers and their families were deported to Siberia. The removal of those farmers, along with forced collectivization, robbed the Forest Brothers of their supporters.[500] These actions constituted deadly blows to the organization.[501] This environment, combined with the patient and ruthless activities of security agencies, practically eradicated the organizations between the autumn of 1949 and spring of 1950. The primary reason for their failure was caused "by the betrayals committed by those who found themselves betrayed."[502]

B. Beginning in the summer of 1945, the vast majority of people killed by the Forest Brothers were civilian collaborators and communists. The resistance portrayed itself as the alternative authority, but doing so also required it to enforce its rule. This meant it had to act upon the civilian population. It therefore punished those who collaborated too enthusiastically. Yet this led to a few ugly and ill-advised actions throughout the Baltics.

C. Eventually, a large part of the populace throughout the Baltics saw the partisans as fighting a lost cause, and simply wanted an end to the violence. As the immediate post-war years went on, the number of Forest Brothers dwindled, but their actions increased against civilian targets in attempts to fight collectivization and collaboration, and their popular support waned. Betrayals became the prevalent method of finding and dismembering the remaining resistance cells, as more people sought some form of accommodation to a situation they could not change.[503]

D. By 1953 the Soviet authorities managed to suppress active armed resistance. In Estonia the resistance claimed the lives of about 2,000 Forest Brothers, while thousands were arrested and sent to Siberian prison camps. After 1953, resistance became increasingly infrequent, although some partisans held out in the forests for decades.[504] The fighting in Estonia was less extensive and less centralized than in Lithuania or Latvia.[505]

CS4.10. The End of Organized Resistance

As the resistance of the Forest Brothers in the Baltics faded, the only remaining hope for freedom from occupation was Western intervention. Envoys were able to surreptitiously leave the Baltic States and arrived in Western capitals to plead for assistance. The most famous of these missions was by the Lithuanian Juozas Luksa. He escaped Lithuania carrying letters to the Pope and the United Nations describing the brutality of the occupation. He received only moral support and a little effort from Western intelligence agencies. He returned to Lithuania in 1950 to a greatly reduced resistance. Thousands of NKVD troops hunted him. He was betrayed, ambushed, and shot dead. Although the last Lithuanian partisan group was destroyed in 1956, some Lithuanian resistance members hid in bunkers or were sheltered by local citizens into the 1960s. Two Estonian Forest Brothers, Hugo and Aksel Mottus were caught by police in 1967 after living for twenty years in underground bunkers in the forests.[506]

CS4.11. Summary

A. Resilience. Estonia and Latvia had suffered lengthy political domination by Russia since the eighteenth century, and gained their independence after World War I, while Lithuania existed intermittently as a duchy of Poland. Each retained a strong sense of nationalism and did not want to be absorbed again into their larger neighbors. This nationalism provided the backbone of resistance to Soviet domination.

B. Pre-Conflict Resistance Planning. The Molotov-Ribbentrop Pact took the Baltics by surprise. They did not have time to organize and develop a resistance capability in the short time before Stalin absorbed them as Soviet Socialist Republics. During that period they developed resistance groups in each country, but without national level organization. Afterwards, they welcomed the Germans in removing the Soviets but suffered Nazi repression, and could not develop a national level organization to face the returning Soviets in 1944.

C. Integration of Resistance During Conflict. In each state, several organizations that existed in 1944 eventually merged into larger umbrella organizations. This may likely have been due to decreasing membership, capability, and hope, making the survival of individual organizations less likely. Communist resistance groups did not develop and compete against independence minded groups as in France and Poland.

D. External Support to Resistance. Despite the stated goals of the Atlantic Charter, the United States and Great Britain ceded the Baltics to the Soviets at the Yalta conference, as they had also done to Poland. Thus Baltic resistance received no material Western support.

E. Resistance Strategy. The Forest Brothers relied on the intervention of the Western powers in order to free them from the Soviets. Their strategy was to maintain themselves as a force, ready to assist allied armies as they fought the Soviets in the Baltics, and ready to reclaim national sovereignty. Toward that end, after costly, large-scale battles against the Red Army in 1944, they conducted only smaller scale operations against smaller Soviet units and attempted to prevent collaboration with the Soviets.

F. Restoration of Sovereignty. Sovereignty was not restored to each nation until the fall of the Soviet Union in December of 1991. Lithuania was the first Soviet republic to break from the Soviet Union in 1990 and was followed in 1991 by Estonia and Latvia.

Appendix E. Cold War Resistance Case Studies

No country can allow its safety to be wholly dependent on faithful observance by other states of rules to which they are obliged. - Arthur Balfour

E1. Introduction

Case studies of resistance organizations established during the Cold War are presented here as examples of what was done to elucidate what can be done. These case studies contain experiences from which society can learn today. They contain examples of mistakes and good ideas that can be adjusted to our present environment. The term stay-behind was used throughout these post-World War II European examples to describe what we refer to in this concept as resistance organizations. This difference in terms can be attributed to the contexts and perspectives of the times. Those stay-behind forces were organized first by individual member states, and then assembled under a NATO umbrella. Each state's stay-behind forces had slightly different original reasons for their establishment and each contains lessons for the present. The first case study focuses on U.S. support to such organizations, while the others focus on national level stay-behind organizations developed in several European countries after World War II. Some of these organizations grew out of existing organizations involved in the fight against the Nazis, while others were established, with no previous experience, by a nation resisting an occupier (e.g., Switzerland).

E2. Case Studies Summaries

A. The first case study focuses on the framework that the United States established to support the stay-behind forces of its NATO allies. This study is different from the others in that it deals with U.S. authorities to assist allied and partner stay-behind organizations. This is significant because the U.S., its allies and partners, must understand each other and have a common basis. Nations ready to assist each other in this effort must establish cooperation through formal agreements and practical exercises in order to form a basis of understanding.

B. Switzerland is the next study and offers both: 1) the perspective of a neutral nation preparing for resistance; and 2) a very complete and transparent model. Uniquely among these studies, Switzerland not only had a very resilient society but also used its willingness to resist a potential occupier as a strong deterrent against that possibility.

C. The case study of the formation of Italian stay-behind forces offers the perspective of a resistance capability born directly from war. The roots of this resistance organization, formed against communist incursion, predate the end of World War II. The Italian capability was initially focused on the cross-border threat from its communist neighbor, Yugoslavia, and fear of infiltration across that border, as well as fear of sympathizers offering cooperation from within Italian borders. It offers lessons of the potential misuse and misdirection of such an organization. It helps to explain why resistance organizations must be formed under strict and suitable legal and policy frameworks and why their focus must be limited to resisting a potential occupier without coming close to resembling an internal intelligence organization, or acting against subversive groups or individuals. Both of those tasks are better performed by actual intelligence and law enforcement agencies. Lastly, the Italian stay-behind effort, eventually known as Gladio offers an examination of the consequences of not establishing clear legal and policy frameworks for such organizations. The negative political revelations of the Italian Gladio operation eventually helped to deconstruct most stay-behind organizations throughout Western Europe in the post-Cold War period, while leaving a very negative impression on the minds of politicians and citizens regarding such organizations.

D. Norway, like most Western European states, established its resistance capability under the threat of advancing communism. This study examines how that resistance capability was developed, piecemeal, into an eventual comprehensive capability. Due to its geography, population, and historical experience, Norway offers yet another perspective on the establishment of resistance organizations. Much like Italy, its resistance capability also later suffered due to inadequate legal and policy frameworks.

E3. Conclusion

In each of these studies, we can discern the importance of legal and policy frameworks, established during peacetime, by both the potential resisting nation and external allies and partners ready to offer assistance. It becomes apparent that a resistance organization should not be used to gather intelligence on political opposition, or for countering a perceived internal insurgency, even if fomented from outside the nation. Furthermore, political support for a nation's resistance organization is inextricably linked to the frameworks established to support that organization, and which convey legitimacy to the organization among the citizenry and can retain legitimacy during an occupation.

Case Study 1. Historical U.S. Support Framework To NATO Stay-Behind Groups[507]

> If I always appear prepared, it is because before entering on an undertaking, I have meditated long and have foreseen what may occur. It is not genius which reveals to me suddenly and secretly what I should do in circumstances unexpected by others; it is thought and preparation. – Napoleon Bonaparte

CS1.1. Background

For the U.S., the concept of stay-behind organizations in Western Europe was part of a panoply of activities, conceived as measures of retaliation against Soviet threats and actions. Stay-behind organizations were kept distinct from other intelligence activities. However, "If you used a local labor union to fight the communists, it is quite possible that you would have asked the same labor leaders into the stay-behind network."[508]

CS1.2. National Security Council (NSC) Document No. 68

A. The most significant U.S. document spelling out the rationale for reacting to the Soviet threat was NSC 68 of 14 April 1950:

> the United States and the other free countries do not now have the forces in being and readily available to defeat local Soviet moves with local action, but must accept reverses or make these local moves the occasion for war–for which we are not prepared. This situation makes for great uneasiness among our allies, particularly in Western Europe, for whom total war means, initially, Soviet occupation.[509]

> [Therefore it is necessary to] strengthen the orientation toward the United States of the non-Soviet nations, and help such of those nations as are able and willing to make an important contribution to U.S. security, to increase their economic and political stability and their military capability.[510]

B. NSC 68 and other NSC enabling documents developed an array of actions considered as justified to be undertaken in order to cope with the Soviet threat. These actions included military assistance and training, economic aid to Western Europe, direct paramilitary operations conducted against the Soviet Union, and stay-behind forces.

CS1.3. The Office of Policy Coordination (OPC)

A. For the U.S., the origins of the stay-behind program go back to the creation of the OPC, authorized in National Security Council directive 10/2 on 18 June 1948.[511] The OPC was the predecessor to the Directorate of Operations at the Central Intelligence Agency (CIA). Though residing within the CIA, it took direction from the Departments of State and Defense. It was charged to organize and support "guerrilla movements" and "underground armies." Its original mission was to assist in liberating Eastern European countries from communist control and pushing back the Soviet border to Russia.[512] NSC 10/2 stated that overt foreign activities of the U.S. government must be supplemented by covert operations because of the "covert activities of the USSR." It also recommended placing responsibility for them in a new entity. This new entity, OPC, would operate independently.[513]

B. Thus, OPC was to be a new and independent body with the mission of carrying out peacetime operations. NSC 10/2 stated that such operations did not include armed conflict by recognized military forces, espionage and counterespionage, nor cover and deception for military operations. It defined such operations as covert activities related to information, economic warfare, preventive direct action (including sabotage, anti-sabotage, demolition, and evacuation measures) and subversion against hostile states (including assistance to underground resistance organizations, guerrillas, and refugee liberation groups).[514] Thus, American justification for support to the stay-behind program was born.

CS1.4. Structure and Organization

OPC was activated on 1 September 1948 with Frank Wisner as its chief. By late October 1948, Wisner categorized "Functional Groups" of activities to be conducted by OPC. This was done with the help of an Advisory Council of representatives from the Army, Navy, Air Force, the Joint Chiefs of Staff, and the State Department. The functional groups encompassed the general areas of: psychological and political warfare, economic warfare, preventive direct action, and miscellaneous activities: [515]

- Functional Group 1: Psychological Warfare
  - Program A—Press (periodical and non-periodical)
  - Program B—Radio
  - Program C—Miscellaneous (direct mail, poison pen, rumors, etc.)
- Functional Group 2: Political Warfare
  - Program A—Support of Resistance (Underground)
  - Program B—Support of DPs [Displaced Persons] and Refugees
  - Program C—Support of Anti-Communists in Free Countries
  - Program D—Encouragement of Defection
- Functional Group 3: Economic Warfare
  - Program A—Commodity Operations (portion redacted)
  - Program B—Fiscal Operations (portion redacted)
- Functional Group 4: Preventive Direct Action
  - Program A—Support of Guerrillas
  - Program B—Sabotage, Counter sabotage and Demolition
  - Program C—Evacuation
  - Program D—Stay-behind
- Functional Group 5: Miscellaneous
  - Program A—Front Organizations
  - Program B—War Plans
  - Program C—Administration
  - Program D—Miscellaneous

CS1.5. Stay-behind Forces, Piercing the Iron Curtain, and NATO

A. Stay-behind operations were also envisioned behind the Iron Curtain. A National Security Council directive of 23 October 1951 ordered that resistance forces be created in areas under Soviet control and be made operationally available in the event of war. As the authors of Battleground Berlin further related: "For OPC Germany, this meant creating 'stay-behind' assets in East Germany— clandestine operatives who would be equipped with radios and would remain inactive unless war actually came."[516]

B. The U.S. military sought repeatedly to take over the stay-behind activity, but the CIA successfully resisted. The European division of OPC became the action arm for the stay-behind program. With the creation of NATO, stay-behind activity became connected with NATO, and eventually fell under a purpose created entity, named the Allied Coordination Committee (ACC). The ACC consisted of intelligence representatives of the member nations. Though all the nations were members of the ACC, not all were participants; for example Turkey, Greece, Portugal, Iceland, and Canada did not participate.[517] Sweden and Switzerland may have been informal, intermittent participants.[518] Specifically, the CIA dealt with Swiss military intelligence on the subject of placing communications equipment in parts of the country.[519]

C. In 1952, OPC was merged with the espionage element of the CIA, the Office of Special Operations. This meant that there was only one clandestine arm of the CIA, the Directorate of Operations (then known as the Directorate of Plans), and it was this element that dealt with the intelligence services in Europe on the matter of stay-behinds, among other things. Stay-behind organization and activity was conducted directly on a country-by-country, or bilateral, basis. The U.S. was thus organized to assist with European national level stay-behind organizations.

CS1.6 Summary

In order to be successful, resistance organizations should plan and coordinate with capable and supportive allies and partners. This planning should be done together and integrated as much as possible. As emphasized in the preceding chapters, though not necessarily contained in the U.S. framework reviewed above, such planning should also account for placement of a government-in-exile, which must receive immediate recognition and material support from allies and partners. Planning should likely include arms, communications equipment, and other material, cached within the threatened state for use by the resistance organization, and possible use by forces from PNs. There should also be planning for additional logistical support during conflict. Critical to legitimacy during both pre-crisis and conflict, this planning must also develop mutually supportive and integrated legal frameworks on both sides, capable of supporting both pre-crisis activities and resistance activities during conflict. Mutually integrated legal, informational, logistical, and operational frameworks are vital to success.

Case Study 2. Switzerland: Total Defense Model[520]

Victory is reserved for those who are willing to pay its price. - Sun Tzu

Shortly before World War I, the German Kaiser was the guest of the Swiss government to observe military maneuvers. The Kaiser asked a Swiss militiaman: "You are 500,000 and you shoot well, but if we attack with 1,000,000 men what will you do?" The soldier replied: "Shoot twice and go home." - Unknown

CS2.1. Background

The Swiss Cold War Total Defense model was a whole-of-society approach to defense. The intent was to involve the whole country (i.e., private sector, local government, and NGOs), and not just the military, in national security. This also included economic and psychological mobilization of the population. The entire populace was subject to call-up for both military and nonmilitary functions, while the national infrastructure and industrial production base were co-opted and tooled for possible defense usage. This extensive civil defense network and wide civic integration achieved a high level of societal resilience. The Cold War Total Defense model was based on the Swiss experience during World War II, when Switzerland was able to deter Nazi Germany from invading this small nation on the border of the Third Reich.

CS2.2. Resilience

At the strategic level, the Swiss viewed the military as only one element of national power. They incorporated the diplomatic, informational, economic, and societal elements of national power into a traditional military domain. To boost resiliency and lower vulnerability to foreign propaganda, they maintained an objective national news service, promoted education, engendered national pride in Swiss institutions, and maintained reserve food supplies at the national, local, and individual levels. The Territorial Service was created as a new military branch shortly after World War II to provide internal security and protect the civilian population with assets such as auxiliary police/guard, hospital groups with medical transport, motor transport, and air defense. The Territorial Service brigades had mixed/interagency headquarters focused on three main tasks: 1) Civil Defense: civilian population defense with Canton (the equivalent to provinces or U.S. states) representatives in charge, along with the Department of Justice and Police; 2) Psychological Defense: information warfare with the Department of the Interior in the lead, and; 3) Economic Defense: led by the Department of National Economy. Civil defense against chemical, biological, radiological, and nuclear (CBRN) threats became a cornerstone of population protection. The strategic objective was to make the society resilient to any form of outside aggression through a holistic Total Defense.

CS2.3. Information Activities

A general principle guiding Swiss efforts was dissuasion, a form of psychological deterrence. This concept was combined with powerful CF, guerrilla resistance, and the planned self-destruction of Switzerland's industrial, communications, and transportation networks to deny their usage to an enemy. This was intended to signal a potential aggressor that the only gain in attacking Switzerland would be the occupation of a hostile area, denuded of economic or transportation value, with continued resistance by a determined and armed population. The objective of Total Defense was to make Switzerland an indigestible hedgehog. The information activities were again heavily influenced by World War II experience when the nation was subjected to Nazi propaganda. The national resilience and defense messages of the time were promulgated during that period via military produced news reels and the satiric acts of cabaret artists. Some self-censorship was imposed to avoid reprisals and comply with the neutrality principles.

CS2.4. Total Defense and Resistance

A. A critical component of Total Defense was the ability to conduct resistance operations in enemy-occupied Swiss territory. A popular misconception is that the plan relied on the writings of Major Hans von Dach, contained in his seven-volume series, *Total Resistance* (printed as one short book in 1965), wherein he propagated a concept of resistance conducted by the entire population, which he termed "partisan warfare."[521] The Swiss General Staff rejected that advice due to concerns with the law of land warfare and the maintenance of governance over a population of partisans, and chose instead a conventional doctrine with an integrated resistance plan.[522] The Swiss military's other major concern was that an overemphasis on von Dach's partisan warfare would neglect other important components of Total Defense.[523] The Territorial Service included a stay-behind force as early as 1948. However, the Soviet invasion of Hungary in 1956 highlighted the operational security limitations of maintaining this stay-behind element in the Territorial Service. By 1967, the requirement to train and equip a stay-behind force was transferred to the Military Intelligence Service.

B. The government's 1973 Swiss Security Policy Report stressed the need for resistance in occupied regions requiring a classical stay-behind UW mission, and an organization to carry it out. Section 426 of the report stated, "The occupation of the country must not mean that all resistance has ended. Even in this case, an enemy shall meet not only with the population's antipathy, but also active resistance."[524] Section 717 highlighted, "Guerrilla warfare and nonviolent resistance in occupied areas are being prepared within the limits of international law and will, if necessary, be carried out."[525] This remained unchanged until the end of the Cold War.

C. Resistance operations would be integrated with the operations of a robust conventional force. Swiss Army 61 became the organizing concept for the military. It designated three field army corps designed to protect the heartland, and one mountain army corps for the alpine regions. These 4 army corps were organized into field, mechanized, mountain, and border divisions.[526] At its peak, Swiss Army 61, based upon a militia concept of universal conscription, encompassed 625,000 personnel.[527] This number stands in relation to a 1962 population of 5.5 million.[528] The main battle doctrine revolved around a defense-in-depth, with static units to channel Soviet forces into destruction zones, and mobile units for counterattacks.[529] An integral part of this plan was resistance in occupied Swiss territory. After the operative collapse of regular military units, their remnants would continue the fight in the occupied regions as guerrillas or partisans. In parallel, the civil population in these areas would practice nonviolent resistance within the parameters of international law. A pre-established resistance cadre organization would support and bring coherence to these efforts. The potential risk of repression and counter violence was noted, and the government called upon the populace to prepare itself for such eventualities.[530] In 1969, the government produced a booklet printed in the three primary languages of Switzerland (German, French, and Italian), entitled Civil Defense, and also known as the Red Booklet. Its purpose was to explain how to prepare for natural disasters and war, and to instill a spirit of patriotism and resistance against communism in an attempt to strengthen national resiliency. It was criticized for its ideological perspective but it contained much practical civil defense advice.

CS2.5. Resistance Organization

A. Switzerland set up covert organizations tasked with the conduct of resistance in the event of a full or partial Soviet occupation. The Swiss Federal Council also established a government-in-exile location in Ireland for such an eventuality.[531] The Military Intelligence Service, which had been tasked with building a stay behind network, created the Special Service to organize popular resistance to the enemy, and supply the government-in-exile with intelligence. The Special Service was made up of three hierarchical levels, with the top level consisting of a small group of directing officers, members of the regular military who always dressed in their military uniforms and were responsible for the administration and training of the secret army. The second level was made up of so-called trusted persons from across Switzerland who were responsible for the recruitment of resistance fighters and supporters. Those individuals formed the third level in their respective parts of the country. Persons recruited by the second level could themselves recruit new members to join the resistance organization.

B. In 1979, an embarrassing spy scandal, the Bachmann-Schilling Affair, led to a parliamentary commission investigation which concluded that there was insufficient oversight and training of the Special Service. The Swiss government transformed and redesignated the effort into Project 26 (P-26); a designation derived from the 26 Swiss cantons.[532] Defense planners conceived it as a top-down, cadre-led structure, rather than a broad, decentralized civilian resistance movement as envisioned by von Dach. Leadership of the effort to establish and develop this organization remained with the Military Intelligence Service, but there was now a 5-member parliamentary oversight committee. Like the Special Service, the P-26 was organized into three levels. The command staff consisted mainly of senior military officials on civilian contracts or secondment. On the second, and core level, the cadre organization formed the secretive and well-trained nucleus of the resistance underground. This formation had a decentralized organizational model with distributed, clandestine cells. The third level would only have been recruited by the cadre if Switzerland had come under foreign occupation. P-26 was tasked with recruiting and training core personnel who could continue the fight after an occupation. P-26 executed this by setting up stay-behind arms caches, storing specialized equipment that would be required by the resistance organization, and organizing the necessary infrastructure for the coordinated command of the resistance from unoccupied parts of Swiss territory or from a potential exile base.[533] In essence, P-26 provided the framework for the creation of both an underground and a partial auxiliary. The underground is understood as a "clandestine cellular organization within the resistance movement that has the ability to conduct operations in areas that are inaccessible to guerrillas such as urban areas under the control of the local security forces," and a partial auxiliary is "that portion of the population that is providing active support to the guerrilla force or the underground."[534]

C. Operationally, the P-26 concept offers four areas for contemporary consideration on how to set up a clandestine organization for the conduct of resistance in the case of occupation. First, the group prepared for four possible and plausible operational scenarios:

- A foreign military transiting Switzerland and occupying only a portion of territory without the goal of full occupation.
- A foreign power attacking Switzerland and occupying a portion of territory with the ultimate goal of full conquest and occupation.
- Full conquest and occupation by a foreign army.
- The overthrow of the Swiss government by external forces resulting in the occupation of Switzerland.[535]

D. Second, the Swiss government placed the organization outside of the traditional military and government bureaucracy, to protect its members from discovery in the event of occupation, and to preclude its surrender as part of an overall capitulation agreement. Its military leader was hired under a private-sector contract, and personnel signed an employment convention via a front company delineating rights and obligations, with members paid and insured discreetly by the federal government. During peacetime, P-26 fell under the direction of the Swiss Chief of the General Staff.[536]

E. Third, P-26 sought members who were balanced, independent, stress-resistant, and trustworthy, but with a low profile from both character and societal perspectives. They were to have regular jobs that would provide cover for periodic training absences. Many had no military service records and there were also a minority of females. Professions included educators, academics, medical field personnel, and engineers.[537] Recruitment was slow, with careful vetting and selection. Once enrolled, members were trained and allocated to one of the approximately eighty resistance regions across the country. The manning for P-26 was set at 800 personnel, about half of whom had been recruited and trained by the time of its deactivation in 1990. The six-to-ten person units found in the 80 resistance regions were autonomous, and each had an active and sleeper cell assigned, with the active cell having no knowledge of the existence of the sleeper cell.[538] A typical cell had an operational chief, communicator, courier, and demolitions/engineering specialist.[539]

F. Fourth, the training was conducted in a centralized manner inside a military bunker near Gstaad, a tourist site in central Switzerland. Most training was conducted in 1- to 2- week increments about 3 to 4 times a year. Persons participating in this training could cover this time away by claiming that they were on vacation or holiday so as not to rouse suspicions or unwanted interest in their activities, and keep their membership secret from co-workers and neighbors. The British provided some of the training based on their expertise. It took about five years for an individual to be fully trained. The bunker, nicknamed *Schweizerhof*, was also the central location where the equipment and supplies were maintained in peacetime. P-26 achieved a high degree of planning, detail, training, secrecy, and operational security for the operationalization of resistance plans during peacetime.

G. A scandal known as the Secret Files Scandal shook the nation in 1989. The Swiss Government had been keeping files on about 900,000 of its citizen, roughly 15 percent of the population, under the codename Project 27. A distinct investigation conducted by a federal judge into P-26 led to a report that is still classified but did not find any wrongdoings by any of the individuals involved. The political climate at that time with the end of the Cold War and the revelations of other NATO stay behind networks in Europe, in particular Gladio in neighboring Italy, led to the Swiss government shutting down P-26 in 1990.

CS2.6. Conclusions for Contemporary Planning

A. Although not actually tested by war and occupation, the Swiss example illustrates a pragmatic approach for a small state in preparing for resistance in the event of full or partial occupation of its national territory by foreign forces. Several lessons for evaluation come to the forefront.

B. First, the Swiss profile as a small country with limited resources has relevance for its equally small European cousins, and for many non-European nations as well. The Swiss P-26 resistance organization would have conducted its operations in the rather flat Swiss Mittelland, which encompasses most of the population centers, as well as the industrial engines of the economy. This pre-alpine region is also not much different than the topography found in the Baltics. Additionally, the Swiss population is highly heterogeneous, having German, French, Italian, and Rhaeto-Romanic regions. The Swiss have successfully meshed these diverse cultural and ethnic groups into a single Swiss identity, providing a foundation for societal resilience and resistance to foreign occupiers. This prerequisite is an important lesson for similar heterogeneous nations.

For example, the Baltic nations have Russian and Polish national minorities with varying but mostly successful degrees of integration.

C. Second, articulating the Total Defense concept and resistance mission in official national security documents provided clear and essential policy guidance for a whole-of-government approach to these efforts by providing national level direction. All elements of national power must be integrated into a defense concept and the psychological/information war component takes a leading position for preparation. Credible media outlets, an educated, critical-thinking population, and a degree of national pride are antidotes to adversarial propaganda campaigns.

D. Third, while guerrillas may come from parts of the armed forces, a clandestine cadre organization can provide one structural model for resistance preparation and clandestine network establishment, with new recruits being brought into the underground and auxiliary forces only after hostilities are initiated. Of particular interest is the recruitment of nonmilitary personnel conducted by the P-26. In an age of biometrics and electronic databases, this approach could provide a resistance organization a greater degree of security against a foreign occupiers' pacification operations.

E. Fourth, resistance planning and operations must be well integrated with an adequate conventional military deterrent. Resistance operations alone are insufficient as a deterrent effect to dissuade an aggressor. The Swiss coupled a resistance concept and organization with a four-corps, 625,000-person conventional military force, which represented almost 12 percent of its population in time of national emergency.

F. Finally, Switzerland did not possess a true SOF capability during the Cold War. Today, SOF are traditionally responsible for UW and resistance missions. SOF can be an important catalyst for resistance planning and preparation by facilitating unified action with their interagency brethren to achieve unity of effort in resistance operations. The Swiss Cold War experience provides a useful starting point for today's national policymakers and their SOF elements in considering how to adapt the Cold War concept of Total Defense to current events—especially its critical resistance element.

CS2.7. Summary

A. **Perceived Threat.** The strategic objective of the Swiss holistic Total Defense approach was to make the society resilient to any form of outside aggression, albeit with the Soviet Union at the forefront of perceived foreign power threats. As part of the preparation against that particular threat, civil defense against CBRN threats became a focus in protecting the population.

1. **Resistance Organization and Preparation.** The Swiss created a hierarchical, cadre-led structure instead of relying on a broad civilian movement. The cadre nucleus was carefully selected, vetted, and trained and the organization even provided for division into underground and auxiliary responsibilities. Recruits mostly held regular jobs to provide cover for periodic training absences, and many had no military records. The lowest tactical organizational level was the typical cell structure. The Swiss also chose a friendly foreign state in which to locate their government-in-exile to continue a legitimate government retaining national sovereignty. Finally, they engaged in dissuasive messaging to ensure a potential invader and occupier that it could not succeed, including the publicized planned destruction of their own industry and facilities to ensure that the enemy gained nothing valuable.

B. **Policy and Legal Frameworks.** The Swiss had a public policy and accompanying legal framework of Total Resistance, to assure both Swiss citizens and any potential occupier that the nation, led by the state, planned to continue resistance even if their CF were defeated and the country was occupied. Conventional troops would become resistance fighters if their formations were defeated, to ensure integration of CF with the resistance organization. The Swiss also established a pre-crisis resistance cadre to lead resistance efforts.

C. **Outstanding Points.** The resistance organization was placed outside of the Swiss military and government bureaucracy to prevent the discovery of its members in case of occupation. Its leader was hired under a private sector contract and cadre members were employed and paid via front companies. The full organization was not intended to be built out in peacetime. During peacetime, only the cadre level of personnel was recruited and trained. This meant that most cell members would be recruited during crisis or occupation. This legal structure was also an attempt to preclude its surrender in case of capitulation, as it was not part of the armed forces.

Case Study 3. Italy: Stay-Behind Group—Operation Gladio[540]

Insurrection by means of guerrilla bands is the true method of warfare for all nations desirous of emancipating themselves from a foreign yoke. It is invincible, indestructible. - Giuseppe Mazzini

CS3.1. Background

A. In August 1990, during a parliamentary debate on terrorism, Italian Prime Minister Giulio Andreotti confirmed rumors of the existence of a secret network run by the Italian intelligence services during the Cold War. The network was established for the purpose of conducting stay-behind operations in case of an invasion by the forces of the Warsaw Pact, similar to structures in other NATO countries. The secret had been well preserved for almost 40 years. This revelation caused inquiries to other West European governments, which gradually admitted to possessing similar organizations.[541]

B. On 8 September 1943, an armistice was declared between Italy and the Allies. However, fighting in Italy continued between the Allies, remnants of the army loyal to Mussolini, and the Germans. Further, the battle began among Italians for the political future of Italy. The next twenty months saw conflict between networks of anti-fascist partisans, Mussolini's supporters, and royalists, fighting each other for the future control of Italy. Further, this fighting occurred against the backdrop of an increasingly likely ideological confrontation between the Soviet Union and Western powers. The partisans, Royalist armed forces, and fascists each possessed their own intelligence networks. The intelligence organizations of the military services each also had inner cores and objectives unknown to each other. Volunteers of the Crown was a paramilitary group fighting to preserve the monarchy.[542] Army Volunteers for the Independence of Sicily fought in quasi-guerrilla warfare style, closely linked to local organized crime.[543] Other organizations abounded throughout Italy until June 1945 when they were officially disbanded, bringing an end to most open conflict.

C. However, tensions remained in northeastern Italy where there was fear of a Yugoslav, communist takeover of the area. The area of Venezia Giulia was contested between Yugoslavia and Italy. Therefore, in November of 1945, former members of the Committee of National Liberation of the city of Gorizia, established a clandestine group of about 1,200 partisans for activation in case of attempted forcible annexation by Yugoslavia.[544] This tension with Yugoslavia also resurrected another wartime resistance organization named Osoppo in early 1946. The Army's Chief of Staff sanctioned the *Osoppo* group, and liaison was established between it and the 5th Territorial Command. In 1947 it was renamed as Voluntary Corps of Freedom (*Corpo Volontaria della Liberta*), with a strength of about 4,500. During the elections of April 1948, it was secretly deployed in the northeast region to protect against possible Yugoslav infiltration in support of a communist coup.[545]

D. In 1950, that unit, derived from the original *Osoppo*, changed again. It was now codenamed "O," and became a cadre organization, designed to expand to 20 battalions of 360 men each, for a total of 6,500 men. It was tasked with securing lines of communication, protecting civil and military installations, conducting guerrilla and counter-guerrilla operations, gathering intelligence, and reporting on enemy activities. Its weapons were stored in the depot of the 8th Alpine Regiment in Udine.

E. A U.S. Army intelligence report from 1947 noted the existence of an anti-communist movement named the Anti-Communist Clandestine Army (*Esercito Clandestino Anticomunista*), or ECA, whose goal seemed to be to prevent a communist takeover. There also existed several neo-fascist organizations, as well as the Italian Communist Party (*Partito Comunista Italiano*), or PCI, which was a communist network. Large numbers of communist partisans refused to surrender their weapons at the end of the war and Italian security forces discovered large weapons caches over the course of the following years. Tensions existed between the PCI communist network which was suspected of colluding with the government of communist Yugoslavia. This suspicion resulted in the sanctioning of or non-interference with other anti-communist secret networks such as ECA, as a balance against the communist network.[546]

F. In 1953, Italian Military Intelligence (*Servizio Informazioni Forze Armate*), or SIFAR, purchased a tract of land in a remote part of Sardinia, at *Capo Marrargiu* for the purpose of establishing a training camp for members of a resistance organization. It became known as the Saboteurs' Training Center (*Centro Addestramento Guastatori*) or CAG, and was partially subsidized by the Americans.[547]

CS3.2 Structure and Organization

A. The structure of the stay-behind network was probably completed by late 1956. By late September, SIFAR authorized the creation of a training branch known as the Training Center (*Sezione Addestramento*) or SAD, inside the R office, the foreign intelligence office (*Ricerche all'estero*) of the intelligence service. The new body, activated by 1 October, included a head of section and two groups: one tasked with general organization and support for two large guerrilla units (codenamed *Stella Alpina* and *Stella Marina*), and one with permanent secretarial work and activation of the operational branches (intelligence, stay-behind, propaganda, escape, and guerrilla) and some smaller units kept in a high state of readiness. Between 1959 and 1964, two more groups were added: one for logistical and operational air activities to support a Light Aircraft Section (*Sezione Aerei Leggeri*), originally created in 1958–59, and one for both long and short range communications. SAD also supervised the training and experimental activities of the CAG, which was defined as the operational and training base for the whole operation. Between 1964 and 1971, a new group was added to maintain liaison with the stay-behind operations in other NATO countries, and with the Alliance's structures in charge of clandestine warfare.[548]

B. The name Gladio first appeared in documents resulting from U.S.-Italian discussions in October of 1956 concerning the possible establishment of a joint organization for the Italian stay-behind effort. Between 1956 and 1958, the tasks and the framework of the new network were further defined by a joint U.S.–Italian body named the Gladio Committee.[549]

C. The old *Osoppo* group was incorporated into the new Gladio operation and basically became the new *Stella Alpina* guerrilla unit, even retaining its old leader. Recruiting personnel for the new, smaller units did not begin until the end of 1958. Before then, the only significant step taken was the training of six members of SAD in the U.S. between October and November 1957. The immediate consequence of the formal creation of the new structure was the end of the previous stay-behind structure of *Osoppo*, which was officially terminated in October 1956. The political backdrop to these events which may have intensified efforts was the Suez Crisis of 1956 and the Soviet invasion of Hungary in the same year. At the same time, the Italian Minister of Defense requested an increased American military presence.[550]

D. In 1959, a report from the SAD section of SIFAR identified the two guerrilla units, *Stella Alpina* and *Stella Marina*, as near fully operational. It also listed the main lines of effort of the organization as: completing the CAG, establishing operational doctrine, establishing a communications center to contact stay-behind units and to jam enemy broadcasting, constituting the 40 special nuclei operational branches, creating arms caches, and training cadres, among others.[551]

E. Within this structure, were 40 small nuclei (30 of which were activated by 1961) which would include two-to-three members and two radio operators to carry out such tasks as intelligence (6 nuclei 'I'), stay-behind

(10 'S'), propaganda (6 'P'), evasion and escape (6 'E'), and guerrilla (12 'G'). They would be based on a grand total of about 172 men, under the leadership of about 32 organizers. Then there would be five large guerrilla units in a high state of readiness, categorized as Rapid Response Units (Unita' di Pronto Impiego), or UPI, namely the *Stella Alpina* (which already a force numbering about 600, but intended to be brought up to a total of 1,000 and be capable of mobilizing another 1,000 men), the *Stella Marina* (another 200 men), the *Rododendro*, *Ginestra*, and *Azalea* (each numbering between 100 and 200 men). Altogether, the five larger units would total a maximum of about 1,500 men, with an additional 1,500 to be mobilized if necessary.

F. According to the original documents on which later parliamentary reports were based, these estimates turned out to be too high: the list of external personnel recruited for the stay-behind operation (i.e., not including the SIFAR personnel acting as instructors, planners, etc.) never amounted to more than 622 with about half selected between 1958 and 1967, and the rest between 1967 and 1990.[552]

G. The organizational plan focused mostly on northern Italy (defined as Zones I and II) but, as a second stage, it also conceived of extending operations to the central and southern parts of the country (Zones III and IV). The units were being prepared to operate either in case of occupation by the enemy or, in case of internal subversion, in order to spur the local population to resist and to maintain the continuity and presence of the national government, either as a shadow government, or representing and loyal to a lawful government-in-exile.

H. Arms caches were deployed throughout the possible zones of operation, with the risk of their being discovered and repurposed. Between 1959 and 1960, most of this material was provided by the U.S., and stored in Naples before being sent to the CAG at Capo Marrargiu. By the early 1960s, there were enough supplies to equip 30 out of the 40 nuclei which should eventually be set up: they included explosives, weapons, ammunition, rifles, hand grenades, daggers, mortars, light machine guns, pistols, rocket launchers, radios, binoculars, and various other devices. When the matter was investigated in 1990, the Italian secret service provided the parliamentary committees with a list of 139 caches. The weapons were usually stored underground and their deployment was done at night, to maintain secrecy of the deployment of the caches and the greater operation. Nevertheless, two of them were accidentally found; one by some workers in 1966 and a second one in 1968 by a Carabinieri patrol. This led to the decision to withdraw the arms caches in 1972, which was completed by June 1973. All the equipment was recovered, with the exception of two caches of light arms.[553]

I. Unfortunately, there was a gradual evolution of the original principles according to which the project should operate. Some of the larger guerrilla units designated to operate near the Yugoslav border were gradually tasked to keep under control, and if necessary neutralize, subversive activities conducted in peacetime in their areas. Additionally, the U.S. representatives on the Gladio committee expressed their desire to intensify the activities of the operation. In particular, they suggested that some of the programs be adjusted to the principles of counterinsurgency (COIN) doctrine. The U.S. military and the Kennedy administration would then promote use of the Italian base for external activities unrelated to Gladio. This would have included COIN training courses for foreign elements (particularly Africans), utilizing Italian instructors outside of Italy for the same purpose. It also would have included activating some members of Gladio on Italian territory to partake in propaganda and counterpropaganda. According to the released documents, these American initiatives were never agreed upon and the only COIN effort occurred during one exercise named *Delfino* in 1966.[554]

CS3.3. Conclusion

Though much information is still classified, the general outlines of Gladio are known. In the immediate aftermath of World War II, Italian security concerns were focused on the northeast Trieste area, due to fear of influence and incursion by communist Yugoslavia. The operation was formed to deal with the fear of enemy occupation by organizing a clandestine resistance in peacetime to deal with such an occurrence, and to maintain state legitimacy in occupied territory. The released documents provide evidence of an extremely high degree of coordination and cooperation with the U.S., including partial American financing and provision of arms and equipment.[555]

CS3.4. Summary

A. **Perceived Threat.** Immediate post-war conflict existed between networks of anti-fascists, Mussolini supporters, royalists, and those who supported integration with the Western Allies of World War II. Additionally, there was concern about communists and their potential to break off a northeast chunk of Italy, due to proximity to communist Yugoslavia.

B. **Resistance Organization and Preparation.** A designated resistance organization was placed within the intelligence service. Its structure contained guerrilla units as well as intelligence, escape and propaganda networks. It sought to establish operational doctrine, and it did eventually establish a small nuclei of cells, supported by caches of arms and communications equipment. Information concerning whether plans existed for an exiled government was not released.

C. **Policy and Legal Frameworks.** A lack of legal framework for the resistance organization and its efforts was a major contribution to the problems it faced when knowledge of its existence was made public. Suspicions surrounding its internal activities against Italian citizens (which may have been criminal), along with strongly perceived foreign (American) influence, ensured that the legacy of Gladio would make the establishment of such organizations in the future extremely difficult.

D. **Outstanding Points.** Eventually, part of this effort was directed near the Yugoslavian border with the task of identifying, and if necessary neutralizing, pro-communist subversive elements that sought to unite with Yugoslavia. The United States almost misdirected this Italian effort by seeking to transition it into a COIN role. Such a role may have assisted in the countersubversion task near Yugoslavia in northeast Italy, but such a task was outside the scope of a resistance organization. The Italian effort suffered from altered and confused intent in its early years, as well as an absence of clear policy and legal frameworks justifying and protecting these activities.

Case Study 4. Norway:[556] Stay-Behind Group—Rocambole[557]

> In the light of our wartime experience, a determined will to fight on even after military defeat and occupation is an essential part of a small country's defense preparedness.[558] - Defence Minister Jens Christian Hauge

CS4.1. Background

A. Norway was propelled to develop a stay-behind capability after World War II based on several events, including its own occupation during the war. In March of 1947, the United States proclaimed what came to be known as the Truman Doctrine.[559] This was a new policy whereby the United States committed to assist nations under communist threat. The initial effort focused on preventing communist takeover in Greece and protecting Turkey from perceived Soviet threats regarding the Dardanelles. In May of 1947, communists took over Hungary, and the Marshall Plan was announced in June 1947. In February 1948, the communists took over Czechoslovakia, followed by a Soviet offer of a non-aggression pact to Finland two days later. In that same month, the Norwegian prime minister publicly warned about and called for vigilance against the internal threat posed by the Norwegian Communist Party.[560] Based on these events, Norway began its stay-behind plans.

B. This stay-behind program began with three classified activities code-named Saturn, Jupiter, and Uranus. Saturn was a project to cache about 50 radio transmitters throughout strategic points in the country. These would be assigned to persons not previously involved in similar activities, to avoid a potential occupier dis-covering the persons, and would be activated only under occupation or internal coup.[561]

C. The next effort was to establish an organization for intelligence collection and sabotage behind enemy lines, prepared well ahead of time, and not improvised during a crisis. This was most likely code-named Jupiter. Its initial purpose was to establish a network to protect important industry and factories against infiltration and sabotage. Interestingly, private initiatives for this purpose were already underway. The intent was to guard against fifth column[562] subversive activities. This evolved into a counterespionage effort to protect industry, while also providing the capability to collect wartime intelligence against an occupying power.[563]

D. Complementing this was the creation of a partisan action organization. This partisan organization was distinguished from guerrilla activity. Guerrilla warfare was to be conducted by special commando type units, left behind by a retreating army to operate behind enemy lines in occupied territory. Partisan units would conduct sabotage and underground activities in occupied territory against targets of military significance. This partisan organization would require a high level of preparedness in peacetime. It was designated as FO 4, which was derived from the name given to Norway's Defense High Command in exile during the war, which planned and executed sabotage and other underground activities in occupied territory. These partisan groups would be based on small action groups of two to four men, with access to pre-positioned caches of radios, weapons, and other supplies. They would be recruited from Army and Home Guard non-commissioned officer ranks with local area knowledge. Veterans from the wartime resistance were only to be used as instructors because they could be too easily identified and eliminated by an occupier through local informers.[564]

E. In the spring of 1949, these code-named, partially filled organizations became a secret organization under the Chief of Defense Staff, named *Rocambole*. It subsumed the previous Saturn, Jupiter, and FO 4 efforts. Its purpose was to perform isolated missions in occupied territory. Its tasks were destruction of material targets, organizing larger secret groups, reception of airlifted personnel and supplies, reconnaissance, special intelligence, guerrilla actions, assassinations, and protection of installations and communications during liberation of territory. Notably, guerrilla warfare was no longer excluded from its missions, and intelligence work was clearly within its tasks.[565]

F. Rocambole was intended to establish fifteen five-man groups. In organizational cadre fashion, only the leader and radio operator for each group would be recruited and trained, with other members being nominated but not activated. If all of Norway fell under occupation, then a wartime headquarters would be established in Great Britain. From its beginning in spring of 1949, Rocambole was primarily occupied with training security officers for important industrial firms and acquiring suitable radio equipment and training radio operators. By the end of 1949, training was completed for nine group leaders and seven radio operators, and Rocambole had become a tripartite, U.S.-U.K.-Norwegian, effort. By the end of 1949, the plan for Rocambole was expanded to between 15 and 45 groups of 5 men each, with a corresponding number of secret caches containing explosives, weapons, ammunition, food, field equipment, and communications equipment. The organization's primary intent was to conduct sabotage against industries within occupied territory that could be of use to the enemy, and its secondary intent was to secure critical sites and communications facilities within free territory if Norway fell under partial occupation.[566]

G. The basic outline was a secret network of small, well-equipped groups, only mobilized after partial or full occupation. There would be an initial period of concealment during which time these groups would find cover and carry out reconnaissance and detailed planning, with only selected groups being switched quickly to the active phase. The caches were established under an assumption that they could not be resupplied for 12 months.[567]

H. Rocambole peacetime leadership consisted of eight men. After mobilization, it would become twice as large, with staff responsibilities divided among communications, operations, intelligence, training, personnel, and supplies. The mobilized leadership would establish itself initially within Norway, with assistance from American and British intelligence agencies, and displace to an allied country if Norway fell under full occupation. During the first few years of establishing the organization, there were staff meetings about once per week, with American and British intelligence services attending.[568]

I. In 1950, the Rocambole staff received a house in the *Smestad* camp in northwest Oslo as its headquarters, while also preparing a disaster plan for rapid mobilization and transfer of cadre staff to an allied country in case of critical developments. A sense of urgency was created by the world political situation, particularly the war in Korea. During this time, Rocambole also conducted "pilot operations" which consisted of groups practicing deployment to cache locations. A joint American-British proposal to carry out a practice mobilization of the whole organization was rejected as jeopardizing the organization and its members.[569]

J. By 1951, several training courses were being conducted. These courses consisted of some sort of basic course, a special industrial course, and radio operator's course. By the end of the year, twenty seven group leaders and twenty seven radio operators were trained and deployed in northern Norway, which was considered the area under the greatest threat.[570]

K. By 1952, information had been collected on "key categories of sabotage targets," which comprised radar installations, airfields, communication lines and nodes, fuel depots, and significant industry. Other High North targets included some major highways and power stations. Planning became intense for the northern part of Norway, especially *Finnmark*, due to its proximity to the Soviet Union. By autumn of 1952, a total of ninety one persons had been trained as cadre and organized into thirty two groups, each with a radio station and radio operator.[571]

CS4.2. Intelligence

A network of intelligence agents tasked to perform general military intelligence work in occupied territory was begun under the cover name of *Lindus*, but later called Group 27. By May of 1950, it had established 40 stations with radio operators in southern Norway. During that summer, two additional networks were established. The first, Group 67, was a stay-behind organization tasked to collect and pass on political and economic information and brief agents about to be infiltrated into occupied territory. The second was Group 134, which was a CI or counterespionage group tasked to infiltrate the enemy's security services in occupied territory to warn of danger. As with Rocambole, these networks, excluding Group 134, were only supposed to be established in skeleton or cadre form during peacetime. Group 134 members had to establish their trusted communist bona fides before a crisis in order to be able to infiltrate the enemy, if necessary. There is insufficient public information available to know whether, or to what degree, this was accomplished. Group 33 was organized to report on movements of enemy planes and ships and trained fourteen stations. Group 45 was established for meteorological data. Group 74 was organized to evacuate certain Norwegian and allied personnel, but was not an escape and evasion organization. Blue Mix was the escape and evasion organization.[572]

CS4.3. Costs

Costs were minimal. The majority of personnel were unpaid volunteers. Their motivation was idealistic, without remuneration beyond very minor gifts such as books or compensation for lost earnings.[573]

CS4.4. NATO

A. In August 1951, Supreme Allied Commander Europe established an ad hoc committee, titled the Clandestine Planning Committee, or CPC, but it was only open to the U.S., Great Britain, and France, with other nations invited only when their interests were involved as the meeting subject. Early on, it was established that stay-behind elements were first and foremost instruments at the disposal of their national governments, with the primary task of forming the nucleus for the re-capture of lost territory.

B. At the start of this coordination, Norway insisted that it would command and control stay-behind forces and not place them under NATO command.[574]

> During the last war, the Norwegian government was located outside the boundaries of the country, but its constitutional powers remained in legal order and it exercised its functions as government throughout the enemy's occupation of Norway. Under the influence of these experiences the Norwegian government views it as self-evident that it should retain responsibility for political leadership in the country–also in occupied parts of the country. Were the leadership of the resistance movement to be subordinated to an American general and his international staff, this would incite a political storm in the country if it became known before an occupation–and after an occupation it would provide an excellent basis for enemy propaganda.[575]

C. However, the relationship between national control and NATO supervision was a recurring topic, until the establishment of the Allied Clandestine Coordination Groups (ACCG) at Supreme Headquarters Allied Powers Europe (SHAPE) and subordinate commands. ACCG documents affirmed that "Command and control will at all times be retained by the respective National Clandestine Services."[576]

D. In furtherance of both national control and organizational safety and continuity, critical information was sealed, and deposited in the Norwegian embassies in London and Washington. These sealed files contained copies of personnel records in both the Rocambole and Blue Mix organizations, and the information necessary to establish and run the radio communications networks for the various services.[577]

CS4.5. NATO Plans for Guerrilla Warfare to Delay and Slow a Soviet Attack

A. By 1961, NATO viewed war in Europe in two phases as a result of potential Soviet attack. The first phase would consist of large-scale maneuver warfare lasting about thirty days. The second phase would last longer and be characterized by efforts to regroup and re-supply. NATO leadership did not expect the Soviets to be able to occupy large parts of Europe. It was seen as desirable for stay-behind activities to be conducted in areas from which NATO forces were forced to temporarily retreat. SHAPE accepted that stay-behind activities were purely national responsibilities, but it still desired coordination for intelligence operations and evasion and escape operations. This effort would go through the ACCG (Renamed Consultative and Coordinating).

B. SHAPE desired to delay an attacking enemy's advance. In line with this, Norway established an organization separate from its stay-behind organization to conduct delaying activities. It planned for twenty groups comprising a total of 375 men. They would be tasked to assist in the early phase of a war by obstructing strategic railway lines, oil pipelines, and ports, while supporting allied air activity, and hindering enemy advance movement for the establishment of tactical air bases. Planning or assessments then extended to the earliest operational stages of such a war and foresaw the launching of such activities as early as possible. This necessitated coordination with friendly neighbors, which meant early operations on Swedish and Finnish territory, though most likely conducted by those nations. As early as 1953, Norwegian-Finnish planning had begun for such operations.[578]

CS4.6. Post 1970

A 1996 report from the Lund Commission, appointed to investigate allegations of illegal surveillance, provides the primary material for information after 1970. According to that report, the elaborate district-by-district organization was abandoned by the 1970s and replaced by a more informal network. Also, in 1974, a network named Argus was established for intelligence missions and evasion and escape activities, and in 1978, it attained a more military structure with leaders appointed from the military. Further, the 1978 revelation of an arms cache in the Oslo residence of a Norwegian ship owner resulted in the re-location of all privately housed caches onto military bases. In 1980, the Norwegian stay-behind effort joined NATO's ACC, which was likely the continuation of the ACCG. Directives affirmed that command and control of stay-behind organizations were a national responsibility in both peace and war with coordination conducted to facilitate joint exercises, economize resources, and establish common terminology. In 1983, Rocambole, the sabotage network, was disbanded. In 1991, after the Italian Gladio revelations, the CPC and ACC were disbanded. As fear of invasion waned, primarily from the 1970s onward, stay-behind forces' raison d'être began to vanish. Further, allegations of improprieties, even if mostly unfounded, contributed to the decline of the stay-behind organizational concept. After the Gladio affair in Italy, stay-behind organizations all gained a bad reputation and were also very difficult to justify after the fall of the Soviet Union.[579]

CS4.7. Summary

A. **Perceived Threat.** External communist threats convinced the Norwegian government to establish a resistance organization. The primary concern seemed to be communist penetration of its society. The nation otherwise did not have significant internal divisions. The organization initially focused on intelligence collection and sabotage behind enemy lines. The purpose was to protect important industry and factories against infiltration and sabotage. This was intended to guard against fifth column activities. It thus evolved into a counterespionage effort.

B. **Resistance Organization and Preparation**

1. Complementing this intelligence effort was a partisan organization (distinguished in the Norwegian lexicon from guerrilla warfare which is conducted by stay-behind special commando units) designed to conduct sabotage and underground activities in occupied territory. These partisans were based on very small teams with access to pre-positioned caches of arms, communications equipment, and other supplies. The whole program was placed under the Chief of Defense Staff, with plans for the displacement of the headquarters in wartime to Great Britain.

2. The outline of the organization was as a secret network of small, well-equipped units mobilized after occupation. Initially they would conceal themselves, then conduct reconnaissance and planning. Then selected groups would conduct actions. The caches were supplied to sustain groups for twelve months. Sabotage would be conducted against industries within occupied territory that could be of use to the enemy. Training courses were established for intelligence and partisan personnel.

C. **Policy and Legal Frameworks.** Though under military leadership, most personnel were unpaid volunteers motivated by idealism. The organization's legal framework was problematical, resulting in allegations of illegal surveillance by it. Further allegations of improprieties after the Italian Gladio revelations, combined with the fall of the Soviet Union, contributed to its further decline.

D. **Outstanding Points.** Security and secrecy concerning the existence and hierarchy of the organization, its cadre members, and potential operations were duly treated as critical to organizational survival and success. Additionally, the government insisted on maintaining national, versus NATO, operational control. For these reasons of national control, and to ensure safety and survivability, copies of personnel records in both the Rocambole and Blue Mix organizations, and the information necessary to establish and run the radio communications networks for the various services, were sealed and deposited in the Norwegian embassies in London and Washington.

Appendix F. Case Studies Lessons Learned

Only a fool learns from his own mistakes. The wise man learns from the mistakes of others. - Otto von Bismarck

F.1. Introduction

A. An examination of the case studies, as compared to the suggested organization and actions laid out in the main body of this concept, can elucidate certain salient points and commonalities. Since this is not a critique of the past actions of others, nor a claim to having created the standard within this volume for the establishment of resistance organizations, this section is not a point by point comparison or an excessively detailed analysis. Rather, it is a general review within selected groupings of topics. These groupings are themselves a summary of the suggestions laid out in this volume, and the information summarized within them reflects the salient points and commonalities brought out within the case studies.

B. First is a review of the World War II case studies. The actual historical occurrence of these events offers the benefit of hindsight, and some ability, at the risk of over-simplification, to rate organizations and activities as successful or unsuccessful. This review includes some causal analysis, while being acutely mindful of the context in which these struggles occurred. The Baltic resistance of the Forest Brothers will again be included in this section, since that resistance, though fought mostly after World War II, arose as a resistance struggle during the war, with strong roots in the first Soviet occupation.

C. The selected Cold War case studies reviewed resistance networks that were established to varying degrees based on the context of the Soviet threat of the time. Except for the Italian networks applied in the immediate post-war period against feared communist penetration from Yugoslavia, these networks were never activated against an enemy, because the Soviet Union never invaded the countries in question. Except for Switzerland, each of these nations was to some degree occupied during World War II, and it can be assumed that these networks were established to a large degree based on lessons learned from World War II resistance organizations. In fact, some of them grew directly from World War II organizations. The studies from this era offer insight mostly into the perceived threat, organizational methodology, strategic messaging, and legal and policy frameworks—rather than success or failure in application.

F.2. Second World War Case Studies

A. Resilience

1. In this concept, we defined resilience as the will of the people to maintain what they have. "The will and ability to withstand external pressure and influences and/or recover from the effects of that pressure or influence."

2. Chapter 1 reviewed several factors contributing to resilient nations. Those factors included a sense of national identity, psychological preparation, vulnerability recognition and reduction, preparation, interoperability with external supporters, and planning for resistance.

3. The World War II case studies in appendix D revealed several strengths and weaknesses regarding resilience within each nation studied, including the Philippines for purposes of this description. The

primary facts or experiences influencing national resilience within the surveyed nations included recent wars and occupations, ethnic diversity or minority ethnic commonality with the aggressor nation, and divergent internal ideological groups (communists) adhering to or seeking unity with a foreign nation. National identities were affected to varying degrees by ethnic identities and how well integrated those identities were in the national identity. The draw of communist ideology was present in each of the four cases, and internally divided nations as resisters sought to either restore pre-war governance ideas or bring about communism. The concerned governments may have conducted extensive pre-war population vulnerability assessments and taken actions to reduce those vulnerabilities, however, we have little evidence of this. Full exploration of this topic is beyond the scope of the present volume, but assessing the success of these activities would be an excellent subject for further study. The World War II case studies revealed an overall lack of adequate preparation to continue the struggle against the occupier through organized national resistance. There were no resistance plans, and thus no interoperability plans or exercises, with other nations. Lack of preparation for resistance during occupation contributed to lower resiliency in these nations.

a. **Wars and Occupations**

01. Relatively recent wars had both positive and negative effects on resiliency. In France, the late nineteenth century war with Prussia demonstrated the costs of guerrilla warfare in terms of French lives lost and physical destruction brought about by reprisals. France's World War I experience left a deep scar on its society from the incredibly massive loss of so much of a generation of its young men. These relatively recent costs of war, though not the only factor, contributed much to its weak national resiliency against the invading Nazi-German forces, and its willingness to come to terms with its enemy while attempting to preserve its honor and sovereignty in the form of Vichy France.

02. In the Philippines, the most recent war was the uprising against the Americans immediately following the U.S. assuming sovereignty over the Philippine islands after the Spanish-American War. That war ended relatively amicably with the surrender of its leader, who then became the leader of the Philippines, but without continuing a military or political war against the United States. The ensuing decades of American occupation involved large degrees of cooperation and economic benefit to the Filipinos. The Americans and Filipinos developed mutual affinity. This made their relationship extremely resilient when faced with Japanese aggression. This degree of adherence, coupled with American assurances of Filipino independence after the war, as opposed to continued domination by Imperial Japan, activated Filipino nationalism. Filipinos worked in concert and cooperation with American goals against the Japanese. Thus, the last war fought by many Filipinos against the United States did not weaken either. In fact, they formed bonds of friendship which sustained their mutual efforts and survived from the beginning of the war through to the end. There was no cleavage for the Japanese to exploit, nor did the Japanese adequately attempt to create such cleavage during their occupation.

03. Poland and the Baltic States of Estonia, Latvia, and Lithuania each experienced degrees of Germanic or Russian domination for centuries. During and after World War I, they each gained their national independence and were able to form states. Each also successfully engaged in wars to protect themselves from Russian communists immediately after World War I. This helped solidify the national identity in each country, and gave them confidence in their survivability as nations against their more powerful neighbors. Poland, in particular, developed many NGOs with civic and charitable focuses, which sometimes operated as quasi-governing institutions in

the period between the two world wars. This concept of self-help assisted it to quickly develop underground organizations. Thus, these recent successful war experiences helped form their national identities and gave them confidence to survive as nations. Those recent successful early twentieth century wars were major factors in the positive resilience of these nations due to their effects on national identity.

b. **Nationality and Ethnic Identities**

    01. In France, the factor of ethnicity was minimal. France had a well formed, strong, historic national identity. The regions of Alsace and Lorraine were mixed between French and Germans, but these border regions with their mixed identity did not have a major impact on the war and were not a major battle ground.

    02. Filipinos, for the most part, identified as a nation of people, with physically isolated minorities. Their presence on an archipelago of islands formed their identity, and allowed for degrees of sub-cultures, but there was a majority common ethnic identity on the major islands which were the primary battleground and prize. Where there were differences which the Japanese could have used to access points of political leverage and turn them into military leverage, they failed to do so by treating all Filipinos as subordinate to them.

    03. Poland had a very large German population in the westernmost part of its new state. When Nazi Germany invaded, it created a separate government for the portion with the large German population thus using ethnic leverage well to seize territory. However, this separation meant that the rest of Poland contained an ethnic Polish majority with less risk of a portion of its population adhering to a foreign nation. The exception was the minority Ukrainian population, which the Soviets in particular were able to leverage against the ethnic Poles. The very large Polish Roman Catholic majority allowed for ethnic solidarity against German and Russian domination.

    04. The Baltic nations were incorporated into larger neighboring states for much of the previous several centuries, but had each formed cultures and individual national identities that were basically European and not Russian. The primary ethnic problems within these nations revolved around the ethnic Russians that were settled into these lands during the Soviet occupations. These Russians presented a problem for national anti-Soviet solidarity. Small portions of these populations continue to form potential leverage or access points within those nations.

c. **Communist Ideology**

    01. Communism was a major factor during both French and Polish resistance efforts, as a competitor to national resistance, with the intent of bringing about a Soviet aligned communist government. It was a lesser factor in Filipino-American resistance. In the Baltics, communist ideology combined with the power of the state to bring about the gradual destruction of

> *The concept of national resilience has many components and is very contextual. Previous wars and occupations strengthened the resilience of some nations but weakened the resilience of others. Ethnic identity was a major factor in each state and an internal ethnic population similar to the occupier typically weakened national resilience, as these populations were used as points of leverage by the occupier. An exception is the Japanese who failed to use internal Filipino ethnic divisions. Communist ideology was a major factor in each case except for the Philippines (though the communists there became a large post-war problem). Communist networks offered internal competition and threatened the restoration of independent governments. They were successful where they were geographically close to the Soviet Union and too distant for support from the Western allies.*

the Forest Brothers. In France, adherents formed networks to support this ideology prior to World War II. During World War II, they saw the opportunity to bring about a communist government for France after the Germans were ejected. Part of their strategy was to engage in violent direct action against the Germans to cause German reprisals against the French population in the hope that these reprisals would cause more French citizens to join the communist cause. The communist networks were in constant competition with non-communist networks. Luckily for France, the Soviets were too geographically distant to assist them. Additionally, de Gaulle, the British, and the Americans adequately supported the non-communist networks and starved the communist networks of enough support to suffer their deserved demise. Yet, this internal competition was a major factor throughout the time of resistance against the Germans.

02. In Poland, communist groups received support from their ideological brethren in the adjoining Soviet Union. This support, along with the physical entry of Soviet forces, allowed the communist cause to win, and crushed the more organized, widespread, and supported PUS. The communist groups won due to external support by a large and powerful neighbor, while the more legitimate forces of a free Poland lacked such support due primarily to geography.

03. In the Baltics, communist ideology was less of a factor than in Poland. This was probably due to the fact that the ideology was so closely associated with their very powerful Russian neighbor that had dominated at least two of the three nations for so much of their history, while the third also recognized the danger associated with this mostly Russian propagated ideology. Therefore, communist fifth columns inside of these nations were less effective.

04. In the Philippines, the communist Huks sought to establish a post-war communist government in the archipelago but they had minimal support among the population, no support by the Americans, and the Japanese occupiers were also anti-communist. Therefore, despite the presence of communist groups in the islands, their ideology was a minimal factor affecting the resistance and the post-war outcome.

d. **Pre-Conflict Resistance Planning**

Prior to World War II, none of the states in the case studies made plans for establishing resistance as a form of warfare, or establishing intelligence or logistical networks. This may have been because establishing such an effort may have appeared to signal impending defeat, though this is speculation. The Polish state engaged in hasty planning immediately prior to being invaded, which laid the basis for the gradual formation of their PUS. The Baltic nations prepared for the second Soviet occupation based on their experiences of the first and on occupation by Nazi Germany, but could not plan as a state entity while under German occupation. Each state had pre-war underground communist networks to some extent.

B. **Establishment of Resistance Networks**

1. Each of the nations in the case studies established resistance networks. The communist networks had advantages in that they were established prior to World War II. These communist organizations were either driven underground by the national governments, or existed as underground networks seeking the overthrow of the national governments and typically unification with the Soviet Union.

> *The existence of pre-conflict networks was not determinative of success. The pre-conflict communist networks began resistance efforts faster, but success was determined by the extent of external assistance. However, pre-crisis establishment of resistance network human and physical infrastructure, within legal and policy frameworks, can lend credibility, legitimacy, and unity to a resistance effort.*

2. The initial French underground networks were communist, formed in 1936. After the German invasion and division of France into occupied and unoccupied zones, other non-communist networks sprang up. Initially, these were not externally directed and each was basically purpose driven, such as gathering intelligence to send out of France to the Allies or escape networks. Later, many of these networks became loyal to de Gaulle and his provisional government. British and American intelligence and special operations agencies also established specific networks to directly assist the coming invasion of France. The Jedburgh teams, formed with American, British and French members, though short lived as a program, were formed to provide direct assistance to the maneuver of Allied CF, using local French networks.

3. The Poles had many organizations or associations that functioned outside of government and became templates for underground networks. Immediately prior to occupation by the Germans and Russians, the Polish government tasked its military to form underground networks. These networks eventually combined with other independent networks to form the PUS. This became the most organized resistance in Europe. Despite its organization and many successful operations, such as smuggling German rocket parts to the west, it lacked Western material support and was eventually crushed by the Soviets.

4. The Baltic States also had organizations and associations independent of government. Their anti-Soviet resistance, both active and passive, began under the first Soviet occupation. They continued underground activity while under German occupation, and as the German occupation ended, their resistance became highly organized. It contained a strong military component to fight the Soviets as they re-entered the Baltic States to reclaim them as Soviet Socialist Republics. Gradually, they were ground down from large, organized, uniformed military units into smaller networks, surviving by hiding in the woods in underground bunkers. Lacking Western material support they eventually failed.

5. In the Philippines, neither the United States—which held sovereignty over the islands—nor local Filipinos established resistance organizations prior to the Japanese invasion. As in the European examples, the communists already had an organization, but without significant outside support such as what the Soviets were able to provide to Polish and Baltic communist organizations. They had little effect on the Japanese and were not able to compete adequately with the resistance groups eventually established by the American military. American and Filipino soldiers escaping Japanese capture fled to the inner parts of the islands and established the first resistance networks. Initially, they did not even know of each other but because they were led by American and Filipino officers, they were ideologically and strategically united in their efforts to oust Japan and restore the status quo. Once the American military outside of the Philippines became aware of these networks, they received support and were used to assist the coming invasion to re-take the islands from Japan.

> Each nation surveyed began with many groups of resisters whose networks and were eventually brought together under the auspices of one or more umbrella organizations linked to an external exiled government, or only an interior shadow government. Each nation also faced the internal problem of competition from communist resistance networks which were supported by external communists to various degrees. Whether supported by the Western allies or the Soviets, the resistance networks with the most outside support, typically based on accessibility, and eventually joined by friendly conventional military forces, were victorious.

C. Governments-in-Exile Maintaining Sovereignty

1. The case of France during World War II is unique. The French government had no exile plans in case of invasion. After the German invasion, and loss of much territory and sovereignty, the French negotiated with the Germans. The result was the splitting of France into two portions. One portion, which included the traditional capital of Paris, was occupied by Germany. The other portion, not occupied by Germany, came about with the Treaty of Vichy and had its capital in that city.

2. After the initial invasion, Colonel Charles de Gaulle escaped to England. There, he sought to gather free French citizens to continue the fight against Germany. He was tried by the Vichy government in absentia and sentenced to death for treason. Over the course of a few short years, he gained the support of the British and then the Americans. He contacted and developed networks inside of France and made himself valuable to the British and Americans by providing information to them from these networks inside of France. He soon established provisional political offices and a government in waiting, ready to take the reins of France. In this way, once France was liberated, he avoided immediate elections which may have given too much power to the communists and avoided the imposition of an Allied military government as was done in Italy. De Gaulle survived and thrived through self-exile, cooperation with and support of British and American military efforts, gradual official British and American proclamations of recognition of him and his organization as the true representatives of France, the abrogation of the Vichy government, and the German invasion of Vichy controlled France in 1942 after the Allied landings in North Africa. De Gaulle had essentially become the de facto, self-exiled, leader of a provisional government and reestablished French sovereignty.

3. The government of Poland exiled itself at the beginning of the conflict and eventually arrived in London. Due to that government's legitimate, legal establishment, it was able to retain international recognition of its sovereignty over Poland. The exception was the Soviet Union which recognized a different Polish government that the Soviets established in Moscow. The exiled government in London had diplomatic recognition and support from both the United States and Britain. This government was able to influence events in Poland through the resistance networks it had ordered established as it fled. These networks eventually became the PUS through which it was able to exert some governance within Poland. Yet, despite the Polish government retaining recognition by the United States and Britain, and despite its acceptance by the majority of Poles, it became much less relevant near the end of the war. This is because the Western Allies ceded Poland to the Soviet sphere at the Yalta conference. After the Soviet invasion the Soviets installed a communist government.

> *Exiled governments, or in the French case, a provisional government, served as rallying points for the occupied nations and offered hope for the return of independence as well as providing a legitimate and credible entity to deal with external allies and partners. Such governments were able to make commitments regarding political decisions, cooperate on intelligence and military matters, and enter into financial support arrangements. Thus, such entities greatly benefited the occupied nation. The exiled government of Poland rendered all these services competently, but lost its status with the Western allies due to their decisions based on what they viewed as political and military realities and necessities.*

4. Each of the three Baltic States attempted to form exiled governments, to varying degrees, but with little success. The Lithuanian president escaped and attempted to form a government-in-exile, but was not successful because he lacked legitimacy with his people and because there was no such allowance in Lithuanian law.

5. The Latvian government authorized its Envoy to Great Britain to represent Latvian national interests under a special powers act under war-time conditions. However, the envoy refused to form a government-in-exile, insisting that it would be an untenable position.

6. An Estonian government-in-exile was not established by the government; however, one was attempted in Sweden by the Estonian diaspora in 1953. It lasted until 1991, but it never achieved international recognition.

7. The Philippine archipelago was ruled locally by a Commonwealth government established in November 1935. However, the Philippines was still an American protectorate over which the United States held sovereignty. Thus, though a local ruling Commonwealth governing entity existed in the capital of Manila, the sovereignty of the islands was held by the United States, through its own governing capital of Washington, D.C. The Philippine Commonwealth government exiled itself from the Philippines during the Japanese invasion and established its headquarters in Washington, D.C. Yet, the United States retained sovereignty over the islands until the scheduled independence of the Philippines, scheduled for 4 July 1946, according to Article XVIII of the 1935 Constitution of the Philippines. That constitution was ratified by popular vote in the islands and by the U.S. government, allowing for a ten year transition period to full independence.

D. **Strategic Communication and Guiding Narratives**

1. The Free French forces of de Gaulle were given five minutes a day to broadcast messages to France through the BBC. These broadcasts, combined with leaflets pasted in public places, the use of the "V for victory" sign which was also broadcast in Morse code (three dots and a dash [...-]), and physical or drawn busts of Marianne, or the Cross of Lorrain, were all powerful symbols used to evoke France and its eventual liberation. De Gaulle and his Free French were consistent in messaging the narrative that the Republic did not cease to exist despite the treaty of Vichy and that France would eventually be free of its occupiers. This consistent and powerful message, evoked in broadcasts by de Gaulle and other Frenchmen outside of France, combined with imagery, facilitated popular hope that the occupiers would be ousted and for the resumption of France as an independent republic. De Gaulle's narrative after the liberation of France was that France had freed herself. This was intended to avoid Allied interference in France through establishment of a military government, or the stationing of forces in the country under any terms other than those agreed to by France.

2. Poland and the Baltic States used the typical underground printing methods for newspapers and journals, and graffiti, to spread the message of resistance against their occupiers. The non-communist movements in each country also acted to spare lives much more so than the communist networks. During Nazi occupation, the non-communist networks desired that the people accept them and cooperate with them, whereas the communist networks relied on provoking Nazi reprisals and on the eventual power of the Soviet Union to impose communism by force. Poland sought acceptance by, and support from, the Western powers and cooperated with them as much as possible. Internally, the PUS relied on its political legitimacy granted by its legal exiled government and its tactics within Poland of mostly sparing Polish live and property. The Baltic resistance against the Soviets relied on much the same message as the PUS did during World War II. The Baltic Forest Brothers held out for Western aid and were also sparing of Baltic lives as much as possible. Both the Polish resistance and Baltic Forest Brothers acted against collaborators and traitors as both vengeance and deterrence.

3. The Philippines, as an American protectorate, relied on eventual American return to reassert U.S. sovereignty over the islands and restoration of the status quo, which meant Philippine independence as a nation in 1946. The American message was best captured in General MacArthur's phrase, "I shall return." As a personality, General MacArthur was very familiar to the Philippine people, through previous service there as well as having been a civilian Military Advisor to the Commonwealth of the Government of the Philippines from his retirement in 1937, until his recall by the U.S. Army in the summer of 1941. The message by the exiled Philippine government, as well as the United States, was that the U.S. will free the islands from the Japanese and the Philippines will gain their independence as scheduled in 1946. This guarantee of independence, based on agreement in 1935, was a powerful motivating force for many Filipinos. Joining the Americans to resist Japan, resulting in their complete independence as scheduled, was a powerful motivator.

> Each nation had a guiding narrative of the restoration of sovereignty and independence. This functioned to maintain the morale of the occupied populace and give them hope for ultimate victory against their occupiers. Resistance networks acted in support of this narrative by psychologically and physically dissuading traitors and collaborators. They also retained credibility amongst the population when they limited deaths and damage and showed cooperation with external allies seeking the restoration of their national independence.

E. Maintaining Political Legitimacy

1. The provisional government established by Charles de Gaulle developed political legitimacy over time though its recognition by the Allied powers, and eventual acceptance by the French people. This began with acceptance by the resistance networks. This legitimacy of the provisional government headquartered in London grew from outside of France and reached into France. It was practically complete when de Gaulle led a small French contingent at the head of a large Allied army parade through *Paris' Arch de Triomphe*. He then set about imposing his government and transitioning power to it.

2. The exiled government of Poland retained its legitimacy among the people of Poland as well as the Western Allies throughout most of the war. Its eventual loss of legitimacy in the international arena was caused not by mistakes attributable to it, but by the United States and Great Britain recognizing the Soviet sphere of influence, though not recognizing the Polish government imposed by the Soviets.

> The political legitimacy of resistance was maintained by the credibility of exiled or provisional governments, as well as through the actions of the resistance forces. France, Poland and the Philippines had the strongest external political support, though this did not translate into adequate military support for Poland from its Western allies. The struggle within the Baltic States was almost exclusively internal. Thus, their political legitimacy relied more on the actions of the Forest Brothers than the resistance efforts of the other nations with external political support.

3. The people of the Baltic States did not have exiled or provisional governments to support. Therefore, the internal shadow governments, as part of their Forest Brothers organizations, fulfilled this requirement. They relied on their actions among the people to speak for them in addition to their internally distributed messages. Due to the lack of an external government, and accompanying resources such as information broadcasts and arrangements with allies, the Forest Brothers probably had the greatest internal struggle with legitimacy.

4. The Philippine Commonwealth government and the United States did not lose political legitimacy in the Philippines among the people, or in the international arena (which together the United States and Great Britain mostly controlled). Upon the ejection of the Japanese from the islands, the Commonwealth government returned, the U.S. continued its sovereignty, and Philippine independence occurred as scheduled on 4 July 1946.

F. Organizational Components of Resistance

Each nation surveyed had the basic components of a resistance organization (guerrillas, underground, auxiliary, and public component in the form of an exiled or provisional government). Each non-communist resistance organization basically also sought to spare as many lives and as much property as possible, as they foresaw the return of independence from their occupiers. The networks of each nation became reasonably well developed. The French and Filipino resistance organizations in particular were guided by Allied militaries to achieve military objectives, especially immediately prior to the invasions of France and the Philippines. The Polish non-communist resistance and Forest Brothers, though they had well developed networks, did not receive such guidance because no invasion by the Western allies was contemplated. Each resistance organization in Poland and the Baltics was gradually ground down under unrelenting Soviet coercion, aided by collaborators and communists.

> *The independence-focused resistance networks each basically became organized along the lines of fighters, underground, auxiliary, and governance functions. People who wanted to somehow resist their occupiers fulfilled one of these functions based on their willingness and abilities. People also moved amongst these functions based on willingness, ability and experience.*

G. Nonviolent Peaceful Protest and Passive Resistance

Passive resistance can be distinguished from nonviolent peaceful protest. Passive resistance can be surreptitious work slow-downs or acts of minor sabotage (not sitting next to a soldier of the occupying army on public transportation, mixing stones into bags of potatoes to increase the weight in an effort to seem to comply with food quotas, spraying graffiti, etc.). Nonviolent peaceful protest is a more encompassing category involving open protestation against the occupier, such as work strikes or carrying placards in front of government buildings. Unlike individual passive nonviolent acts, nonviolent, open protest clearly identifies the protester and the issue. During the occupation of the reviewed countries by Nazi Germany or the Soviet Union, there were innumerable instances of passive resistance as well as open nonviolent protests. In particular, several instances of such resistance occurring as open, nonviolent protests in France. Yet, these instances were not successful. The typical early reaction of the German occupiers was to allow the local authorities time to put an end to such activities. If such activities continued into weeks, and particularly if they affected the German war effort, the Germans stepped in directly to arrest and deport people in order to bring an end to the activities. Thus, in the context of resistance against either the Nazi

> *Passive resistance occurred as a matter of opportunity and risk analysis. When an opportunity presented itself to cheat the occupier of labor, money or goods and the risk was low, then many people often did so. However, open protests against Nazi and communist occupiers was treated as defiance and eventually resulted in the application of forceful coercion resulting in arrests, deportations and deaths. Open resistance against an occupier willing to use the degree of force necessary to coerce and maintain control was not successful.*

Germans or communist Soviets, peaceful nonviolent protests were not effective due to the willingness of the occupier to apply the level of force necessary, often lethal, to end the protest.

H. Restoration of Sovereignty

1. France and the Philippines had their sovereignty returned at the end of World War II. In France, Charles de Gaulle was probably the single most important individual in ensuring this outcome when the Allies liberated the country. His provisional government received internal recognition from a significant proportion of the French citizenry, as well as international recognition, most importantly by the Allied victors. French sovereignty was restored as Allied troops moved through France, liberating each village, town, and city. American sovereignty over the Philippines was restored in the same manner. American and Filipino troops moved through the islands to liberate them and resume actions toward the scheduled transfer of sovereignty from the United States to the people of the Philippines in 1946.

> *The restoration of sovereignty and independence occurred in those nations with significant external support to their resistance effort and the eventual entry of large scale conventional military forces of the Western Allies. Orderly post-war legal processes to handle charges of treason and collaboration with as much fairness as possible assisted those nations employing those processes. The nations unfortunate enough to come under Soviet domination also suffered political purges based on inadequate adherence to communist ideology.*

2. The nations of Estonia, Latvia, Lithuania, and Poland suffered very different fates. Each was overrun by the Soviet Red Army with no intention by the government of the Soviet Union to restore full sovereignty to those nations. The Baltic States were re-incorporated into the Soviet Union as Soviet Socialist Republics, even before the defeat of the Forest Brothers resistance movement. The Soviets recognized Poland as an independent nation but only after arranging for unchallenged communist supremacy within it. This was accompanied by the stationing of large numbers of Soviet troops and eventually incorporating Poland into what became the Soviet controlled Warsaw Pact of communist controlled nations.

3. Within each nation, processes were put in place to deal with wartime traitors and collaborators, with accompanying rules attempting to eliminate the settling of scores resulting from personal feuds. Many people in each nation suffered numerous extra-judicial actions, ranging from physical assaults, to property destruction, and killings. The Soviets also took this opportunity to purge as many non-communists as possible from government offices and positions of authority in order to bring about as much communist ideological purity as possible, while also using these methods to intimidate the population into compliance.

F3. Cold War Resistance Case Studies

A. Perceived Threat

1. During the early part of the Cold War, the U.S. focus was on potential Soviet invasion of European countries, which included providing support to "indigenous anticommunist elements in threatened countries." Therefore, the U.S. established a policy and legal framework that was ready to assist, and in fact encouraged, peacetime efforts against Soviet communist infiltration.

2. Stay-behind or resistance networks of this era were originally developed to defend against Soviet infiltration, subversion, and sabotage. The Italian and Norwegian cases are good examples. The U.S., through its policy and legal frameworks designed to support the efforts of military and intelligence organizations, assisted those efforts. In the examples of those two nations, however, the focus of efforts was not limited to planning for a crisis period preceding and following an actual invasion, but also included attempts to prevent communist infiltration. Problems arose because many communists were citizens of those countries. This meant that anti-communist efforts by stay-behind or resistance forces, actually became ill-trained and ill-equipped counterespionage or counter subversion, with the attendant political and legal problems.

3. Of three cases in this sample, the Swiss were the only ones who established a resistance capability focused solely on an actual physical invasion by the Soviets, whether temporary, partial, or complete. The Swiss objective was to harden the resilience of their society against outside aggression through a holistic Total Defense. The Swiss model maintained a clear focus on a potential invading enemy, and avoided the negative political ramifications that occurred in other countries, such as Italy (Gladio).

B. Resistance Organization and Preparation

1. Resistance organizations in the sampled nations were organized variously under established military or intelligence organizations. The Swiss created a Special Service under military auspices and, uniquely among the sampled nations, expanded their inchoate resistance organization to encompass civilian institutions and organizations. They intentionally picked resistance leaders who had no military connections to lead the organization during occupation, in order to better prevent their discovery by the occupying power.

2. Generally, the organizations were not fully manned for reasons of financial expense, as well as to not reveal members. Thus, nations generally conducted some pre-vetting of prospective individuals, and planned further recruitment only when finally necessary to resist against an occupier. Nations did create caches for weapons, ammunition, communications devices, and other material, but some learned only gradually that such activity required legal and policy framework authority.

3. Each nation established specific functions based skeletal networks focused on activities such as intelligence gathering, sabotage, propaganda or IO, and movement of persons for reasons such as escape. They also engaged in site surveys for planned sabotage of facilities and equipment useful to the invader. The Swiss, in particular, even prepared tunnels and bridges for collapse in case they were threatened with invasion. The Swiss not only focused on specific networks, but also on preparing most of the population to engage in some form of resistance, creating deep and thorough opposition to the invaders as part of their Total Defense concept. The Swiss model publicized these efforts to engage in dissuasive messaging against potential aggression.

4. Since these organizations were never tested by an occupier, it is difficult to draw insights based on experience. However, positive lessons can be learned. The Swiss model of dissuasive messaging, separating the resistance organization from the military for security purposes, and preparing targets for sabotage to prevent their use by the enemy can provide guidance for resistance. The cadre nature of the organizations is also important for security and resource preservation. Involving the whole-of-society to increase the depth and breadth of potential resistance is a Total Defense concept advocated within this document for nations that lack strategic depth.

C. **Policy and Legal Frameworks**

1. Of the nations surveyed, Switzerland seems to have had the most detailed legal and policy frameworks to support the establishment and conduct of their resistance networks. A great deal of control over resistance actions was held at the most senior levels, because sabotage actions were designed to maintain national morale which is inherently a political calculation.

2. Italy and Norway seemed to have lesser frameworks to support their resistance efforts. This inadequacy in Norway led to accusations of illegal surveillance. In Italy, it led to strong public suspicions of illegality and wrongful American clandestine influence within the country. These perceptions resulted in making the Italian stay-behind plan code word of Gladio infamous and bringing negative attention to other similar European networks.

3. Though these networks were not used, we know from the political and legal backlash spurred by the Gladio revelations, that strong policy and legal frameworks are absolutely essential to the establishment of national resistance networks. Such frameworks can protect the actors from false political and legal accusations, while granting legitimacy and credibility to the effort to establish and conduct resistance.

D. **Outstanding Points**

1. The Swiss resistance organization was placed outside of the Swiss military and government bureaucracy to prevent the discovery of its members in case of occupation. Additionally, its leader was hired under a private contract while cadre members were employed and paid via front companies. This legal structure also functioned to legally preclude its surrender in case of capitulation, since the resistance organization was not part of the armed forces.

2. The Italian effort was initially directed near the Yugoslavian border with the task of countering pro-communist subversive elements seeking to unite with Yugoslavia. It suffered early from altered and confused intent.

3. The Norwegians deposited sealed copies of personnel records and the information necessary to establish and run resistance networks in the Norwegian embassies in London and Washington. This protected identities and operational intent, increased the survivability of crucial information, and allowed access to this information by a legitimate government-in-exile.

Appendix G. Assessing Resilience

A Country is not a mere territory; the particular territory is only its foundation. The Country is the idea which rises upon that foundation; it is the sentiment of love, the sense of fellowship which binds together all the sons of that territory. - Giuseppe Mazzini

G.1. Introduction

A. As stated in chapter 1, a society's resilience enables deterrence and supports national defense planning, including regaining national sovereignty. Resilience is a society's survivability and durability. It is the will of the people to maintain their status quo—the will and ability to withstand external pressure and influences and/or recover from the effects of that pressure or influence.

B. Identifying and reducing vulnerabilities is a critical aspect of resilience. Reducing vulnerabilities requires whole-of-government and whole-of-society approaches that cover all the elements of the operational environment. For example, proactively countering adversary messaging, diversifying and protecting the national economy and critical industries/infrastructure, facilitating a common operating picture and information sharing, protecting basic standards of living, securing borders, promoting national unity, adopting data and cyber protection and information assurance measures, reducing vulnerabilities of key populations, and maintaining existing military advantages. These must be assessed as potential vulnerable points of political leverage (influence) by an aggressor.

C. Governments must understand and recognize the manner in which an adversary can, or is, exploiting its advantages in the domestic operational environment. Governments inform and educate the population about the threat, particularly those elements of the population that are vulnerable to activities of the adversary.

D. Governments can use assessment tools to measure malign foreign influence within their borders. This can be accomplished by categorizing areas of human activity to provide perspective on potential areas of malign influence within society. At the end of this appendix is a sample of how such activity can be categorized and rated.

G.2. Resilience and Hybrid Warfare[580]

A. Hybrid warfare is inextricably linked to the concept of resilience, because it exploits the vulnerabilities of a state instead of directly confronting that state's armed forces. The concept of resilience is better understood if it is associated with hybrid warfare as the threat to be protected against.

B. In broad terms, hybrid warfare can be understood as a creative combination of civil and military ways and means that are deployed in a synchronized manner.[581] Hybrid warfare allows a tailored combination of efforts to explore and exploit a target state's internal weaknesses, or political and social cleavages, for the purpose of gaining access that can be used as internal leverage against that state.

C. A state's vulnerabilities create strategic options for the actor using hybrid warfare. A hostile state actor, or even sub-state group, can use asymmetric options to identify, access, and exploit cleavages, and gain internal leverage within the targeted state. As presented in appendix C of this document, Russia is a prime example of a state that has developed asymmetric tactics combined into hybrid warfare against its targets very effectively, and continues to enhance its ability to use hybrid means to achieve its international political objectives.

D. Hybrid warfare is based on the internal characteristics of the target state. It thus begins with discovering weaknesses or cleavages within the society of the target state that can be exploited.[582] This can mean the instrumentalization of a segment of the target population, by causing them to act in ways that support the political objectives of the state engaging in hybrid warfare. This is a form of instigation of "war among the people."[583] During its execution or implementation, the state employing this means will likely engage in a form of maskirovka (see appendix C) to conceal itself as the actor. This maintains political deniability for the state, helping to avoid overt international responses, and reduces risk by reducing political vulnerability in case the actions are not successful. This is all aimed at preventing the target state from clearly identifying these methods and successfully countering them.[584]

G.3. Resilience Enhances Deterrence by Reducing Adversary Leverage Points

A. With the rise of hybrid warfare, much discussion of the term resilience has associated it with the increasing complexity of the modern security environment, the changing threats, and the unpredictability of attacks. However, in protecting their own societies, democratic states cannot guarantee complete security for their people without becoming threats to their own values. State institutions and societal organizations must be able to identify their internal weaknesses and cleavages that can be leveraged against them by an outside power.[585] Based on this understanding, resilience contributes to deterring hybrid attacks. This concept of resilience differs somewhat from the approach of NATO.[586]

B. An element of resilience is to focus on preventing or identifying asymmetric attacks that are designed to destabilize states, polarize societies, and spread distrust. The focus should be on discovering and assessing weaknesses or societal cleavages which are most dangerous when they affect the critical juncture where Clausewitz's "fascinating trinity"[587] of government, people and the military interact in strategy-making.

G.4. Engaging in Continuous Resilience Assessment

A. The state and society seeking to protect itself must perform critical self-assessments to honestly identify internal weaknesses. This is a continuing and permanent process for as long as an external power, or combination of powers, seeks to leverage weakness within the targeted state. As with resistance, this too is a whole-of-society, whole-of-government approach. Political leaders should address security questions with their electorates. Externally originating propaganda campaigns targeting specific groups within a society should be countered by public diplomacy campaigns using truthful and independent media.[588] Resilience also requires improving educational programs to help build civil awareness and critical thinking.

B. The Foreign Malign Influence and Summary Assessment rationale charts, figures 19 and 20, identify categorized sections of a society that a state should continually assess for vulnerabilities.[589] These are historically typical areas in which weaknesses or cleavages exist that an outside power can use as leverage, and therefore should be protected.

G.5. Individual Resilience Preparation

Each free and sovereign state has its own contextual historical development, legal and policy frameworks, and cultural traditions and customs. Most of these are more than adequate to guide individual behavior during times of adversity imposed by a foreign aggressor state. During a state of formal peace, adversity and heightened internal pressure on the population brought by that foreign state may still exist. Many states have informal codes of behavior that are not legally enforceable, but rather serve as common guides for its citizens. The U.S. has such a guide in the form of its Pledge of Allegiance. Citizens should commonly understand, as also guided by law, the bounds of acceptable behavior during times of adversity and crisis (see appendix H, fig. 22).

Foreign Malign Influence Assessment

Influence areas & threat rating	Summary	Risk mitigation
Political (3)	**Threat level: Medium** • Organizations: • Leaders:	**Resilience: Medium** • Analysis:
Economic (3)	**Threat level: Medium** • Industries and trade:	**Resilience: Medium** • Analysis:
Social (3)	**Threat level: Medium** • Demographics: • Organizations:	**Resilience: Medium** • Analysis:
Ideological (4)	**Threat level: High** • Information environment: • Anti-national/Western narratives:	**Resilience: Low** • Analysis:

Figure 19. Foreign Malign Influence Assessment Chart Sample.

SOURCE: PRODUCED BY NAVANTI GROUP FOR SOCEUR/USED WITH PERMISSION

Foreign Malign Influence Summary Assessment Rationale

Definitions of threat level, resilience, degree of Foreign malign influence

Threat ratings are based on a qualitative assessment of influence factors and exploitable social cleavages combined with a level of resilience to those factors. Resilience is the ability of communities to mitigate and recover from shocks and stresses in a manner that reduces vulnerability. **Low resilience heightens threat, while high resilience reduces overall threat.** Influence factors and resilience analysis are summarized in each box below. Arrows indicate an assessment of whether the threat level is currently increasing or decreasing.

🟩	1-2	**Low threat**: Foreign malign influence remains nonexistent or is minimal, and presents only minor risks.
🟨	3	**Medium threat**: Foreign malign or pro-Foreign groups, individuals, or political parties present several avenues for a malign actor to exploit local divisions or weaknesses. Resilence is moderate or high.
🟧	4-5	**High threat**: Multiple ties to foreign malign entity and active promotion of pro-Foreign, anti-Western groups and individuals or activity present a significant security risk. Resilence is moderate to low.

Influence areas	Summary	Risk mitigation
Political ⬆️	**Threat level:** • **Organizations:** Key parties and political groups • **Leaders:** Influential leaders	**Resilience:** • Competing political influences and potential counters to foreign influence
Economic ⬆️	**Threat level:** • **Industries and trade:** Economic ties to Foreign malign businesses and market	**Resilience:** • Economic ties to the EU/US allies and ability to withstand a fall in trade with Foreign malign businesses and market
Social ⬆️	**Threat level:** • **People or demographics:** Influencial figures in civil sphere or important demographic data • **Organizations:** Major civil society groups	**Resilience:** • Depth of social ties to Foreign malign actor; competing social groups; opportunities to build ties to Western groups
Ideological ⬆️	**Threat level:** • **Information environment:** Media access and typical media consumption • **Narrarives:** Key ideas in major media	**Resilience:** • **Analysis:** Access and consumption of Western media sources; opportunities to expand and build trust toward national and Western information sources

Figure 20. Foreign Malign Influence Assessment Chart Legend Sample.

SOURCE: PRODUCED BY NAVANTI GROUP FOR SOCEUR/USED WITH PERMISSION

Appendix H. Population Interaction With Foreign Occupier

War is an ugly thing, but not the ugliest of things. The decayed and degraded state of moral and patriotic feeling which thinks that nothing is worth war is much worse. - John Stuart Mill

H.1. Introduction

A. Throughout the history of conflict, societies have had to deal with individual and group level interactions between the occupied and the occupiers. These interactions cover the entire range of human behavior. However, in societies under occupation by a foreign power, not all interaction with the occupier is deemed acceptable by the majority of the occupied population that seeks to reestablish its independence and sovereignty. These modes of unacceptable interaction have typically led to violence, and often death of the persons whose behavior was deemed by the majority as unacceptable.

B. Behavior deemed as unacceptable by the majority of people seeking to regain their freedom from a foreign occupier has typically been qualified along a range of cooperative behaviors. Depending on the circumstances, these behaviors can range from simple actions, such as minor actions of politeness, to obedience to newly imposed laws or regulations, to sexual fraternization, and on the farthest end of the spectrum, voluntarily betraying one's society, nation, or fellow citizens to the advantage of the occupying regime.

C. Focusing on twentieth century conflicts covered in the case studies in this volume, some examples of the punishments meted out to such collaborative individuals run the gamut from public shaming to killings. These ranged from shaving the heads of women in France at the end of World War II for sexual fraternization with the occupiers, to physical beatings for similar offenses, and to outright extra-judicial killings for suspected egregious offenses.

D. Some scholarly work has been conducted to examine the behavior of groups and individuals under repressive regimes and their behavior toward each other when the political situation devolves into a civil war. This issue is one which will hopefully be minimized in the context of pre-planned resistance and societal fortification through resilience, as well as a common societal focus on the restoration of the status quo ante.

H.2. Collaboration, Accommodation, and Resistance (C-A-R)

A. Professor Marek Jan Chodakiewicz has put forth an understanding of popular resistance which divides attitudes of the population into three methods by which they cope with foreign invaders; collaboration, accommodation, and resistance. His concept is graphically represented here with the ROC's creation of the C-A-R bar in figure 21. These methods are generally defined as:[590]

- **Collaboration**: An active relationship with the occupier for reasons of self-interest and to the detriment of the occupied population
- **Accommodation**: Multilevel gradational compliance with the occupiers, with its character depending on its relative proximity to either collaboration or resistance
- **Resistance**: Passive and Active opposition to the occupiers

B. In this understanding, when the occupier gradually introduces policies aimed at increasing control over the population, the scope of officially permissible behavior quickly narrows, eventually forcing increased resistance.[591] Passive resistance is a response to oppressive policies such as economic exploitation. Active resistance is a response to the lack of traditional and accepted social institutions and translates into organized resistance.[592] When an occupier increases the level of control and possibly terror, armed resistance to it is reduced due to the lethal danger to the actors and to those around them from collective reprisals. Blatant resistance invites death. Armed resistance and its accompanying risk, in such conditions, increases only when a degree of accommodation is not an option.[593]

C. In this model, population behavior can be seen as a spectrum, with collaboration and resistance occupying opposite ends. Accommodation encompasses the graduated area between them. Accommodation spans a plethora of attitudes that borders resistance on one end and stops short of collaboration on the other end. Accommodation is a matter of finding a modus vivendi with the occupying powers while surrendering as little as possible on the individual and group levels, and is thus distinguishable from collaboration.

D. **Collaboration** is a conscious and consistent collusion with the occupiers, for reasons of self-interest that causes grave harm to the population as a whole, a group within it, or one of its individual members.

E. **Accommodation** is directly related to resistance, or fused to it, and is a complementary function or mode of behavior. Participation by the citizenry in institutions and organizations sanctioned by the occupiers is a necessary prerequisite for successful resistance.[594] In fact, in this concept's construct of the components of resistance, based on historic experience captured in U.S. doctrine and other professional writings, a measure of accommodation is necessary to facilitate the existence and functioning of an auxiliary and even some members of the underground. The Resistance, as an organization, draws its members from the accommodation-resistance range of population interactive modes. Accommodationists provide the bulk of the auxiliary.

F. **Resistance** is an action taken to adversely affect the occupier. Most narrowly defined, it is armed struggle. However, its most widespread form is cheating those authorities acting on behalf of the occupiers. For example, in the Polish case study a form of resistance included intentionally poor execution of the occupiers' orders. Resistance can be spontaneous or planned and active or passive. Resistance is most often shaped by nationalism and the national conscience increases as the occupier increases oppression. An increase in occupier oppression is what the communist resistance groups attempted to bring about through acts intended to bring the consequences of harmful reprisals against the people, in an attempt to draw more people to those communist groups.[595]

H.3. Relationship between the C-A-R Spectrum and Components of the Resistance Organization

A. Persons inclined to actively or passively resist tend to fill the underground and guerrilla ranks. Collaborators present the greatest security threat to the organization. Figure 21 is a model graphic created for this concept to represent these relationships, which are based on the collaboration, accommodation, and resistance spectrum, which we will term as the C-A-R bar.

B. The C-A-R Bar contains the spectrum of behavior among the population. It is not meant to demonstrate equal division of the population into thirds. A person can move along the spectrum depending on their personal experiences through time. It also demonstrates linkage between members of the population engaging in accommodation and resistance modes and the organized resistance. Accommodationists are essential and tend to serve as auxiliary members. People most inclined to engage in active or armed resistance are more likely to be members of the underground or guerrillas. The resistance organization components bar is below the behavior modes bar to demonstrate that resistance organization members are under, or within

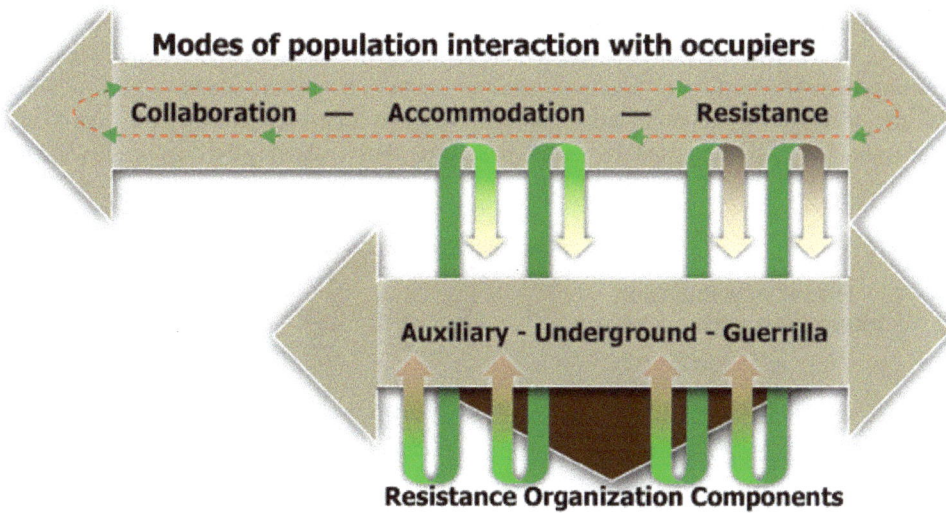

Figure 21. The spectrum of Collaboration, Accommodation and Resistance (C-A-R) related to Resistance Organizational Components.

SOURCE: THIS MODEL, CONSTRUCTED BY OTTO C. FIALA AND ASSISTED BY CATHEY SHELTON, IS BASED ON THE WORK OF PROFESSOR MAREK JAN CHODAKIEWICZ

the population, masked by it. Its placement also demonstrates the portions of the population which typically contribute to the different resistance components. The establishment of a shadow government entity (usually an adjunct of the underground) by resistance organizations has been used historically to discourage collaboration through application of its judicial resources, sometimes rendering harsh results. Such a shadow government interacts with the population as part of the underground depicted in this graphic. Additionally, the establishment of a government-in-exile has also been used to restrain collaboration.[596]

H.4. Effects of Significant Coercion or Terror

A. **General Effect.** When an occupier increases terror, larger segments of the population can be terrorized into accommodation. Some people can even be forced into collaboration as a form of self-preservation. Paradoxically, terror can also cause an increase in the resistance instinct. Extreme coercion or terror can force an accommodative, passive attitude while also creating outlaws with no choice but to resist. In these conditions, survival depends on successful alternation between accommodation and resistance.[597]

B. **Divisive Effect.** The effect of terror on an occupied population also expedites national consciousness. When that population has different identity groups, and they each identify as separate from the majority nation, there is a reflex for a defensive outlook and self-preservation. This attitude can reduce and even preclude cooperation between these groups. With an increase in terror, each group undertakes self-defense in various forms of resistance that often adversely affects its relations with the other groups. Thus, the occupier can apply a strategy of divide and conquer to the population to exacerbate differences among them and undermine the effectiveness of accommodation and resistance of each group.[598]

C. **Effect on Links between the Population and Organized Resistance.** When the occupier applies a high degree of coercion on the population, the mode of active or armed resistance becomes more difficult and dangerous, thus reducing the number of people willing to engage in such resistance. Though some forms of passive resistance can still be disguised, active resistance is met with harsher countermeasures. This causes varying degree increases in accommodation and collaboration. It also reduces the links or connections between the resistance organization and the general population, thus weakening the organization (see fig. 22).

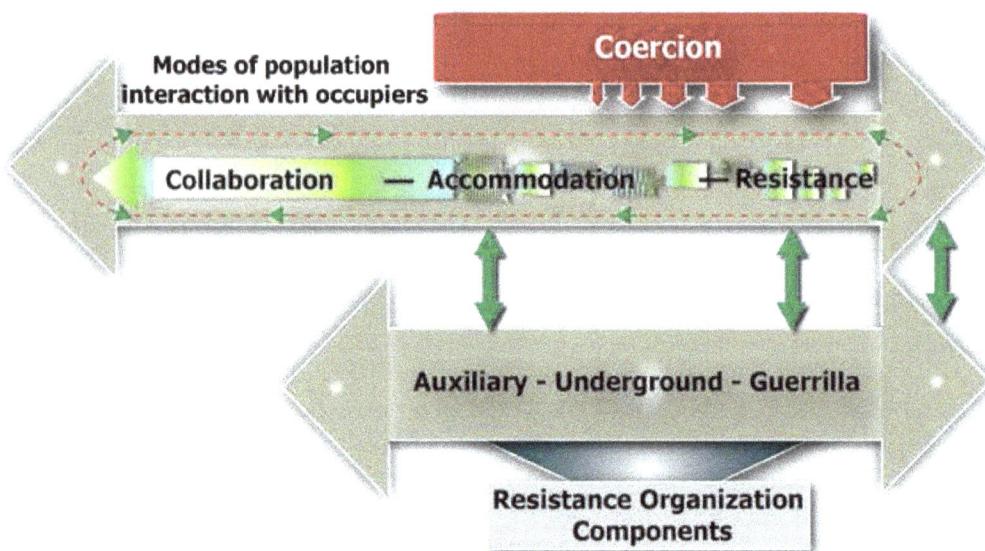

Figure 22. The Effect of Coercion on the C-A-R Bar.

SOURCE: THIS MODEL, CONSTRUCTED BY OTTO C. FIALA AND ASSISTED BY CATHEY SHELTON, IS BASED ON THE WORK OF PROFESSOR MAREK JAN CHODAKIEWICZ

H.5. Context of Today's Threat

The above elucidations taken from historical analysis should be applied to the context of the likely actual threats posed today. Based on twenty first century mores, rapid and accurate information exchanges among most advanced nations, and various forms of international integration and inter-dependence, an occupier from among these nations is not likely to apply widespread terror in the forms analyzed in the above case studies. However, the previous analysis is still valid in demonstrating relationships and the effects of coercion on those relationships. This is because today's aggressive state actors are likely to use more subtle means of coercion and terror. In accord with the previously discussed hybrid warfare section, they are more likely to use asymmetric methods that can publicly be denied while bringing about very similar effects to the methods of the occupiers examined.

H.6. Individual Conduct under Occupation

Historically, collaboration has resulted in judicial or extra-judicial actions against collaborating individuals or groups. This occurs in each of three conditions: 1) while under occupation; 2) during the turbulent ouster of the occupier, which may be characterized by some chaos and lawlessness, and; 3) later, when sovereignty has been restored with a fully operating judicial system. Due to the historical issues of collaboration, and at times confusion with accommodation, states may consider educating their populace during peacetime on such modes of cooperation and the bounds of acceptable behavior during occupation. For example, states can publish guides of behavior for their populations prior to a crisis as part of the fortification of their societal resilience. To that end, figure 23 is a suggestion of general possible points for consideration and dissemination, which should be modified for compliance with national traditions, social acceptability and legal frameworks of each state.

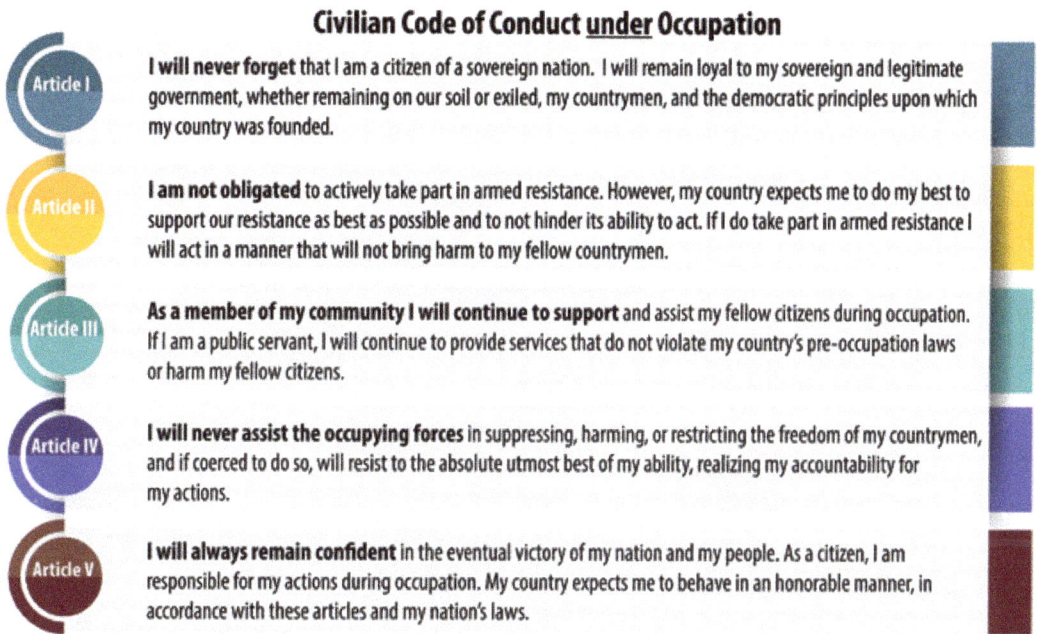

Civilian Code of Conduct under Occupation

Article I

I will never forget that I am a citizen of a sovereign nation. I will remain loyal to my sovereign and legitimate government, whether remaining on our soil or exiled, my countrymen, and the democratic principles upon which my country was founded.

Article II

I am not obligated to actively take part in armed resistance. However, my country expects me to do my best to support our resistance as best as possible and to not hinder its ability to act. If I do take part in armed resistance I will act in a manner that will not bring harm to my fellow countrymen.

Article III

As a member of my community I will continue to support and assist my fellow citizens during occupation. If I am a public servant, I will continue to provide services that do not violate my country's pre-occupation laws or harm my fellow citizens.

Article IV

I will never assist the occupying forces in suppressing, harming, or restricting the freedom of my countrymen, and if coerced to do so, will resist to the absolute utmost best of my ability, realizing my accountability for my actions.

Article V

I will always remain confident in the eventual victory of my nation and my people. As a citizen, I am responsible for my actions during occupation. My country expects me to behave in an honorable manner, in accordance with these articles and my nation's laws.

Figure 23. Civilian Code of Conduct under Occupation Sample.

SOURCE: DEVELOPED BY DANIEL BOWLER WITH ASSISTANCE FROM CATHEY SHELTON, ADAPTED FROM THE AMERICAN CODE OF CONDUCT FOR PRISONERS OF WAR. USED WITH PERMISSION FROM DANIEL BOWLER

Appendix I.[599] Swedish Government Swedish Civil Contingencies Agency (MSB): *If Crisis or War Comes* Pamphlet

IMPORTANT INFORMATION FOR THE POPULATION OF SWEDEN

IF **CRISIS** OR **WAR** COMES

Contents

This brochure is available to download in several different languages at dinsäkerhet.se.

Questions and answers about the brochure can be found at dinsäkerhet.se.

MSB
Swedish Civil
Contingencies
Agency

PEFC

MSB is a central government agency that works to improve Sweden's ability to prevent and manage accidents and emergencies. In the event of a serious accident or emergency, we provide support to those who are responsible.

Swedish Civil Contingencies Agency (MSB)
651 81 Karlstad
www.msb.se

Graphic design and production: Kreab AB
Illustrations: Arvid Steen
Printed by: Stibo Graphic A/S
Publ. no.: MSB1214 · May 2018
ISBN: 978-91-7383-836-8

Emergency preparedness

Total defence

Warning systems

For the population of Sweden

This brochure is being sent to all households in Sweden at the behest of the Swedish Government. The Swedish Civil Contingencies Agency (MSB) is responsible for its content. The purpose of the brochure is to help us become better prepared for everything from serious accidents, extreme weather and IT attacks, to military conflicts.

Many people may feel a sense of anxiety when faced with an uncertain world. Although Sweden is safer than many other countries, there are still threats to our security and independence. Peace, freedom and democracy are values that we must protect and reinforce on a daily basis.

Public authorities, county councils and regions, municipalities, companies and organisations are responsible for ensuring that society functions. However, everyone who lives in Sweden shares a collective responsibility for our country's security and safety. When we are under threat, our willingness to help each other is one of our most important assets.

If you are prepared, you are contributing to improving the ability of the country as a whole to cope with a major strain.

KEEP THIS BROCHURE!

3

What would you do if your everyday life was turned upside down?

An emergency can result in society not functioning in the way we are used to. Climate change may mean that flooding and forest fires become more common. Incidents in the rest of the world may result in shortages of certain foodstuffs. Disruptions to important IT systems may have an impact on the electricity supply. In just a short time, your everyday life can become problematic:

- The heating stops working.
- It becomes difficult to prepare and store food.
- The shops may run out of food and other goods.
- There is no water coming from the taps or the toilet.
- It is not possible to fill up your car.
- Payment cards and cash machines do not work.
- Mobile networks and the internet do not work.
- Public transport and other means of transport are at a standstill.
- It becomes difficult to obtain medicines and medical equipment.

Think about how you and people around you will be able to cope with a situation in which society's normal services are not working as they usually do.

212

Your emergency preparedness

Your municipality is responsible for ensuring that services including care of the elderly, the water supply, the fire and rescure service and schools continue to function, even in the event of a societal emergency. As a private individual, you also have a responsibility. Preparing correctly can enable you to cope with a difficult situation, regardless of what has caused it.

In the event of a societal emergency, help will be provided first to those who need it most. The majority must be prepared to cope on their own for some time. The better prepared you are, the greater the opportunity you will also have to help others who do not have the same prerequisites.

What is most important is that you have water, food and warmth and are able to obtain information from the authorities and the media. You also need to be able to make contact with relatives. There are check-lists on pages 10 and 11 with foodstuffs and items that are good to have at home.

> Think about what risks may affect you and your local area. Do you live in an area that is sensitive to landslides or flooding? Is there some sort of hazardous industry or something else in your area that may be good to know about?

5

Be on the lookout for false information

States and organisations are already using misleading information in order to try and influence our values and how we act. The aim may be to reduce our resilience and willingness to defend ourselves.

The best protection against false information and hostile propaganda is to critically appraise the source:

- Is this factual information or opinion?
- What is the aim of this information?
- Who has put this out?
- Is the source trustworthy?
- Is this information available somewhere else?
- Is this information new or old and why is it out there at this precise moment?

- Search for information – the best way to counteract propaganda and false information is to have done your homework.

- Do not believe in rumours – use more than one reliable source in order to see whether the information is correct.

- Do not spread rumours – if the information does not appear trustworthy, do not pass it on.

6

You can find more information at dinsäkerhet.se

In the event of a terror attack

Terror attacks may be targetted against individual people or groups, against the general public or against vital societal functions such as the electricity supply or the transport system. Even though there are many different ways to carry out a terrorist attack, there are some pieces of advice that may be applicable in most situations:

Emergency preparedness

- Move to a safe place and avoid large groups of people.
- Call the police on 112 and inform them if you see something important.
- Warn those who are in danger and help those who are in need of assistance.
- Put your mobile on silent and do not call anyone who may be in the danger area. The sound of their phone ringing may reveal the location of someone who is hiding.
- Do not call anyone with your mobile unless you have to. If the network is overloaded, it may be difficult for vital calls to get through.
- Comply with requests from the police, the fire and rescue service and the authorities.
- Do not share unconfirmed information online or in any other way.

7

215

Sweden's defences

Sweden's combined defences are in place to protect the country, our freedom and our right to live as we ourselves choose to. All of us have a duty to act if Sweden is threatened.

Total defence

Total defence

The term 'total defence' denotes all activities that are needed in order to prepare Sweden for war. Sweden's total defence consists of military defence and civil defence.

Military defence

Sweden's military defence consists of the Swedish Armed Forces, including the Home Guard, and a number of other public authorities whose main duty is to support Sweden's military defence. The Armed Forces defend our territory and our borders.

Civil defence

Civil defence deals with the whole of society's resilience in the event of the threat of war and war. Civil defence is the work that is carried out by central government agencies, municipalities, county councils and regions, private companies and voluntary organisations. This work aims to protect the civilian population and to ensure that, for example, healthcare and the transport system continue functioning in the event of the threat of war and war. In the event of the threat of war and war, Sweden's civil defence also has to be able to support the Armed Forces.

8

216

Duty to contribute to Sweden's total defence

In Sweden there is a duty to contribute to total defence. This means that everyone who lives here and is between the ages of 16 and 70 can be called up to assist in various ways in the event of the threat of war and war. Everyone is obliged to contribute and everyone is needed.

The duty to contribute to total defence has three forms:

- **Conscription** into the Armed Forces.
- **Civil conscription** into organisations controlled by the Government.
- **General national service** involves serving in organisations that must function even in the event of the threat of war and war. This means that you continue to do your normal job, work in a voluntary organisation or are tasked by Arbetsförmedlingen with performing work that is of particular importance to Sweden's total defence.

Those compelled to contributed to Sweden's total defence can be given wartime postings. If you are given a wartime posting, you will have received wartime posting orders or another form of confirmation from your employer about this.

Total defence

For many years, the preparations made in Sweden for the threat of war and war have been very limited. Instead, public authorities and municipalities have focused on building up the level of preparedness for peacetime emergencies such as flooding and IT attacks. However, as the world around us has changed, the Government has decided to strengthen Sweden's total defence. That is why planning for Sweden's civil defence has been resumed. It will take time to develop all parts of it again. At the same time, the level of preparedness for peacetime emergencies is an important basis of our resilience in the event of war.

Follow what is happening at **dinsäkerhet.se**

9

Home preparedness tips

Your prerequisites and needs vary, for example, depending on whether you live in the countryside or in a built-up area, in a house or in an apartment. Here are some general home preparedness tips. Use that which is appropriate for you and those close to you. It is a good idea to share certain things and borrow from one another.

Food

It is important to have extra food at home that provides sufficient calories. Use non-perishable food that can be prepared quickly, requires little water or can be eaten without preparation.

- [ ] potatoes, cabbage, carrots, eggs
- [ ] bread with a long shelf-life, e.g. tortillas, hard bread, crackers, rusks
- [ ] cheese spread, soft whey cheese and other spreads in tubes
- [ ] oat milk, soy milk, milk powder
- [ ] cooking oil, hard cheese
- [ ] quick-cook pasta, rice, grains, instant mashed potatoes
- [ ] precooked lentils, beans, vegetables, hummus in tins
- [ ] chopped tomatoes to, for example, cook pasta in
- [ ] tins of bolognese sauce, makerel, sardines, ravioli, salmon balls, boiled meat, soup
- [ ] fruit purée, jam, marmelade
- [ ] prepared blueberry and rosehip soup, juice or another drink that can be stored at room temperature
- [ ] coffee, tea, chocolate, energy bars, honey, almonds, nuts, nut butter, seeds

Water

Clean drinking water is vital. Allow for at least three litres per adult per day. If you are uncertain about its quality, you need to be able to boil the water.

If the toilet is not working, you can take strong plastic bags and place them in the toilet bowl. Good hand hygiene is important for avoiding infection.

- [ ] bottles
- [ ] buckets with lids
- [ ] Plastic bottles to freeze water in (do not fill to the top as the bottle will crack if you do)
- [ ] mineral water
- [ ] jerry cans, ideally with a tap, to collect water in. You can also have a couple of clean jerry cans that are filled with water as a reserve. These are to be stored in a cool, dark place.

Learn more about home preparedness at **dinsäkerhet.se**

10

Warmth

If the electricity goes off at a cold time of the year, your home will quickly become cold. Gather together in one room, hang blankets over the windows, cover the floor with rugs and build a den under a table to keep warm. Think about the risk of fire. Extinguish all candles and alternative heating sources before you go to sleep. Air the room regularly to let in oxygen.

- [ ] woolen clothes
- [ ] warm all-weather outdoor clothing
- [ ] hats, gloves, scarves
- [ ] blankets
- [ ] sleeping mats
- [ ] sleeping bags
- [ ] candles
- [ ] tea lights
- [ ] matches or fire-lighter
- [ ] alternative heat sources, e.g. LPG heaters, paraffin heaters.

Other

- [ ] spirit stove and fuel
- [ ] torch, head torch
- [ ] batteries

Communications

In the event of a serious incident, you need to be able to receive important information from the authorities, primarily Sveriges Radio's radio station P4. You also need to be able to follow how the media are reporting events, remain in contact with relatives and friends and be able to reach the emergency services in the event of an emergency.

- [ ] a radio powered by batteries, solar cells or winding
- [ ] a car radio
- [ ] a list of important telephone numbers on paper
- [ ] extra batteries/power bank for devices such as mobile phones
- [ ] mobile phone charger that works in the car.

- [ ] cash in small denominations
- [ ] medicine cabinet and extra medicines
- [ ] wet wipes
- [ ] hand sanitiser
- [ ] nappies and menstrual products
- [ ] paper printouts of information such as insurance policies, bank details, registration certificates
- [ ] fuel in the tank.

11

If Sweden is attacked, resistance is required

We must be able to resist various types of attacks directed against our country. Even today, attacks are taking place against our IT systems and attempts are being made to influence us using false information. We may also be affected by conflicts in our region. Potential attacks include:

- Cyberattacks that knock out important IT systems.
- Sabotage of infrastructure (e.g. roads, bridges, airports, railways, electricity cables and nuclear power stations).
- Terror attacks that affect a large number of people or important organisations.
- Attempts to influence Sweden's decision makers or inhabitants.
- Severed transport links that result in a shortage of foodstuffs and other goods.
- Military attack, for example airstrikes, rocket attacks or other acts of war.

If Sweden is attacked by another country, we will never give up. All information to the effect that resistance is to cease is false.

Total defence

12

Heightened state of alert

The Government can decide to put the country on a
heightened state of alert in order to improve Sweden's
chances of defending itself. In a heightened state of alert,
peacetime laws apply, but other laws may also be used. For
example, the state can requisition private property that is of
particular importance to Sweden's total defence.

In a heightened state of alert, the whole of society has to
gather its collective forces in order to ensure that which is
most important functions. In a heightened state of alert, you
may be called up to help in various ways.

Information about the heightened state of alert will be
broadcast on radio and TV. Sveriges Radio's radio station P4
is the emergency channel.

Total defence

13

221

🔊 **Important public announcement**
Signal 7 seconds – break 14 seconds   `7` 14 `7` 14 `7`

🔊 **Danger over**
Unbroken signal 30 seconds   `30`

Warning systems

Important public announcement

The warning and information system IPA (important public announcement) is used in emergency situations – for example in the event of emissions of hazardous substances, fires where there is a risk of explosion, forest fires and other natural disasters.

Important public anouncements are broadcast primarily on Sveriges Radio's radio stations, Sveriges Television's TV channels and SVT's teletext system. IPAs can also be sent as text messages to mobile phones within a specific area.

Warning systems

14

14

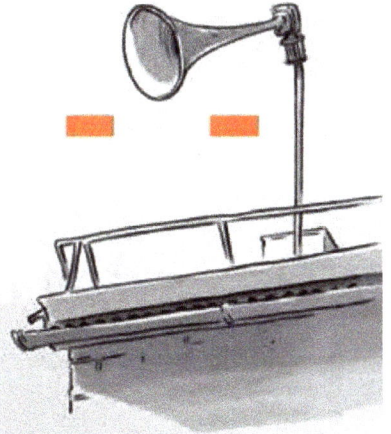

Outdoor warning

On rare occasions, the outdoor warning system is used ("Hesa Fredrik"). Facilities for the outdoor warning system are located in the majority of large built-up areas and around Sweden's nuclear power stations.

If you hear the signal: go indoors, close windows, doors and ventilation and listen to Sveriges Radio's radio station P4, which is tasked with providing public information.

The outdoor warning system is tested at 15:00 on the first non-public holiday monday in March, June, September and December.

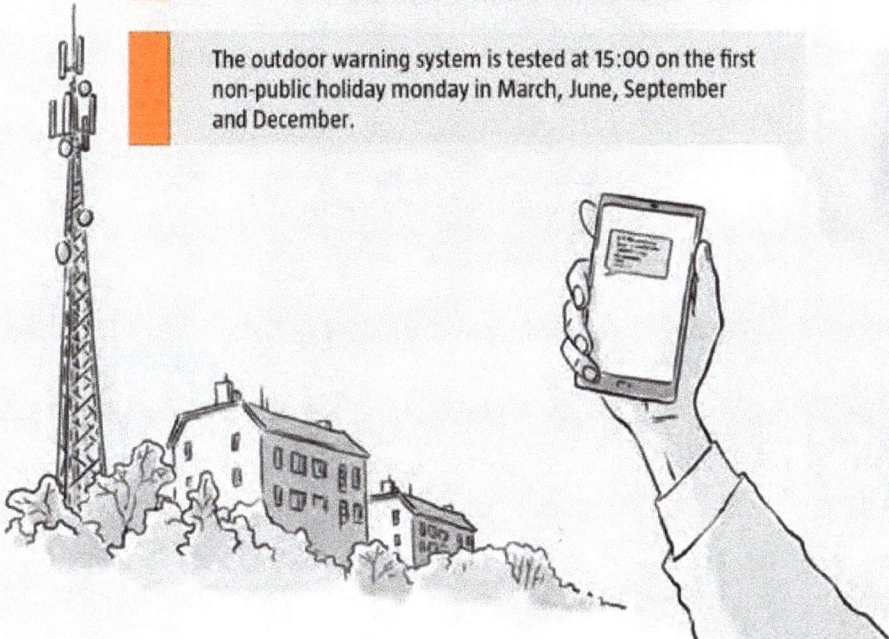

Warning systems

15

🔊 Emergency alarm
Signal 30 seconds – break 15 seconds

🔊 Air raid warning
Signal with short bursts for one minute

🔊 Danger over
Unbroken signal 30 seconds

Emergency alarm and air raid warning

The emergency alarm is a way for the Government to announce that there is the imminent threat of war, or that the country is at war.

If you hear the signal, you have to go indoors immediately and listen to Sveriges Radio's radio station P4. Get ready to leave home with that which is most important, warm clothes, something to eat and drink and identification documents. If you have been given a wartime posting, you are to proceed immediately to the place you have been instructed to go.

The air raid warning means that you are to find shelter immediately, for example an air raid shelter or the cellar of the building in which you are located.

New ways to warn the population may be applicable.

Keep yourself up to date by visiting **dinsäkerhet.se**

Warning systems

16

Shelters and other protective spaces

Shelters can provide protection to the population in the event of war. All shelters and buildings that contain shelters are marked with a sign. You do not belong to any specific shelter, you use whichever is nearest.

SKYDDSRUM

Find out the location of the shelters that are nearest to where you live and where you are during the daytime. In the event of an air raid alarm, go immediately to a shelter or, in an emergency, to another protective space such as a celler, tunnel or metro station.

Warning systems

17

Educate yourself!

Learn to provide first aid. Your knowledge can save lives. If you are the first on the scene after an accident or other serious incident, call SOS Alarm on the emergency number 112. Even if your pay-as-you-go SIM card has no credit, or your mobile has no SIM card, you can still call 112. SOS Alarm can provide advice about what to do at the site of the accident.

Get involved!

Many non-profit organisations and faith communities make important contributions to our collective security and preparedness. The voluntary defence organisations have specific duties as part of Sweden's total defence and offer both courses and training programmes. In the event of emergencies and heightened states of alert, their tasks include distributing important information to Sweden's population. You are needed and your contribution makes a difference!

You can find more information at dinsäkerhet.se

18

Important notes

Write down important telephone numbers, addresses and the closest shelter or other protective space.

The purpose of the brochure is to help us become better prepared for everything from serious accidents, extreme weather and IT attacks, to military conflicts. It is a good idea to talk about its contents with people around you.

19

IMPORTANT TELEPHONE NUMBERS AND WEBSITES

112

In an emergency situation that requires the immediate assistance of an ambulance, the fire and rescue service or the police.

113 13

To provide or obtain information about serious accidents or emergency situations.

114 14

All police matters that are not about crimes or incidents that are ongoing.

1177

Healthcare advice.

Dinsäkerhet.se

More detailed information about the contents of this brochure.

Krisinformation.se

Emergency information from Sweden's public authorities collected in one place.

Appendix J. Government Interagency Planning and Preparation Chart

Government Interagency Planning and Responsibilities			
	Pre Crisis/Deterrence	Crisis	Enemy Occupation
Ministry of Defense (MOD)	• Resistance focused training • Lead resistance planning and preparation • Engage in joint training and exercises • Purchase specialized equipment and supplies • Designate cache site and store equipment and supplies	• Activate resistance • Stock caches • Distribute equipment • Disperse select leadership to external, pre-planned locations	• Prepare for incoming allied forces and conduct activities against occupier guided by political leadership of exiled government • Conduct sabotage, subversion, intelligence gathering • Recruit, train and equip additional resistance underground and guerrilla members
Ministry of Interior (MOI) / Ministry of Justice (Internal Security and Law Enforcement)	• Assist the national legislative body in creating a national legal framework for resistance • Assist the national legislative body in writing laws • Identify, monitor, and disrupt subversive elements within the population	• Conduct raids and arrests of select known subversive elements that are assisting the foreign adversarial power • Increase surveillance of groups and individuals suspected of assisting the adversarial power • Employ any authorized emergency powers	• Gather and pass intel • Support and mask the activities of local law enforcement agencies • Maintain law enforcement against traditional criminal activities
Agency for Disaster Response or Civil Emergency	• Educate and inform the general public regarding individual and organizational responses to natural or man-made disasters • Assist the general public with the earliest stages of preparation and response	• Prepare the nation for shortages and execute most disaster protocols	• Continue functioning as a disaster response agency • Assist and prepare the nation, as possible, for sustenance and power shortages during combat
Ministry of Foreign Affairs (MFA)*	• Agreements with allies and partners to ensure legal recognition of resistance • Foster relationships with diaspora communities • Engage international organizations and international nongovernmental organizations	• Coordinating bilateral and multilateral support in key institutions • Execute plan for exiling key members of government • Execute strategic communication plan	• Ensure continued bilateral and multilateral recognition of the legitimate government • Continue pre-planned strategic communication* • Promote cause of international sanctions against occupying state with governments and international organizations*
Ministry of Communication	• Develop and distribute strategic communication messages • Prepare crisis strategic communication • Identify and disrupt hostile communication penetration of domestic networks • Coordinate government and non-government entities to leverage existing cyberspace capabilities • Develop alternative communication distribution systems and methods • Assist in development and acquisition of communication capability for resistance • Increase resiliency of communication networks	• Distribute national strategic communication messages • Attack adversary communication capability • Engage partner and other international media outlets to gain international support	• Restore and repair cyber and telecommunications infrastructure, especially emergency systems • Distribute pre-planned resistance information operations products • Surreptitiously monitor occupier activities
Ministry of Education/Culture	• Oversee the conduct of patriotic education and events • Promote a national culture available to all citizens • Engage with neighboring nations and international organizations to conduct cultural and educational events • Involve domestic civil society organizations in such activities to strengthen domestic bonds • Educate population on peaceful and other resistance methods	• Communicate message of strong national bonds and national resilience • Promote national cooperation against adversary	• Continue general education as far as allowed by occupier • Promote surreptitious schooling in cooperation with resistance organization • Engage in peaceful culturally based resistance activities against the occupier • Assist in maintaining popular morale • Disseminate clandestine information to inform population of occupier activities
Ministry of Transportation	• Formulate plan for priority transport of military essential items in time of national crisis • Coordinate and support development of alternate and clandestine transport networks • Identify weaknesses of transportation infrastructure for purpose of sabotage	• Execute national defense priority transportation plan as required by MoD • Prepare transportation networks for national resistance activities • Conduct sabotage as required for national defense and directed by MoD	• Restoration and recovery of emergency transportation infrastructure • Disrupt enemy use of domestic transportation infrastructure • Support resistance use of transportation networks domestically and for border infiltration and exfiltration • Assist sabotage of transportation systems as directed
Ministry of Finance (MOF)	• Develop plan to financially support acquisition of materials, supplies and services • Develop alternate methods of revenue generation, collection and distribution	• Activate contingency finance measures • Execute diaspora financial support network	• Continue engagement with diaspora communities and international banks and donors*

* Designates functions are performed by displaced or exiled ministry members outside of occupied territory

Table 4. Government Interagency Planning and Responsibilities for Resistance Operations.

SOURCE: THE INTERAGENCY RESPONSIBILITY CHART IS ROOTED IN THE INITIAL WORK OF PROFESSOR DOOWAN LEE, SENIOR LECTURER, UNITED STATES NAVAL POSTGRADUATE SCHOOL, AND MODIFIED WITH ADDITIONS BY GORDON (JIM) JAMES WORRALL AND OTTO C. FIALA

Acronyms

A2AD	anti-access, area denial
ACC	Allied Coordination Committee
ACCG	Allied Clandestine Coordination Groups
ARIS	Assessing Revolutionary and Insurgent Strategies
ASCOPE	area, structures, capabilities, organizations, people, and events
ASU	active service unit
AK	*Armia Krajowa* (The Home Army)
BBC	British Broadcasting Corporation
BCRA	*Bureau Central de Renseignements et d'Action* (World War II French Central Bureau of Intelligence and Action)
CAG	*Centro Addestramento Guastatori* (Saboteurs' Training Center)
C-A-R	Collaboration, Accommodation, and Resistance
CBRN	chemical, biological, radiological, and nuclear
CF	conventional forces
CFLN	*Comité Français de la libération nationale* (World War II French Committee of National Liberation)
CI	counterintelligence
CIA	Central Intelligence Agency
CIL	critical information list
CNR	*Conseil national de la résistance* (World War II French National Council of Resistance)
COIN	counterinsurgency
DIMEFIL	diplomacy, information, military, economic, financial, intelligence, law/law enforcement
DDoS	distributed denial of service
DOD	Department of Defense
FARC	*Fuerzas Armadas Revolucionarias de Colombia*
FFI	*Forces Françaises de l'Interieur* (French Forces of the Interior)
FID	foreign internal defense
FSF	foreign security forces
FTP	*Francs-Tireurs et Partisans*

HN	host nation
HUMINT	human intelligence
IAC	international armed conflict
ICRC	International Committee of the Red Cross
IDAD	internal defense and development
IHL	international humanitarian law
IHRL	international human rights law
IO	information operation
IRGC	Iranian Revolutionary Guard Corps
ISIS	Islamic State in Iraq and Syria
IW	irregular warfare
JORSS	Jerlet, Oras, Raadik, Saaliste, Saalists
KLO	Kosovo Liberation Army
LAP	Lithuanian Activist Front
LOAC	law of armed conflict
LOW	law of war
MOI	ministry of interior
MOD	ministry of defense
MSB	*Myndigheten för samhällsskydd och beredskap* (Swedish Government's Civil Contingencies Agency)
NATO	North Atlantic Treaty Organization
NGO	nongovernmental organization
NKVD	People's Commissariat for Internal Affairs (USSR)
NSC	National Security Council
OPC	Office of Policy Coordination (U.S.)
OPSEC	operations security
OSINT	open-source intelligence
OSS	Office of Special Services
PIRA	Provisional Irish Republican Army
PN	partner nation
POW	prisoner of war

PUS	Polish Underground State
ROC	Resistance Operating Concept
SAD	*Sezione Addestramento* (Italian Training Center)
SC	security cooperation
SFA	security force assistance
SHAPE	Supreme Headquarters Allied Powers Europe
SIFAR	*Servizio Informazioni Forze Armate* (Italian Military Intelligence)
SIGINT	signals intelligence
SOCEUR	Special Operations Command-Europe
SOE	special operations executive
SOF	Special Operations Forces
SSRs	Soviet Socialist Republics
UNCC	*Union Nationale des Combattants Coloniaux*
USASOC	United States Army Special Operations Command
USSR	Union of Soviet Socialist Republics
UW	unconventional warfare

Glossary

auxiliary. The support element of the irregular organization whose organization and operations are clandestine in nature and whose members do not openly indicate their sympathy or involvement with the irregular movement.

cache. A source of subsistence and supplies, typically containing items such as food, water, medical items, and/or communications equipment, packaged to prevent damage from exposure and hidden in isolated locations by such methods as burial, concealment, and/or submersion, to support isolated personnel.

cadre. A small group of people specially trained for a particular purpose or profession.

fifth column. A group within a country at war who are sympathetic to or working with its enemies. The term dates from the Spanish Civil War, when General Mola, leading four columns of troops towards Madrid, declared that he had a fifth column inside the city.

general war. Armed conflict between major powers in which the total resources of the belligerents are employed, and the national survival of a major belligerent is in jeopardy.

government-in-exile. A government that has been displaced from its country, but remains recognized as the legitimate sovereign authority.

guerrilla. A combat participant in guerrilla warfare.

guerrilla base. A temporary site where guerrilla installations, headquarters, and some guerrilla units are located. A guerrilla base is considered to be transitory and must be capable of rapid displacement by personnel within the base.

guerrilla warfare. Military and paramilitary operations conducted in enemy-held or hostile territory by irregular, predominantly indigenous forces.

insurgency. The organized use of subversion and violence to seize, nullify, or challenge political control of a region. Insurgency can also refer to the group itself. (JP 3-24).

intelligence preparation of the operational environment. The analytical process used by intelligence organizations to produce intelligence estimates and other intelligence products in support of the commander's decision-making process. It is a continuous process that includes defining the operational environment, describing the impact of the operational environment, evaluating the adversary, and determining adversary courses of action.

irregular warfare. A violent struggle among state and non-state actors for legitimacy and influence over the relevant population(s). Also called IW. (JP 1).

limited war. Armed conflict just short of general war, exclusive of incidents involving the overt engagement of the military forces of two or more nations.

operations security. A process of identifying critical information and subsequently analyzing friendly actions attendant to military operations and other activities. Also called OPSEC. (JP 3-13.3).

partisan. A member of an armed group formed to fight secretly against an occupying force, in particular in German-occupied Yugoslavia, Italy, and parts of eastern Europe in the Second World War (The Oxford

Dictionary). The term is used in this book limited to this historical context, unless used to describe a political supporter of a cause in a non-warfare context.

proxy. An agent or organization authorized to act on behalf of another. A proxy is an agent or substitute authorized actor. The necessary element is authorization.

resilience. The will and ability to withstand external pressure and influences and/or recover from the effects of that pressure or influence.

resistance. A nation's organized, whole-of-society effort, encompassing the full range of activities from nonviolent to violent, led by a legally established government (potentially exiled/displaced or shadow) to reestablish independence and autonomy within its sovereign territory that has been wholly or partially occupied by a foreign power.

shadow government. Governmental elements and activities performed by the irregular organization that will eventually take the place of the existing government. Members of the shadow government can be in any element of the irregular organization (underground, auxiliary, or guerrilla force).

strategic communication. Focused efforts to understand and engage key audiences to create, strengthen, or preserve conditions favorable for the advancement of the governments' interests, policies, and objectives through the use of coordinated programs, plans, themes, messages, and products synchronized with the actions of all instruments of national power.

subversion. Actions designed to undermine the military, economic, psychological, or political strength or morale of a governing authority.

surrogate. A person or organization acting as substitute for another. A surrogate is used to replace or substitute someone or some organization with another. A surrogate is a substitute when the primary actor is for some reason unable to act in that capacity.

unconventional warfare. Activities conducted to enable a resistance movement or insurgency to coerce, disrupt, or overthrow a government or occupying power by operating through or with an underground, auxiliary, and guerrilla force in a denied area. Also called UW. (U.S. DOD Dictionary of Military and Associated Terms, as of April 2019).

underground. A covert unconventional warfare organization established to operate in areas denied to the guerrilla forces or conduct operations not suitable for guerrilla forces.

Endnotes

1. This concept reflects the synthesis of ideas, discussions, and intellectual capital contributed in five seminars and two writing workshops, with the most recent event occurring in October 2018. SOCEUR facilitated the seminars and workshops, which included representation from Estonia, Latvia, Lithuania, Sweden, Poland, Norway, Denmark, Finland, and the NATO Special Operations Headquarters. The participants included subject matter experts, government and military practitioners, and representatives from academia and industry. Those initial ideas were built upon within several previous limited distribution versions of this work, adding U.S. Joint and Army doctrine, select portions of the ARIS series developed by the USASOC and the Johns Hopkins University Applied Physics Laboratory, articles from professional journals, historic material and material from other relevant publications, entities, and individual professionals.

2. This comprehensive defense strategy has been instituted or is being reinstituted in Sweden, Switzerland, Singapore, Finland, Denmark, and Austria. This summary was based on the following sources: George J. Stein, "Total Defense: A Comparative Overview of the Security Policies of Switzerland and Austria," *Defense Analysis* vol. 6, no. 1 (1990): 17–33; "The Five Pillars of Total Defence," Singapore Ministry of Defence, accessed July 2018, https://www.mindef.gov.sg/oms/imindef/mindef_websites/topics/totaldefence/about_us/5_Pillars.html; Swedish Defence Commission Report, "Resilience-The Total Defence Concept and the Development Of Civil Defence 2021-2025," 20 December 2017, a summary issued by the Swedish Defence Commission Secretariat, https://www.government.se/4afeb9/globalassets/government/dokument/forsvarsdepartementet/resilience---report-summary---20171220ny.pdf; "Sweden's Defence Policy, 2016 to 2020," Government Offices of Sweden, 1 June 2016, https://www.government.se/globalassets/government/dokument/forsvarsdepartementet/sweden_defence_policy_2016_to_2020.

3. Finland expresses this as its Societal Security Strategy. See Charly Salonius-Pasternak, "An Effective Antidote: The Four Components that Make Finland More Resilient to Hybrid Campaigns," *FIIA Comment*, 3 October 2017, https://storage.googleapis.com/upi-live/2017/10/comment19_finland.pdf.

4. Finland, Norway, Sweden, and Denmark are members of the Nordic Defence Cooperation and Norway and Denmark are also members of the NATO.

5. One priority may be the governmental establishment of large cash reserves to continue economic transactions in case automated credit systems fail due to power failure or https://storage.googleapis.com/upi-live/2017/10/comment19_finland.pdf.

6. MPK is a voluntary organization overseen by public law, promoting national defense through safety and security training and education to Finnish citizens. It also trains the territorial forces with oversight from the Finnish Defence Forces.

7. The Swedish Home Guard, Singapore Armed Forces Volunteer Corps, Estonian Defense League, Latvian Zemessardze (National Guard), Danish Home Guard, Finnish Territorial Forces.

8. Office of the Chairman of the Joint Chiefs of Staff, *DOD Dictionary of Military and Associated Terms* (Washington D.C.: The Joint Staff, October 2016), 119.

9. *Security Cooperation*, Joint Publication (JP) 3-20 (Washington, D.C.: The Joint Staff, May 2017), GL-5.

10. *Security Cooperation*, JP 3-20, GL-6, vii.

11. *Special Operations*, JP 3-05, (Washington, D.C.: The Joint Staff, 2014), GL-12.

12. *Foreign Internal Defense*, JP 3-22 (Washington, D.C.: The Joint Staff, 2014), I-1.

13. Modified from After Action Report, Resistance Seminar, Krakow, Poland, 26-27 April 2016, 4.

14. Modified from After Action Report, Resistance Seminar, Krakow, Poland, 26-27 April 2016, 5.

15. Donald J. Trump, *National Security Strategy of the United States of America* (Washington, D.C.: White House, 2017), 27, https://www.whitehouse.gov/wp-content/uploads/2017/12/NSS-Final-12-18-2017-0905.pdf.

16. Common operating picture is defined as, "A single identical display of relevant information shared by more than one command that facilitates collaborative planning and assists all echelons to achieve situational awareness. Also called COP." Office of the Chairman of the Joint Chiefs of Staff, *DOD Dictionary*.

17. Mark Grdovic, *A Leader's Handbook to Unconventional Warfare* (Fort Bragg: U.S. Army John F. Kennedy Special Warfare Center and School, November 2009), 13, 16-17.

18. Aleksa, Karolis, ed., *Prepare to Survive Emergencies and War: A Cheerful Take on Serious Recommendations* (Vilnius: Ministry of National Defence, 2015).

19. Nic Robertson, Antonia Mortensen, Elizabeth Roberts and Woj Treszczynski, "Lithuania Issues Manual on What to do if Russia Invades," *CNN online*, 28 October 2016, http://edition.cnn.com/2016/10/28/europe/lithuania-war-manual/index.html.

20. See H. von Dach, *Total Resistance: Swiss Army Guide to Guerrilla Warfare and Underground Operations* (Boulder: Paladin Press, 1992), 127-140. Von Dach provides myriad examples of how individuals, from teachers to clergy to farmers to caterers and more, can disrupt the daily operations of the enemy.

21. Mao Tse-Tung, *On Guerrilla Warfare* (New York: Frederick A. Praeger, Inc., 1965), 71-87.

22. David Kilcullen, *Out of the Mountains* (Oxford: Oxford University Press, 2015).

23. Grdovic, *A Leader's Handbook to Unconventional Warfare*, 13-14.

24. *Mission Command*, Army Doctrine Publication (ADP) 6-0 (Washington D.C.: Headquarters, Department of the Army. 2014).

25. "Sweden Mobilises Entire Home Guard for First Time Since 1975," *The Local*, 6 June 2018, https://www.thelocal.se/20180606/sweden-mobilises-entire-home-guard-for-first-time-since-1975.

26. Carl von Clausewitz, *On War*, edited and translated by Michael Howard and Peter Paret (Princeton: Princeton University Press, 1984), 481.

27. International Conferences (The Hague), *Hague Convention (IV) Respecting the Laws and Customs of War on Land and Its Annex: Regulations Concerning the Laws and Customs of War on Land* (The Hague: October 18, 1907), Section III, Article 42, 1907.

28. This model is consistent with both seminar series findings and U.S. Joint Doctrine.

29. Grdovic, *A Leader's Handbook to Unconventional Warfare*, 12; "satellite television" added to original.

30. Grdovic, *A Leader's Handbook to Unconventional Warfare*, 12.

31. Leonhard, Robert, ed., *Assessing Revolutionary and Insurgent Strategies: Undergrounds in Insurgent, Revolutionary, and Resistance Warfare*, 2nd Edition (Fort Bragg: United States Army Special Operations Command, 2013), 153.

32. Leonhard, *Undergrounds in Insurgent, Revolutionary, and Resistance Warfare*, 188.

33. Office of the Chairman of the Joint Chiefs of Staff, *DOD Dictionary*, 221.

34. Leonhard, *Undergrounds in Insurgent, Revolutionary, and Resistance Warfare*, 171.

35. Leonhard, *Undergrounds in Insurgent, Revolutionary, and Resistance Warfare*, 175.

36. *Merriam-Webster's Dictionary of Law by Merriam-Webster* (Springfield, MA: Merriam-Webster, Inc., 2016).

37. Leonhard, *Undergrounds in Insurgent, Revolutionary, and Resistance Warfare*, 177-178.

38. United States Army Institute for Military Assistance, *The Underground*, ST 31-202 (Fort Bragg: U.S. Army Institute for Military Assistance, 1978), 190.

39. Leonhard, *Undergrounds in Insurgent, Revolutionary, and Resistance Warfare*, 180.

40. United States Department of Defense, *Commander's Communication Synchronization*, Joint Doctrine Note 2-13, (Washington D.C.: Department of Defense, 2013).

41. *Public Affairs*, JP 3-61 (Washington D.C.: The Joint Staff, 2016); *Commander's Communication Synchronization*, Joint Doctrine Note 2-13, (Washington D.C.: Department of Defense, 2013).

42. *Public Affairs*, JP 3-61; *Commander's Communication Synchronization*.

43. Leonhard, *Undergrounds in Insurgent, Revolutionary, and Resistance Warfare*, 140.

44. See for example Jānis Bērziņš, *Russia's New Generation Warfare in Ukraine: Implications for Latvian Defence Policy* (Riga: National Defence Academy of Latvia, Center for Security and Strategic Research, April 2014), 10. "Since language is a strong factor determining one's cultural identity, the result is that many ethnic groups are now culturally closer to Russia than to Latvia or to their real heritage. This is the case with many Poles, Byelorussians, Ukrainians, Tartars, Uzbeks, Kazakhs, and others, who make up approximately 13% of Latvia's population. Since these countries have their own political issues with Russia, the Latvian government should stimulate ethnic division between Russian-speakers by increasing their cultural self-awareness, thus making them proud of their real heritage. A concomitant step should be to apply the same strategy to the ethnic groups that supposedly form the Russian nation,

in other words, Bashkirs, Chuvashs, Chechens, Mordvins, Kazakhs and Avars, just to name a few. It is important to establish policies to increase their awareness of being unique, and thus not part of the Russian nation."

45. Leonhard, *Undergrounds in Insurgent, Revolutionary, and Resistance Warfare*, 7-9.

46. Paul A. Jureidini, Norman A. La Charite, Bert H. Cooper, and William A. Lybrand, *Casebook on Insurgency and Revolutionary Warfare, Volume I: 1933-1962*, Revised Edition (Fort Bragg: United States Army Special Operations Command, 2013), 70-100.

47. Bryan Gervais, "Hutu-Tutsi Genocides," in *Casebook on Insurgency and Revolutionary Warfare, Volume II: 1962-2009*, ed. Charles Crossett (Fort Bragg: United States Army Special Operations Command, 2013).

48. Leonhard, *Undergrounds in Insurgent, Revolutionary, and Resistance Warfare*, 9-10.

49. Office of the Chairman of the Joint Chiefs of Staff, *DOD Dictionary*, 157.

50. Crossett, Chuck, ed., *Casebook on Insurgency and Revolutionary Warfare, Volume II: 1962-2009*, 285-287.

51. Grdovic, *A Leader's Handbook to Unconventional Warfare*, 12.

52. Chuck Crossett and Summer Newton, "Solidarity," in *Casebook on Insurgency and Revolutionary Warfare, Volume II: 1962-2009*, ed. Chuck Crossett (Fort Bragg: United States Army Special Operations Command, 2013). See also appendix B.

53. Leonhard, *Undergrounds in Insurgent, Revolutionary, and Resistance Warfare*, 45-46.

54. Leonhard, *Undergrounds in Insurgent, Revolutionary, and Resistance Warfare*, 47.

55. Leonhard, *Undergrounds in Insurgent, Revolutionary, and Resistance Warfare*, 45.

56. Leonhard, *Undergrounds in Insurgent, Revolutionary, and Resistance Warfare*, 45-46.

57. *Joint Intelligence*, JP 2-0 (Washington D.C.: The Joint Staff, 2013), B4-B9; *Special Operations*, JP 3-5, (Washington D.C.: The Joint Staff, 2014), IV-2.

58. *Joint Intelligence*, JP 2-0, B9.

59. JP 3-13.3, *Operations Security*.

60. JP 2-0, *Joint Intelligence*, B7.

61. Leonhard, *Undergrounds in Insurgent, Revolutionary, and Resistance Warfare*, 111-114, 189.

62. *JP 2-0, Joint Intelligence*, IV-20, V1-13.

63. *Joint Intelligence, JP 2-0*, IV-19; Leonhard, *Undergrounds in Insurgent, Revolutionary, and Resistance Warfare*, 47.

64. *Joint Intelligence*, JP 2-0, IV-20, I-19.

65. Anthony R. Molnar, *Undergrounds in Insurgent, Revolutionary and Resistance Warfare* (Washington, D.C.: Special Operations Research Office, The American University, 1963), 108.

66. Molnar, *Undergrounds in Insurgent, Revolutionary and Resistance Warfare*, 108; Leonhard, *Undergrounds in Insurgent, Revolutionary, and Resistance Warfare*, 46-47.

67. Molnar, *Undergrounds in Insurgent, Revolutionary and Resistance Warfare*, 108.

68. Leonhard, *Undergrounds in Insurgent, Revolutionary, and Resistance Warfare*, 48.

69. Leonhard, *Undergrounds in Insurgent, Revolutionary, and Resistance Warfare*, 48.

70. Leonhard, *Undergrounds in Insurgent, Revolutionary, and Resistance Warfare*, 49.

71. Leonhard, *Undergrounds in Insurgent, Revolutionary, and Resistance Warfare*, 49.

72. Olav Riste and Berit Nokleby, *Norway 1940-45: The Resistance Movement* (Otta: Arthur Vanous Co., 1970), 59-61.

73. Leonhard, *Undergrounds in Insurgent, Revolutionary, and Resistance Warfare*, 49.

74. Margen Nix and Dru Daubon, "Kosovo Liberation Army, 1996-1999," in *Casebook on Insurgency and Revolutionary Warfare, Volume II: 1962-2009*, ed. Chuck Crossett (Fort Bragg: United States Army Special Operations Command, 2013).

75. Leonhard, *Undergrounds in Insurgent, Revolutionary, and Resistance Warfare*, 51.

76. Jerome Conley, "Orange Revolution (Ukraine): 2005-2005," in *Casebook on Insurgency and Revolutionary Warfare, Volume II: 1962-2009*, ed. Chuck Crossett (Fort Bragg: United States Army Special Operations Command, 2013).

77. Leonhard, *Undergrounds in Insurgent, Revolutionary, and Resistance Warfare*, 51.

78. Leonhard, *Undergrounds in Insurgent, Revolutionary, and Resistance Warfare*, 57.

79. Leonhard, *Undergrounds in Insurgent, Revolutionary, and Resistance Warfare*, 57; Molnar, Undergrounds in Insurgent, Revolutionary and Resistance Warfare, 61.

80. Leonhard, *Undergrounds in Insurgent, Revolutionary, and Resistance Warfare*, 58.

81. Leonhard, *Undergrounds in Insurgent, Revolutionary, and Resistance Warfare*, 58.

82. Leonhard, *Undergrounds in Insurgent, Revolutionary, and Resistance Warfare*, 58-59.

83. Leonhard, *Undergrounds in Insurgent, Revolutionary, and Resistance Warfare*, 58-59.

84. Leonhard, *Undergrounds in Insurgent, Revolutionary, and Resistance Warfare*, 59.

85. Leonhard, *Undergrounds in Insurgent, Revolutionary, and Resistance Warfare*, 60.

86. Leonhard, *Undergrounds in Insurgent, Revolutionary, and Resistance Warfare*, 60-61.

87. Molnar, *Undergrounds in Insurgent, Revolutionary and Resistance Warfare*, 63.

88. Leonhard, *Undergrounds in Insurgent, Revolutionary, and Resistance Warfare*, 61-64.

89. The Telegraph, "Nazi Fake Banknote 'Part of Plan to Ruin British Economy,'" *The Telegraph*, 29 September 2010, https://www.telegraph.co.uk/history/world-war-two/8029844/Nazi-fake-banknote-part-of-plan-to-ruin-British-economy.html.

90. Leonhard, *Undergrounds in Insurgent, Revolutionary, and Resistance Warfare*, 62.

91. Leonhard, *Undergrounds in Insurgent, Revolutionary, and Resistance Warfare*, 65.

92. Leonhard, *Undergrounds in Insurgent, Revolutionary, and Resistance Warfare*, 65.

93. Leonhard, *Undergrounds in Insurgent, Revolutionary, and Resistance Warfare*, 65-66.

94. Leonhard, *Undergrounds in Insurgent, Revolutionary, and Resistance Warfare*, 67-71.

95. Leonhard, *Undergrounds in Insurgent, Revolutionary, and Resistance Warfare*, 77.

96. Leonhard, *Undergrounds in Insurgent, Revolutionary, and Resistance Warfare*, 77-82.

97. Leonhard, *Undergrounds in Insurgent, Revolutionary, and Resistance Warfare*, 82.

98. Leonhard, *Undergrounds in Insurgent, Revolutionary, and Resistance Warfare*, 82-83.

99. Leonhard, *Undergrounds in Insurgent, Revolutionary, and Resistance Warfare*, 82-85.

100. Leonhard, *Undergrounds in Insurgent, Revolutionary, and Resistance Warfare*, 85-86.

101. Leonhard, *Undergrounds in Insurgent, Revolutionary, and Resistance Warfare*, 85-86.

102. Leonhard, *Undergrounds in Insurgent, Revolutionary, and Resistance Warfare*, 87.

103. Leonhard, *Undergrounds in Insurgent, Revolutionary, and Resistance Warfare*, 87.

104. Leonhard, *Undergrounds in Insurgent, Revolutionary, and Resistance Warfare*, 88.

105. Leonhard, *Undergrounds in Insurgent, Revolutionary, and Resistance Warfare*, 89.

106. Leonhard, *Undergrounds in Insurgent, Revolutionary, and Resistance Warfare*, 89.

107. Leonhard, *Undergrounds in Insurgent, Revolutionary, and Resistance Warfare*, 93.

108. Leonhard, *Undergrounds in Insurgent, Revolutionary, and Resistance Warfare*, 93-94.

109. Leonhard, *Undergrounds in Insurgent, Revolutionary, and Resistance Warfare*, 97.

110. Leonhard, *Undergrounds in Insurgent, Revolutionary, and Resistance Warfare*, 99.

111. Leonhard, *Undergrounds in Insurgent, Revolutionary, and Resistance Warfare*, 101.

112. Leonhard, *Undergrounds in Insurgent, Revolutionary, and Resistance Warfare*, 102-103.

113. Leonhard, *Undergrounds in Insurgent, Revolutionary, and Resistance Warfare*, 109.

114. Leonhard, *Undergrounds in Insurgent, Revolutionary, and Resistance Warfare*, 111-114.

115. Leonhard, *Undergrounds in Insurgent, Revolutionary, and Resistance Warfare*, 109-110.

116. U.S. Army Institute for Military Assistance, *The Underground*, 56.

117. U.S. Army Institute for Military Assistance, *The Underground*, 69.

118. Al J. Venter, *War Dog: Fighting Other People's Wars* (Drexel Hill: Casemate, 2003), 90.

119. Leonhard, *Undergrounds in Insurgent, Revolutionary, and Resistance Warfare*, 133-138.

120. Michael Fitzsimmons, *Governance, Identity, and Counterinsurgency: Evidence from Ramadi and Tal Afar* (Carlisle: Strategic Studies Institute and U.S. Army War College Press, 2013).

121. Chuck Crosset and Summer Newton, "The Provisional Irish Republican Army: 1969-2001" in *Casebook on Insurgency and Revolutionary Warfare, Volume II: 1962-2009*, ed. Chuck Crossett (Fort Bragg: United States Army Special Operations Command, 2013).

122. Leonhard, *Undergrounds in Insurgent, Revolutionary, and Resistance Warfare*, 138-142.

123. Erin H. Hahn, ed., *Special Topics in Irregular Warfare: Understanding Resistance, Assessing Revolutionary and Insurgent Strategies* (Fort Bragg: United States Army Special Operations Command, 2018), 143-145.

124. Leonhard, *Undergrounds in Insurgent, Revolutionary, and Resistance Warfare*, 140.

125. Bairbre De Brun, *The Road to Peace in Ireland* (Berlin: Berghof-Forschungszentrum fur Konstruktive Konfliktbearbeitung, 2008), 14-15.

126. Nathan Bos, ed., *Assessing Revolutionary and Insurgent Strategies: Human Factors Considerations of Undergrounds in Insurgencies* 2nd Edition. (Fort Bragg: United States Army Special Operations Command, 2013) 275-276.

127. A Japanese defensive martial art that uses locks, holds, throws, and the opponent's own movements.

128. Bos, *Human Factors Considerations of Undergrounds in Insurgencies* 2nd ed., 276-278.

129. Bos, *Human Factors Considerations of Undergrounds in Insurgencies* 2nd ed., 279.

130. Bos, *Human Factors Considerations of Undergrounds in Insurgencies* 2nd ed., 283-284.

131. Bos, *Human Factors Considerations of Undergrounds in Insurgencies* 2nd ed., 287-288.

132. Bos, *Human Factors Considerations of Undergrounds in Insurgencies* 2nd ed., 288-289.

133. Bos, *Human Factors Considerations of Undergrounds in Insurgencies* 2nd ed., 289-290.

134. Primarily based on: United States Department of Homeland Security, *National Response Framework* (Washington D.C.: Department of Homeland Security, 2008); and supplemented by Frederick M. Kaiser, *Interagency Collaborative Arrangements and Activities: Types, Rationales, Considerations* (Washington D.C.: Congressional Research Service, 2011); Towell, *Organizing the U.S. Government for National Security: Overview of the Interagency Reform Debates* (Washington, D.C.: Congressional Research Service, 2008); Veronica de Allende and Alison Vernon, *Coordination in a JTF Formation* (Arlington, Center for Naval Analysis, 2018); *Interorganizational Cooperation*, JP 3-08, (Washington, D.C.: The Joint Staff, 2017); *Homeland Defense*, JP 3-27, (Washington: The Joint Staff, 2018).

135. Clausewitz, *On War*, 480.

136. *If Crisis or War Comes (Om krisen eller kriget kommer)* (Karlstad: Swedish Civil Contingencies Agency (MSB), May 2018), https://www.msb.se/Upload/Forebyggande/Krisberedskap/Krisberedskapsveckan/Fakta%20om%20broschyren%20Om%20krisen%20eller%20Kriget%20kommer/om-krisen-eller-kriget-kommer---engelska.pdf.

137. Preparing to exile the national government while maintaining maximum control within occupied territory has elements similar to continuity of government operations. See *Continuity of Operations Plan Template for Federal Departments and Agencies*, Department of Homeland Security, Washington, D.C., 2013, http://www.fema.gov/about/org/ncp/coop/templates.shtm.

138. An example of this is the Swedish Rakel national digital communication system established for the purposes of national emergencies and managed by the MSB.

139. *Interorganizational Cooperation*, JP 3-08, (Washington: The Joint Staff, 2017) I-9–I-10, I-13.

140. *National Response Framework*, 74-75.

141. We are very grateful to Dr. Marcin Marcinko, Chair of Public International Law, Jagiellonian University in Krakow, for his presentation at one of our Resistance seminars in Riga, Latvia, on 30 October 2018. His presentation was used to "fine-tune" several aspects of this appendix and adding other sections.

142. "*Founding and Early Years of the ICRC (1863-1914)*," 12 May 2010, International Committee of the Red Cross, https://www.icrc.org/en/document/founding-and-early-years-icrc-1863-1914.

143. 1949 Geneva Convention, common Article 2 states "the present Convention shall apply to all cases of *declared war* or of any other *armed conflict* which may arise between two or more of the High Contracting Parties, even if the state of war is not recognized by one of them." Convention (III) relative to the Treatment of Prisoners of War art. 2, 12 August 1949, 6 U.S.T. 3116, 75 U.N.T.S. 135 (emphasis added).

144. U.S. Department of Defense, *Law of War Manual*, June 2015 (as updated in December 2016) Para. 1.3.

145. U.S. Department of Defense, *Law of War Manual*, Para. 1.3.1.2.

146. Gary D. Solis, *The Law of Armed Conflict: International Humanitarian Law in War* (New York: Cambridge University Press, 2012), 20.

147. Jean Pictat, ed., *Commentary, Convention Relative to the Treatment of Prisoners of War* (Geneva: International Committee of the Red Cross, 1960), 23.

148. International Committee of the Red Cross, *Protocol Additional to the Geneva Conventions of 12 August 1949, and relating to the Protection of Victims of International Armed Conflicts* (Protocol I), 8 June 1977, Commentary of 1987.

149. United States Army Judge Advocate General's Legal Center and School, *Law of Armed Conflict Deskbook* (Charlottesville: International and Operational Law Department, 2015); Lee et al., *Law of Armed Conflict Deskbook*, 133.

150. United States Army Judge Advocate General's Legal Center and School, *Law of Armed Conflict Deskbook*, 135–151.

151. Convention (IV) Respecting the Laws and Customs of War on Land, and its Annex: Regulations Concerning the Laws and Customs of War on Land art. 23, 18 October 1907, 36 Stat. 2277, 205 Consol. T.S. 277.

152. Common Article 3 was included to emphasize the "principle of respect for the human personality" regardless of whether the conflict occurs between two recognized states (i.e., is international in nature). The drafters determined that "the same logical process could not fail to lead to the idea of applying the principle to all cases of armed conflicts, including those of an internal character." See; Jean Pictat, ed., *Commentary, Convention Relative to the Protection of Civilians in Time of War* (Geneva: International Committee of the Red Cross, 1958), 26.

153. Jean Pictat, ed., *Commentary, Convention Relative to the Protection of Civilians in Time of War* (Geneva: International Committee of the Red Cross, 1958), 34.

154. International Committee of the Red Cross, *Geneva Convention for the Amelioration of the Condition of the Wounded and Sick in Armed Forces in the Field (First Geneva Convention)*, 12 August 1949, 75 UNTS 31.

155. Nils Melzer, *Interpretive Guidance on the Notion of Direct Participation in Hostilities Under International Humanitarian Law* (Geneva: International Review of the Red Cross, 2009), 20–40.

156. Richard R. Baxter, "So-Called 'Unprivileged Belligerency': Spies, Guerillas, and Saboteurs," *British Yearbook of International Law* 28 (1951): 340, 343.

157. Richard R. Baxter, "So-Called 'Unprivileged Belligerency': Spies, Guerillas, and Saboteurs," *British Yearbook of International Law* 28 (1951): 338, 343.

158. Protocol Additional to the Geneva Conventions of 1949, and relating to the Protections of Victims of International Armed Conflicts (Protocol I), 12 December 1977, 1125 U.N.T.S. 609 [hereinafter Protocol I]; Protocol Additional to the Geneva Conventions of 1949, and relating to the Protections of Victims of Non-International Armed Conflicts (Protocol II) arts. 4, 13, 14, 12 December 1977, 1125 U.N.T.S. 609 [hereinafter Protocol II].

159. International Committee of the Red Cross, *Geneva Convention Relative to the Treatment of Prisoners of War (Third Geneva Convention)*, Art. 4(A)(2)(a)-(d), 12 August 1949, 75 UNTS 135.

160. Yoram Dinstein, *The Conduct of Hostilities Under the Law of International Armed Conflict* (New York: Cambridge University Press, 2010), 45.

161. This comes from judicial interpretation and Commentary to GC III of 1960: "For partisans a distinctive sign replaces a uniform; it is therefore an essential factor of loyalty in the struggle and must be worn constantly, in all circumstances. If it is to be distinctive, the sign must be the same for all the members of any one resistance organization, and must be used only by that organization. This in no way precludes the wearing of additional emblems indicating rank or special functions."

162. Two of the most prominent cases and decisions on point follow: a) Judicial Committee of the Privy Council, *Bin Haji Mohamed Ali and Another v. Public Prosecutor*, Appeal Judgement, Case No. 20/1967. During the conflict between Indonesia and Malaysia (1963-1966) two members of the Indonesian armed forces dressed as civilians placed a bomb in an office building in Singapore. The explosion killed three civilians. The perpetrators were found guilty of murder and sentenced to death. The case was finally decided in 1968 by the Privy Council in London, which, inter alia, cited the following statement in Section 96 of the Military Manual of the United Kingdom: "Members of the armed Forces caught in civilian clothes while acting as saboteurs in enemy territory are in a position analogous to that of spies". Furthermore, the Privy Council emphasized the importance of the uniform to mark the difference between combatants and peaceful civilians. b) *Military Prosecutor v. Omar Mahmud Kassem and others* (Ramallah Military Tribunal, Case No. 6/49, 1969): The Tribunal underlined, that some PFPL members had infiltrated Israeli occupied territory wearing dark green uniforms and mottled peaked caps. The Tribunal admitted that "civilians resident in the area of the encounter with the Israeli forces do not usually wear green clothes or mottled peaked caps and that the accused, therefore, fulfilled the condition under reference."

163. A prominent decision on point is *Trial of Wilhelm List and others*, United States Military Tribunal, Nuremberg, 1947-1948 (The Hostages Trial): "[The partisans] generally wore civilian clothes although parts of German, Italian and

Serbian uniforms were used to the extent they could be obtained. The Soviet Star was generally worn as insignia. The evidence will not sustain a finding that it was such that it could be seen at a distance. Neither did they carry their arms openly except when it was to their advantage to do so."

164. International Committee of the Red Cross, *Protocol Additional to the Geneva Conventions of 12 August 1949, and relating to the Protection of Victims of International Armed Conflicts* (Protocol I), 8 June 1977, Article 46 (3).

165. International Committee of the Red Cross, *Protocol Additional to the Geneva Conventions of 12 August 1949, and relating to the Protection of Victims of International Armed Conflicts (Protocol I)*, 8 June 1977, Article 46 (4).

166. Yoram Dinstein, *The International Law of Belligerent Occupation* (New York: Cambridge University Press, 2009), 97-98.

167. International Committee of the Red Cross, *Protocol Additional to the Geneva Conventions of 12 August 1949, and relating to the Protection of Victims of International Armed Conflicts (Protocol I)*, 8 June 1977, Commentary of 1987, Combatants and Prisoners of War, para 1704; "It seems to be accepted nowadays that a 'levée en masse' can take place in any part of the territory which is not yet occupied, even when the rest of the country is occupied, or in an area where the Occupying Power has lost control over the administration of the territory and is attempting to regain it. (54) On the other hand, it does not seem conceivable with regard to a retreating enemy, as this is manifestly no longer an invading army."

168. International Committee of the Red Cross (ICRC), *Protocol Additional to the Geneva Conventions of 12 August 1949, and relating to the Protection of Victims of International Armed Conflicts (Protocol I)*, 8 June 1977, Commentary of 1987, Combatants and Prisoners of War, para 1705: Further, related to the concept of Total Defense, this paragraph states; "According to the traditional interpretation, it is not necessary for the population engaged in the 'levée en masse' to be surprised by the invasion; the provision on the 'levée en masse' is also to the advantage of a population which has been forewarned, provided that it has not had time to organize itself in accordance with the requirements laid down in Article 43 (Armed forces). This benefit is also accorded to a population acting on the orders of its government, for example, obeying orders given by radio or through the media. The requirement of carrying arms openly must probably be understood as an obligation to always carry them visibly."

169. Nils Melzer, "Interpretive Guidance on the Notion of Direct Participation in Hostilities under International Humanitarian Law," *International Review of the Red Cross* 90, no. 872 (2008), 36.

170. International Committee of the Red Cross, *Protocol Additional to the Geneva Conventions of 12 August 1949, and relating to the Protection of Victims of International Armed Conflicts (Protocol I)*, Art. 51(1)-(3), 8 June 1977, 1125 UNTS 3.

171. Geneva Convention, Protocol I, Art. 51(1)-(3).

172. Melzer, "Interpretive Guidance," 46.

173. Melzer, "Interpretive Guidance," 49.

174. Melzer, "Interpretive Guidance," 53.

175. Melzer, "Interpretive Guidance," 60-63.

176. In a review before the United States Supreme Court, in *Ex Parte Quirin et al.* (1942), No. 100, decided 31 July 1942; In June 1942, Quirin and seven other German soldiers in uniform landed in the United States for the purpose of committing sabotage. They buried their uniforms and brought along explosives and proceeded into the country dressed in civilian clothes. They were taken prisoners shortly thereafter. The court held: "(1) That the charges preferred against petitioners on which they are being tried by military commission appointed by the order of the President of July 2, 1942, allege an offense or offenses which the President is authorized to order tried before a military commission. (2) That the military commission was lawfully constituted. (3) That petitioners are held in lawful custody, for trial before the military commission, and have not shown cause for being discharged by writ of habeas corpus. The motions for leave to file petitions for writs of habeas corpus are denied. The orders of the District Court are affirmed. The mandates are directed to issue forthwith."

177. George Fletcher and Jens David Ohlen, *Defending Humanity* (New York: Oxford University Press, 2008), 183.

178. Geneva Convention, Protocol I, Art. 37.

179. *The United Nations War Crimes Commission, Law Reports of Trials of War Criminals*, vol. VIII, 1949, 34-76. The Trial of Wilhelm List and others in *United States of America v. Wilhelm List, et al.*, United States Military Tribunal, Nuremberg, 1947-1948 (The Hostages Trial) elucidates this: "International Law makes no distinction between a lawful and an unlawful occupant in dealing with the respective duties of occupant and population in occupied territory. There is no reciprocal connection between the manner of the military occupation of territory and the rights and duties of

the occupant and population to each other after the relationship has in fact been established. Whether the invasion was lawful or criminal is not an important factor in the consideration of this subject."

180. A case on point is *Military Prosecutor v. Omar Mahmud Kassem and others* (Ramallah Military Tribunal, Case No. 6/49, 1969): At trial, POW status was refused captured members of the Popular Front for the Liberation of Palestine (PFLP). The main reasons were that PFLP members did not belong to a Party to the conflict, and the mere membership in the resistance organization was not sufficient for recognition as POWs according to the GC III.

181. Case in point is: ICTY, *Prosecutor v. Dusko Tadić*, Appeals Chamber Judgement, Case No. IT-94-1-A, 1999: "In order for irregulars to qualify as lawful combatants, it appears that international rules and State practice (...) require control over them by a Party to an international armed conflict and, by the same token, a relationship of dependence and allegiance of these irregulars vis-à-vis that Party to the conflict. These then may be regarded as the ingredients of the term *belonging to a Party to the conflict*."

182. *Nonviolent Civic Action, A Study Guide Series on Peace and Conflict* (Washington D.C.: United States Institute of Peace, 2009), 8-9.

183. *Nonviolent Civic Action*, 13.

184. *Nonviolent Civic Action*, 14.

185. Gene Sharp, *Sharp's Dictionary of Power and Struggle: Language of Civil Resistance in Conflicts* (New York: Oxford University Press, 2011).

186. Feliks Gross, *The Seizure of Political Power in a Century of Revolutions* (New York: Philosophical Library, 1958), 51.

187. Gene Sharp, *The Politics of Nonviolent Action, Part 2: The Methods of Nonviolent Action* (Boston: Porter Sargent, 1980), 357.

188. H. von Dach, *Total Resistance*, 127-140.

189. Bos, *Human Factors Considerations of Undergrounds in Insurgencies* 2nd ed., 290-293.

190. Bos, *Human Factors Considerations of Undergrounds in Insurgencies* 2nd ed., 293-295.

191. Bos, *Human Factors Considerations of Undergrounds in Insurgencies* 2nd ed., 295-296.

192. Bos, *Human Factors Considerations of Undergrounds in Insurgencies* 2nd ed., 296-297.

193. Mike Elkin, "Tunisia's Internet Chief Gives Inside Look at Cyber Uprising," *Wired*, 28 January 2011, http://www.wired.com/dangerroom/2011/01/as-egypt-tightens-its-internet-grip-tunisia-seeks-to-open-up/.

194. Charles Levinson and Matt Bradley, "Egypt's Regime on the Brink," *Wall Street Journal Online*, 29 January 2011, http://online.wsj.com/article/SB10001424052748703956604576109323492986438.html?mod=WSJ World LeadStory.

195. Bos, *Human Factors Considerations of Undergrounds in Insurgencies* 2nd ed., 297-299.

196. Ernest K. Bramstedt, *Dictatorship and Political Police: The Technique of Control by Fear* (New York: Oxford University Press, 1945), 210.

197. Bos, *Human Factors Considerations of Undergrounds in Insurgencies* 2nd ed., 299-300.

198. Ruaridh Arrow, "Gene Sharp: Author of the Nonviolent Revolution Rulebook," *BBC News*, 21 February 2011, http://www.bbc.co.uk/news/world-middle-east-125228-48.

199. Issued by the Heads of State and Government participating in the meeting of the North Atlantic Council in Warsaw, 8-9 July 2016.

200. Frank G. Hoffman, "Hybrid Warfare and Challenges," *Joint Forces Quarterly* 52, no. 1 (2009): 34-39.

201. William J. Nemeth, "Future Wars and Chechnya: A Case for Hybrid Warfare," (thesis, Naval Postgraduate School, 2002), http://calhoun.nps.edu/bitstream/handle/10945/5865/02Jun_Nemeth.pdf?sequence=1.

202. Frank Hoffman and James N. Mattis, "Future Wars: The Rise of Hybrid Wars," *Proceedings* vol. 131/II/1,233 (November 2005).

203. Liang, Qiao and Xiangsui, Wang, *Unrestricted Warfare*, trans. Ian Strauss (Brattleboro: Echo Point Books and Media, 2015).

204. Frank G. Hoffman, *Conflict in the 21st Century: The Rise of Hybrid Wars* (Arlington: Potomac Institute of Policy Studies, 2007): 8. http://www.potomacinstitute.org/images/stories/publications/potomac_hybridwar_0108.pdf.

205. "Wales Summit Declaration" issued by the Heads of State and government participating in the meeting of the North Atlantic Council in Wales, 5 September 2014, paragraph 13. In the same paragraph, NATO established the

NATO-accredited Strategic Communications Centre of Excellence in Latvia "as a meaningful contribution to NATO's efforts in this area."

206. Andrés B. Muñoz Mosquera and Sascha-Dominik Bachmann, "Understanding Lawfare in a Hybrid Warfare Context," *NATO Legal Gazette*, Issue 37 (October 2016): 23.

207. Sun Tzu, *The Art of War, The Book of Lord Shang* (Hertfordshire: Wordsworth Editions Limited, 1998), 25, as quoted in Mosquera and Bachmann "Understanding Lawfare in a Hybrid Warfare Context," 23.

208. Fulvio Poli, "An Asymmetrical Symmetry: How Convention has become Innovative Military Thought" (thesis, U.S. Army War College, 2010), 2.

209. Williamson Murray and Peter R. Mansoor, *Hybrid Warfare, Fighting Complex Opponents from the Ancient World to the Present* (New York: Cambridge University Press, 2012), 104-150.

210. Matt M. Matthews, *We Were Caught Unprepared: The 2006 Hezbollah-Israeli War* (Fort Leavenworth: Combat Studies Institute Press, 2008), 20.

211. Hoffman, *Conflict in the 21st Century*, 4.

212. Timothy McCulloh and Richard Johnson, *Hybrid Warfare* (Tampa: JSOU Press, 2013), 3.

213. The Islamic State is also known as the Islamic State in Iraq and Syria (ISIS) and Islamic State of Iraq and the Levant (ISIL).

214. Franklin D. Kramer and Lauren M. Speranza, *Meeting the Russian Hybrid Challenge, A Comprehensive Strategic Framework* (Washington D.C.: Atlantic Council, Brent Scowcroft Center on International Security, 2017).

215. Dmitry Adamsky, *Cross-Domain Coercion: The Current Russian Art of Strategy* (Paris: Institut Francais des Relations Internationales, November 2015) http://www.ifri.org/sites/default/files/atoms/files/pp54adamsky.pdf.

216. Merle Maigre, "Nothing New in Hybrid Warfare: The Estonian Experience and Recommendations for NATO," The German Marshall Fund of the United States (12 February 2015): 2, http://www.gmfus.org/publications/nothing-new-hybrid-warfare-estonian-experience-and-recommendations-nato.

217. Julio Miranda Calha, "Hybrid Warfare: NATO's New Strategic Challenge?" Draft General Report (Brussels: NATO Parliamentary Assembly (7 April 2015), http://www.nato-pa.int/documents.

218. Vladislava Vojtiskova, Vit Novotny, Hubertus Schmid-Schmidsfelden, and Kristina Potapova, *The Bear in Sheep's Clothing: Russia's Government Funded Organizations in the EU* (Brussels: Wilfried Martens Centre for European Studies, 2016), 20, https://www.martenscentre.eu/sites/default/files/publication-files/russia-gongos_0.pdf.

219. This term was coined by Colonel Charles J. Dunlap Jr., in 2001. See Charles Dunlap, "Law and Military Interventions: Preserving Humanitarian Values in 21st Conflicts," *Humanitarian Challenges in Military Intervention Conference* (November 2001).

220. Charles J. Dunlap, Jr., "Lawfare: A Decisive Element of 21st Century Conflicts?" *Joint Forces Quarterly* 54 (July 2009), 34-39, http://scholarship.law.duke.edu/cgi/viewcontent.cgi?article=6034&context=faculty_scholarship.

221. Victoria Barber, Andrew Koch, Kaitlyn Neuberger, "Russian Hybrid Warfare" (capstone project, The Fletcher School of Law and Diplomacy at Tufts University, June 2017), 28-29, 40-41.

222. Victoria Nuland, Assistant Secretary, Bureau of European and Eurasian Affairs, Statement Before the Senate Committee on Foreign Relations, *Ukraine: Countering Russian Intervention and Supporting Democratic State*, 113th Cong., 2d sess., 6 May 2014.

223. John Kerry, Secretary of State, Opening Statement Before the Senate Committee on Foreign Relations, *National Security and Foreign Policy Priorities in the FY 2015 International Affairs Budget*, 113th Cong., 2d sess., 8 April 2014.

224. See Valery Gerasimov, "The Value of Science in Prediction," *Military-Industrial Kurier* (27 February 2013); cited in "The 'Gerasimov Doctrine' and Russian non-Linear War," http://inmoscowsshadows.wordpress.com/2014/07/06/the-gerasimov-doctrine-and-russian-non-linear-war/.

225. Andreas Schmidt, "The Estonian Cyberattacks" in *The Fierce Domain—Conflicts in Cyberspace 1986-2012*, ed. Jason Healey (Washington, D.C.: Atlantic Council, 2013), https://www.researchgate.net/publication/264418820_The_Estonian_Cyberattacks.

226. Emily Tamkin, "Ten Years After the Landmark Attack on Estonia, Is the World Better Prepared for Cyber Threats?" Foreign Policy, 27 April 2017, accessed 17 November 2018.

227. Mohan B. Gazula, "Cyber Warfare Conflict Analysis and Case Studies" (partial fulfillment of the requirements for master of science degree in engineering and management, Cybersecurity Interdisciplinary Systems Laboratory Sloan School of Management, Massachusetts Institute of Technology, May 2017).

228. Andreas Schmidt, "The Estonian Cyberattacks," https://www.researchgate.net/publication/264418820_The_Estonian_Cyberattacks.

229. Max Gordon, "Lessons From the Front: A Case Study of Russian Cyber Warfare" (submitted in partial fulfillment of master's degree requirements, Air Command and Staff College, Air University, Maxwell AFB, Alabama, December 2015), http://www.dtic.mil/dtic/tr/fulltext/u2/1040762.pdf.

230. Gordon, "Lessons From the Front."

231. Robert A. Miller and Daniel T. Kuehl, "Cyberspace and the 'First Battle' in 21st-Century War" *Defense Horizons*, no. 68, September 2009, National Defense University, https://ndupress.ndu.edu/Media/News/Article/1006254/cyberspace-and-the-first-battle-in-21st-century-war/.

232. Maksym Beznosiuk, "Russia's Military Reform: Adapting to the Realities of Modern Warfare," *New Eastern Europe* (October 2016).

233. Beznosiuk, "Russia's Military Reform."

234. Roger McDermott, "Russia Activates New Defense Management Center," *Eurasia Daily Monitor* vol. 11, Issue 196 (November 2014).

235. Margarete Klein, "Russia's Military: On the Rise?" *Transatlantic Academy*, 2015-16 Paper Series, No. 2 (March 2016): 17.

236. Klein, "Russia's Military: On the Rise?" 2.

237. The remainder of this appendix borrows heavily from Andras Racz, *Russia's Hybrid War in Ukraine, Breaking the Enemy's Ability to Resist* (Helsinki: The Finnish Institute of International Affairs, FIIA Report 43, n.d.).

238. Robert Johnson, "Hybrid War and its Countermeasures: A Critique of the Literature," *Small Wars and Insurgencies* vol. 29, no. 1 (2018): 141-163.

239. Valery Gerasimov, "The Value of Science is in the Foresight: New Challenges Demand Rethinking the Forms and Methods of Carrying out Combat Operations," translated by Robert Coalson, *Military Review* (January-February 2016).

240. Racz, *Russia's Hybrid War in Ukraine*, 43.

241. Sergey G. Chekinov and Sergey A. Bogdanov, "The Nature and Content of a New-Generation War," *Military Thought* (October-December 2013): 12–23, http://www.eastviewpress.com/Files/MT_FROM%20THE%20CURRENT%20ISSUE_No.4_2013.pdf.

242. Racz, *Russia's Hybrid War in Ukraine*, 37-39.

243. Maria Snegovaya, *Putin's Information Warfare in Ukraine: Soviet Origins of Russia's Hybrid Warfare* (Washington D.C.: Institute for the Study of War, September 2015), 7.

244. Barber, "Russian Hybrid Warfare."

245. Zane M. Galvach, Thomas B. Everett, Matthew J. Mesko, Jeffrey V. Dickey, and Anton V. Soltis, "Russian Political Warfare: Origin, Evolution, and Application" (thesis, Naval Postgraduate School, 2015).

246. The following phases and steps are summarized from Racz, *Russia's Hybrid War in Ukraine*, 57-70.

247. Racz, *Russia's Hybrid War in Ukraine*, 57-70.

248. Racz, *Russia's Hybrid War in Ukraine*, 58-60.

249. Racz, *Russia's Hybrid War in Ukraine*, 60-64.

250. Racz, *Russia's Hybrid War in Ukraine*, 67-70.

251. Racz, *Russia's Hybrid War in Ukraine*, 64-67

252. Johnson, "Hybrid Warfare and its Countermeasures," 158-159.

253. This case study relies heavily on Robert Gildea, *Fighters in the Shadows, A New History of the French Resistance* (Cambridge: The Belknap Press, 2015).

254. Julian Jackson, *France: The Dark Years, 1940-1944* (New York: Oxford University Press, 2001), 406.

255. Benjamin F. Jones, *Eisenhower's Guerrillas, The Jedburghs, the Maquis, and the Liberation of France* (New York: Oxford University Press, 2016), 60.

256. Robert Gildea, *Fighters in the Shadows, A New History of the French Resistance* (Cambridge: The Belknap Press, 2015), 182.

257. Julian Jackson, *France: The Dark Years*, 97-111.

258. Paul Reynaud was a French politician and lawyer prominent in the interwar period. Reynaud was prime minister of France in June 1940, he persistently refuse to support an armistice with Germany and resigned on 16 June.

259. Gildea, *Fighters in the Shadows*, 23.

260. Gordon Wright, "Reflections on the French Resistance (1940-1944)," *Political Science Quarterly*, vol. 77, no. 3 (September 1962): 336-349. http://www.jstor.org/stable/2146309

261. Francois Kersaudy, *Churchill and De Gaulle* (London: Collins, 1981), 83. See also Gildea, *Fighters in the Shadows*, 25.

262. Gildea, *Fighters in the Shadows*, 26-27.

263. Gildea, *Fighters in the Shadows*, 110.

264. Gildea, *Fighters in the Shadows*, 32-34.

265. Gildea, *Fighters in the Shadows*, 111.

266. Gildea, *Fighters in the Shadows*, 242-243.

267. Gildea, *Fighters in the Shadows*, 241.

268. Gildea, *Fighters in the Shadows*, 277.

269. Gildea, *Fighters in the Shadows*, 257.

270. Gildea, *Fighters in the Shadows*, 182.

271. Gildea, *Fighters in the Shadows*, 37, 59.

272. Gildea, *Fighters in the Shadows*, 43-45.

273. Gildea, *Fighters in the Shadows*, 66.

274. Gildea, *Fighters in the Shadows*, 62-63.

275. Gildea, *Fighters in the Shadows*, 65.

276. Gildea, *Fighters in the Shadows*, 83-84.

277. Gildea, *Fighters in the Shadows*, 96.

278. Gildea, *Fighters in the Shadows*, 82.

279. Wright, "Reflections on the French Resistance (1940-1944)," 337.

280. Gildea, *Fighters in the Shadows*, 73, 119, 126.

281. Gildea, *Fighters in the Shadows*, 294.

282. Gildea, *Fighters in the Shadows*, 73, 98, 158.

283. Gildea, *Fighters in the Shadows*, 74.

284. Gildea, *Fighters in the Shadows*, 105, 176.

285. Gildea, *Fighters in the Shadows*, 88-91.

286. Gildea, *Fighters in the Shadows*, 67-69, 155.

287. Gildea, *Fighters in the Shadows*, 81.

288. Gildea, *Fighters in the Shadows*, 70.

289. Robert B. Asprey, *War in the Shadows, The Guerrilla in History* (New York: William Morrow and Co., 1994), 316.

290. Gildea, *Fighters in the Shadows*, 67.

291. Gildea, *Fighters in the Shadows*, 265.

292. Jones, *Eisenhower's Guerrillas*, 67.

293. The term "*maquis*" is a reference to a type of Mediterranean vegetation, roughly translated as thicket, scrub or bush, and to a Corsican expression "prendre le *maquis*" signifying to take refuge in inhospitable terrain covered by low lying scrubs to avoid the law or to fight a vendetta against a foe. The term is used to describe for both the terrain and the resistance group. H. R. Kedward. "Refusal and Revolt, Spring 1943," in *In Search of the Maquis: Rural Resistance in Southern France* (New York: Oxford University Press, 1993), 30.

294. Gildea, *Fighters in the Shadows*, 265, 327.

295. Jones, *Eisenhower's Guerrillas*, 93.

296. Gildea, *Fighters in the Shadows*, 298.

297. Gildea, *Fighters in the Shadows*, 262-264.

298. Gildea, *Fighters in the Shadows*, 171-172.

299. Asprey, *War in the Shadows*, 311-313, 316.

300. Gildea, *Fighters in the Shadows*, 161.

301. Gildea, *Fighters in the Shadows*, 156.

302. Gildea, *Fighters in the Shadows*, 167-168.

303. Gildea, *Fighters in the Shadows*, 164.

304. Gildea, *Fighters in the Shadows*, 275.

305. Gildea, *Fighters in the Shadows*, 277, 285.

306. This republican slogan was replaced by the Vichy government, with "Work, Family, and Fatherland." See "France Since 1940, Wartime France," Encyclopedia Britannica, accessed 10 January 2018, https://www.britannica.com/place/France/France-since-1940#ref465491.

307. Gildea, *Fighters in the Shadows*, 112-119.

308. Gildea, *Fighters in the Shadows*, 286-291.

309. Gildea, *Fighters in the Shadows*, 302.

310. Gildea, *Fighters in the Shadows*, 312.

311. Gildea, *Fighters in the Shadows*, 303-304.

312. Gildea, *Fighters in the Shadows*, 307-311.

313. Gildea, *Fighters in the Shadows*, 141.

314. Gildea, *Fighters in the Shadows*, 63-64.

315. Gildea, *Fighters in the Shadows*, 142. See also Jon Zimmerman, "The 'V for Victory' Campaign," DefenseMediaNetwork, 24 July 2011, https://www.defensemedianetwork.com/stories/the-v-for-victory-campaign/.

316. Gildea, *Fighters in the Shadows*, 177.

317. Gildea, *Fighters in the Shadows*, 3.

318. Jones, *Eisenhower's Guerrillas*, 2.

319. Jones, *Eisenhower's Guerrillas*, 2.

320. Jones, *Eisenhower's Guerrillas*, 2.

321. Jones, *Eisenhower's Guerrillas*, 7, 31. This was despite an internally estimated 72 hours required to organize an operation by a team.

322. Jones, *Eisenhower's Guerrillas*, 7, 271, 280. Of 265 team members deployed, 13 were KIA, and 9 out of 10 returned.

323. Gildea, *Fighters in the Shadows*, 314.

324. Gildea, *Fighters in the Shadows*, 341.

325. Gildea, *Fighters in the Shadows*, 120.

326. Gildea, *Fighters in the Shadows*, 113.

327. Gildea, *Fighters in the Shadows*, 324.

328. Gildea, *Fighters in the Shadows*, 316-317.

329. Gildea, *Fighters in the Shadows*, 343-345.

330. Gildea, *Fighters in the Shadows*, 351.

331. Asprey, *War in the Shadows*, 318-319.

332. Gildea, *Fighters in the Shadows*, 366-368.

333. Gildea, *Fighters in the Shadows*, 375-377.

334. Gildea, *Fighters in the Shadows*, 347.

335. Gildea, *Fighters in the Shadows*, 379-383.

336. Gildea, *Fighters in the Shadows*, 395.

337. Gildea, *Fighters in the Shadows*, 420.

338. Gildea, *Fighters in the Shadows*, 3, 445.

339. Gildea, *Fighters in the Shadows*, 421.

340. Gildea, *Fighters in the Shadows*, 425-427.

341. "France Since 1940, Wartime France," Encyclopedia Britannica, accessed 11 January 2018, https://www.britannica. com/place/France/France-since-1940#ref465491.

342. Gildea, *Fighters in the Shadows*, 447-448.

343. Encyclopedia Britannica, "France Since 1940, Wartime France."

344. John E. Sawyer, "The Reestablishment of the Republic in France: The De Gaulle Era, 1944-1945," *Political Science Quarterly*, vol. 62, no. 3 (September 1947): 354-367, http://www.jstor.org/stable/2144294.

345. This case study relies heavily on Marek Jan Chodakiewicz, *Between Nazis and Soviets, Occupation Politics in Poland, 1939-1947* (Lanham: Lexington Books, 2004).

346. Quoted from Colonel Harold B. Perkins of the British SOE, Jan Nowak, *Courier From Warsaw* (Detroit: Wayne State University Press, 1982), 236.

347. Chester M. Nowak, "The Polish Resistance Movement in the Second World War," *Bridgewater Review* vol. 4, Issue 1 (1986): 4-7, http://vc.bridgew.edu/br_rev/vol4/iss1/6.

348. Marek Jan Chodakiewicz, *Between Nazis and Soviets, Occupation Politics in Poland, 1939-1947* (Lanham: Lexington Books, 2004), 106-111.

349. Nowak, "The Polish Resistance," 4-5.

350. Chodakiewicz, *Between Nazis and Soviets*, 40-41.

351. Chodakiewicz, *Between Nazis and Soviets*, 44-46, 50, 54.

352. Chodakiewicz, *Between Nazis and Soviets*, 47-48. See also note 3 on page 56.

353. Chodakiewicz, *Between Nazis and Soviets*, 42-43.

354. Robert B. Asprey, *War in the Shadows, The Guerrilla in History* (New York: William Morrow and Co., 1994), 305-306.

355. Eduards Bruno Deksnis, "Latvian Exile Government Proposals," *Journal of the University of Latvia, Law, Riga*, no. 9 (2016): 79.

356. Deksnis, "Latvian Exile Government Proposals," 79-80.

357. Deksnis, "Latvian Exile Government Proposals," 80.

358. "The Polish Resistance," C.N. Truman, *The History Learning Site*, accessed 12 January 2018, historylearningsite. co.uk.

359. Chodakiewicz, *Between Nazis and Soviets*, 186.

360. Though instances of heroic rescue of Jews by Polish Christians did occur, the horrible plight of the Jews was not within the everyday experience of most people and their intended fate at the hands of the Nazis was only gradually realized by most of the population. Ultimately, Nazi terror, with horrific penalties for aiding Jews determined Polish-Jewish relations after 1941; Chodakiewicz, *Between Nazis and Soviets*, 68, 114, 119, 144-157.

361. The Order Police (Ordungspolizei–Orpo) was the constabulary force, while the Security Police and Security Service (Sicherheitspolizei and Sicherheitsdienst–Sipo and SD) were the investigative branch. The Orpo had the urban and rural branches: the Guard Police (Schutzpolizei–Schupo) and the gendarmerie respectively. The Sipo and SD were composed of the Secret State Police (Geheimestaatspolizei–Gestapo), the Criminal Police (Kriminalpolizei–Kripo), and the Security Office (Sicherheitsdienst–SD), Chodakiewicz, *Between Nazis and Soviets*, note 17, 92.

362. Chodakiewicz, *Between Nazis and Soviets*, 70-71.

363. Chodakiewicz, *Between Nazis and Soviets*, 82.

364. Marek Jan Chodakiewicz, *Intermarium, The Land Between the Black and Baltic Seas* (New Brunswick NJ: Transaction Publishers, 2016), 137.

365. Chodakiewicz, *Between Nazis and Soviets*, 71-73.

366. Chodakiewicz, *Between Nazis and Soviets*, 76-82.

367. Chodakiewicz, *Between Nazis and Soviets*, 86.

368. Chodakiewicz, *Between Nazis and Soviets*, 186.

369. Nowak, "The Polish Resistance," 4.

370. Nowak, "The Polish Resistance," 4.

371. Chodakiewicz, *Between Nazis and Soviets*, note 86, 102.

372. Chodakiewicz, *Between Nazis and Soviets*, 82.

373. Chodakiewicz, *Between Nazis and Soviets*, 105.

374. Chodakiewicz, *Between Nazis and Soviets*, 120-122.

375. Chodakiewicz, *Between Nazis and Soviets*, 106, 189.

376. Nowak, "The Polish Resistance," 5.

377. Chodakiewicz, *Between Nazis and Soviets*, 337.

378. Nowak, "The Polish Resistance," 5.

379. Chodakiewicz, *Between Nazis and Soviets*, 186-188.

380. Trueman, "The Polish Resistance." The AK was founded in July 1942 as the only legal underground armed force. Only by 1944 was the AK generally recognized by the Poles and all other non-communist resistance organizations as the official armed force of the PUS, Chodakiewicz, *Between Nazis and Soviets*, note 17, 217.

381. Nowak, "The Polish Resistance," 5.

382. Chodakiewicz, *Between Nazis and Soviets*, 182.

383. Chodakiewicz, *Between Nazis and Soviets*, 195-197.

384. Chodakiewicz, *Between Nazis and Soviets*, 138-144.

385. Named for Józef Piłsudski, a Polish revolutionary, statesman, and the first chief of state (1918–22) of the newly independent Poland established in November 1918. Professor Marek Jan Chodakiewicz has coined the term "independentists" to refer to these groups, since independence was the goal of each, excluding the communists; Chodakiewicz, *Between Nazis and Soviets*, 19.

386. Nowak, "The Polish Resistance," 6.

387. Chodakiewicz, *Between Nazis and Soviets*, 181-194.

388. Nowak, "The Polish Resistance," 6.

389. Nowak, "The Polish Resistance," 6. See also Asprey, *War in the Shadows*, 306-307.

390. Nowak, "The Polish Resistance," 7.

391. Michael Sontheimer, "When We Finish, Nobody is Left Alive," *Spiegel Online*, 27 May 2011, http://www.spiegel.de/international/europe/germany-s-wwii-occupation-of-poland-when-we-finish-nobody-is-left-alive-a-759095-3.html.

392. Chodakiewicz, *Intermarium*, 146-147.

393. Anita Jean Prazmowska , "Anticipation of Civil War: The Polish Government-in-Exile and the Threat Posed by the Communist Movement During the Second World War," *Journal of Contemporary History*, vol. 48, no. 4 (October 2013): 717-741, http://www.jstor.org/stable/24671829.

394. Chodakiewicz, *Between Nazis and Soviets*, 17.

395. David Joel Steinberg, *Philippine Collaboration in World War II* (Ann Arbor: University of Michigan Press, 1967), 57.

396. Sam McGowan, "Guerrilla War on Luzon During WWII," *Warfare History Network*, 12 August 2015, http://warfarehistorynetwork.com/daily/wwii/guerrilla-war-on-luzon-during-world-war-ii/.

397. Larry S. Schmidt, "American Involvement in the Filipino Resistance Movement on Mindanao During the Japanese Occupation, 1942-1945" (thesis, United States Army Command and General Staff College, 1982), 23-24.

398. Schmidt, "American Involvement," 32.

399. Schmidt, "American Involvement," 45-47.

400. Schmidt, "American Involvement," 47-50.

401. McGowan, "Guerrilla War on Luzon During WWII."

402. McGowan, "Guerrilla War on Luzon During WWII."

403. McGowan, "Guerrilla War on Luzon During WWII."

404. Asprey, *War in the Shadows*, 384.

405. Fertig was a mining engineer and Army Reserve officer. Asprey, *War in the Shadows*, 381-382. See also Ronald H. Spector, *Eagle Against the Sun, the American War with Japan* (Norwalk: Easton Press, 1985), 467, and John Keats, *They Fought Alone* (Philadelphia: Lippincott, 1963), and Russell W. Volkmann, *We Remained* (New York: Norton, 1954).

406. Li Yuk-wai, "The Chinese Resistance Movement in the Philippines During the Japanese Occupation," *Journal of Southeast Asian Studies* vol. 23 issue 2 (September 1992): 308.

407. McGowan, "Guerrilla War on Luzon During WWII."

408. Schmidt, "American Involvement," 206-207.

409. Schmidt, "American Involvement," 34.

410. Spector, *Eagle Against the Sun*, 468.

411. Schmidt, "American Involvement," 65.

412. Estimates are not precise, but somewhere between 100 and 200 American military personnel avoided capture and became part of resistance groups either as leaders or were cared for due to illness or wounds. Schmidt, "American Involvement," 73.

413. Schmidt, "American Involvement," 70-71.

414. Asprey, *War in the Shadows*, 379.

415. Schmidt, "American Involvement," 77.

416. Spector, *Eagle Against the Sun*, 467.

417. Schmidt, "American Involvement," 108-109.

418. Asprey, *War in the Shadows*, 383.

419. McGowan, "Guerrilla War on Luzon During WWII."

420. Schmidt, "American Involvement," 2. These were the numbers recognized and approved after the war to confer veteran's benefits and other qualifications. In the southern islands, about 1 percent of the population was recognized as former guerrillas.

421. Schmidt, "American Involvement," 22-23.

422. McGowan, "Guerrilla War on Luzon During WWII."

423. Edward M Kuder, "The Moros in the Philippines." *The Far Eastern Quarterly*, vol. 4, no. 2, 1945, *JSTOR*, 119–126, www.jstor.org/stable/2048961.

424. James S. Corum, "A View from Northeastern Europe: The Baltic States and the Russian Regime," *South Central Review*, vol. 35, no. 1 (spring 2018): 129-130.

425. Corum, "A View from Northeastern Europe: The Baltic States and the Russian Regime," 130.

426. Corum, "A View from Northeastern Europe: The Baltic States and the Russian Regime," 131.

427. Roger D. Petersen, *Resistance and Rebellion, Lessons from Eastern Europe* (Cambridge: Cambridge University Press, 2001) 90.

428. Deksnis, "Latvian Exile Government Proposals," 79.

429. Deksnis, "Latvian Exile Government Proposals," 79, 82.

430. Deksnis, "Latvian Exile Government Proposals," 83.

431. Deksnis, "Latvian Exile Government Proposals," 79.

432. Deksnis, "Latvian Exile Government Proposals," 81, 85.

433. Uldis Neiburgs, "Latvia during World War II," accessed 4 April 2018, http://okupacijasmuzejs.lv/en/history/nazi-occupation/latvia-during-world-war-ii.

434. Mart Laar, *War in the Woods: Estonia's Struggle for Survival, 1944-1956* (Washington D.C.: Compass Press, 1992), xiii-ix.

435. Daniel J. Kaszeta, "Lithuanian Resistance to Foreign Occupation 1940-1952," *Lituanus, Lithuanian Quarterly Journal of Arts and Sciences* vol. 34, no. 3 (fall 1988).

436. Laar, *War in the Woods*, xiv-xv.

437. Kaszeta, "Lithuanian Resistance to Foreign Occupation 1940-1952."

438. Uldis Neiburgs, "Latvia During World War II," accessed 4 April 2018, http://okupacijasmuzejs.lv/en/history/nazi-occupation/latvia-during-world-war-ii.

439. Olavi Punga, "Estonia's Forest Brothers in 1941: Goals, Capabilities, and Outcomes," *Combating Terrorism Exchange* vol. 3, no. 3 (August 2013), https://globalecco.org/estonias-forest-brothers-in-1941-goals-capabilities-and-outcomes.

440. Punga, "Estonia's Forest Brothers in 1941: Goals, Capabilities, and Outcomes."

441. Punga, "Estonia's Forest Brothers in 1941: Goals, Capabilities, and Outcomes."

442. Uldis Neiburgs, "Latvia During World War II," accessed 4 April 2018, http://okupacijasmuzejs.lv/en/history/nazi-occupation/latvia-during-world-war-ii.

443. Thomas Remeikis, *Opposition to Soviet Rule in Lithuania, 1945-1980* (Chicago: Institute of Lithuanian Studies Press, 1980), 43-44.

444. Kaszeta, "Lithuanian Resistance to Foreign Occupation 1940-1952."

445. Vylius M. Leskys, "'Forest Brothers,' 1945: The culmination of the Lithuanian Partisan movement" (thesis, United States Marine Corps Command and Staff College, 2009), 9-11.

446. Joseph Pajaujis-Javis, *Soviet Genocide in Lithuania* (New York: Maryland Books Inc., 1980), 89.

447. Laar, *War in the Woods*, 41.

448. Laar, *War in the Woods*, 42-43.

449. Uldis Neiburgs, "Latvia during World War II," accessed 4 April 2018, http://okupacijasmuzejs.lv/en/history/nazi-occupation/latvia-during-world-war-ii.

450. Uldis Neiburgs, "Latvia during World War II," accessed 4 April 2018, http://okupacijasmuzejs.lv/en/history/nazi-occupation/latvia-during-world-war-ii.

451. Marlies Voelzke, ed., *War After War, Armed Anti-Soviet Resistance in Lithuania, 1944-1953* (Vilnius: The Museum of Genocide Victims of the Genocide and Resistance Research Centre of Lithuania, n.d.).

452. Marlies Voelzke, ed., *War After War.*

453. Marlies Voelzke, ed., *War After War.*

454. Marlies Voelzke, ed., *War After War.*

455. Marlies Voelzke, ed., *War After War.*

456. Marlies Voelzke, ed., *War After War.*

457. Pajaujis-Javis, *Soviet Genocide in Lithuania*, 94.

458. Keith Lowe, *Savage Continent, Europe in the Aftermath of World War II* (New York: St. Martin's Press, 2012), 344.

459. Kaszeta, "Lithuanian Resistance to Foreign Occupation 1940-1952."

460. Kaszeta, "Lithuanian Resistance to Foreign Occupation 1940-1952."

461. Pajaujis-Javis, *Soviet Genocide in Lithuania*, 94.

462. Lowe, *Savage Continent*, 341.

463. Martin Herem, "The Strategy and Activity of the Forest Brothers: 1947-1950," *Combatting Terrorism Exchange* vol. 3. no. 3 (August 2013), https://globalecco.org/226, accessed 23 May 2018.

464. "The post-WW II armed resistance to Soviet power in Estonia," Estonica, Encyclopedia about Estonia, accessed May 2018, http://www.estonica.org/en/The_post-WW_II_armed_resistance_to_Soviet_power_in_Estonia/.

465. Laar, *War in the Woods*, 23.

466. Many former members of the Latvian Legion—a Waffen-SS unit and part of the German formation—formed the core of the resistance guerrillas.

467. Laar, *War in the Woods*, 24.

468. Laar, *War in the Woods*, 24.

469. Laar, *War in the Woods*, 25-26.

470. Marlies Voelzke, ed., *War After War.*

471. Marlies Voelzke, ed., *War After War.*

472. Specifically, it stated; "the Council of the LLKS shall be the supreme political body of the nation during the occupation period. The governing of Lithuania shall be exercise by the Seimas (Parliament) elected through free, democratic,

general, equal elections by secret ballot." See: Dalia Kuodyte and Rokas Tracevskis, *The Unknown War; Armed Anti-Soviet Resistance in Lithuania in 1944-1953* (Vilnius: Genocide and Resistance Research Center of Lithuania, 2013), 33.

473. Marlies Voelzke, ed., *War After War.*

474. Estonica, "The Post-WWII Armed Resistance to Soviet Power in Estonia."

475. Herem, "The Strategy and Activity of the Forest Brothers: 1947-1950."

476. Herem, "The Strategy and Activity of the Forest Brothers: 1947-1950."

477. Herem, "The Strategy and Activity of the Forest Brothers: 1947-1950."

478. Estonica, "The Post-WWII Armed Resistance to Soviet Power in Estonia."

479. Estonica, "The Post-WWII Armed Resistance to Soviet Power in Estonia."

480. Estonica, "The Post-WWII Armed Resistance to Soviet Power in Estonia."

481. Herem, "The Strategy and Activity of the Forest Brothers: 1947-1950."

482. Herem, "The Strategy and Activity of the Forest Brothers: 1947-1950."

483. The Atlantic Charter was a joint statement by the United States and Great Britain, issued on 14 August 1941, at the conclusion of a meeting between Churchill and Roosevelt aboard the U.S.S. Augusta in Newfoundland, Canada, though the U.S. was officially neutral at the time and not a belligerent. The document contained eight "common principles." The reference here is to their commitment to the principle of the restoration of self-governments for all countries having been occupied and for all people to choose their form of government.

484. Herem, "The Strategy and Activity of the Forest Brothers: 1947-1950."

485. Pajaujis-Javis, *Soviet Genocide in Lithuania*, 95.

486. Lowe, *Savage Continent*, 345.

487. Lowe, *Savage Continent*, 343.

488. Herem, "The Strategy and Activity of the Forest Brothers: 1947-1950."

489. Herem, "The Strategy and Activity of the Forest Brothers: 1947-1950."

490. Herem, "The Strategy and Activity of the Forest Brothers: 1947-1950."

491. Herem, "The Strategy and Activity of the Forest Brothers: 1947-1950."

492. Herem, "The Strategy and Activity of the Forest Brothers: 1947-1950."

493. Lowe, *Savage Continent*, 354-356.

494. Lowe, *Savage Continent*, 346-347.

495. Lowe, *Savage Continent*, 347.

496. Lowe, *Savage Continent*, 348-349.

497. Lowe, *Savage Continent*, 349.

498. Lowe, *Savage Continent*, 349.

499. Lowe, *Savage Continent*, 350.

500. Estonica, "The Post-WWII Armed Resistance to Soviet Power in Estonia."

501. Laar, *War in the Woods*, 17.

502. Herem, "The Strategy and Activity of the Forest Brothers: 1947-1950."

503. Lowe, *Savage Continent*, 350-354.

504. August Sabbe, the last known Estonian forest brother, was killed while being arrested in Võrumaa in 1978. Estonica, "The post-WW II armed resistance to Soviet power in Estonia."

505. Estonica, "The post-WW II armed resistance to Soviet power in Estonia."

506. Lowe, *Savage Continent*, 356-357.

507. The following is condensed from: Charles Cogan, "'Stay-behind' in France: Much Ado about Nothing?" Journal of Strategic Studies, 30:6, (2007): 937-954, DOI: 10.1080/01402390701676493. Though the original title mentions France, the article lays a solid foundation for understanding early U.S. authorities under which activities were conducted, while the article possesses very little information on the actual establishment of stay-behind efforts in post-war France.

508. Jonathan Kwitny, "The CIA's Secret Armies in Europe," *The Nation*, 6 April 1992, 445.

509. Ernest May, ed., *American Cold War Strategy: Interpreting NSC 68* (Boston: Bedford Books of St. Martin's Press, 1993), 72.

510. May, *American Cold War Strategy: Interpreting NSC 68*, 79.

511. Michael Warner, ed., *CIA Cold War Records: The CIA under Harry Truman* doc. 30 (Washington D.C.: CIA History Staff, Center for the Study of Intelligence, 1993), 134.

512. Warner, *CIA Cold War Records*, doc. 43, 134.

513. Warner, *CIA Cold War Records*, doc. 43, 214.

514. Warner, *CIA Cold War Records*, doc. 43,215.

515. Warner, *CIA Cold War Records*, doc. 47, 241–42.

516. David E. Murphy, Sergei A. Kondrashev and George Bailey, *Battleground Berlin: CIA vs. KGB in the Cold War* (New Haven: Yale University Press 1997), 123. Reference is to Warner, *CIA Cold War Records*, doc. 73.

517. Jean Guisnel, "Chevènement balance le glaive," *Libération*, 13 November 1990, 15.

518. Interview of Franklin Lindsay (CIA) by Charles Cogan, February 2005. The activity he described would have taken place in the period 1949–1952.

519. Interview of Franklin Lindsay (CIA) by Charles Cogan, February 2005. The activity he described would have taken place in the period 1949–1952.

520. Condensed from Kevin D. Stringer, "Building a Stay-Behind Resistance Organization; The Case of Cold War Switzerland Against the Soviet Union," *Joint Forces Quarterly* 85 (2nd Quarter 2017): 109-114. That article had several additional significant points added here based on the later research of Mr. Georges T. Egli.

521. Marc Tribelhorn, "Terror-Rezepte aus der Schweiz," *Neue Zürcher Zeitung*, 26 July 2013, 12.

522. *"Bundesrates an die Bundesversammlungbetreffend die Organisation des Heeres"* (Truppenordnung) (Bern: Bundesrat, 30 June 1960), 329.

523. Tribelhorn, "Terror-Rezepte aus der Schweiz," 12.

524. *"Bericht des Bundesrates ueber die Sicherheitspolitikder Schweiz. Konzeption der Gesamtverteidigung"* (Berne, 27 June 1973), 16.

525. *"Bericht des Bundesrates ueber die Sicherheitspolitikder Schweiz. Konzeption der Gesamtverteidigung"* (Berne, 27 June 1973), 16, 38.

526. *"Bundesrates an die Bundesversammlung betreffend die Organisation des Heers,"* 367–368, and Heinz Haesler, "Grundsaetzliche Ueberleguungen eines ehemaligen Generalstabschefs," in *Erinnerungen an die Armee 61*, ed. Franz Betschon and Louis Geiger (Frauenfeld, Switzerland: Verlag Huber, 2009), 96.

527. Haesler, 96.

528. World Bank data, 2016.

529. *"Kleine Heereskunde"* (Bern: EMD, 1992), 13–14.

530. Mauro Mantovani, "Der 'Volksaufstand': Vorstellungen und Vorbereitungen der Schweiz im 19. und 20. Jh," in *Military Power Revue der Schweizer Armee*, r. 1 (2012), 52–60.

531. Daniele Ganser, "The British Secret Service in Neutral Switzerland: An Unfinished Debate on NATO's Cold War Stay-Behind Armies," *Intelligence and National Security* 20, no. 4 (December 2005), 553–580.

532. Lucien Fluri, "Ehemalige Geheimarmee, P-26," *Solothurner Zeitung*, 14 July 2012.

533. *"Bericht der Parlamentarischen Untersuchungskommission zur besonderen Klaerung von Vorkommnissen von grosser Tragweite im Eidgenossischen Militaerdepartement,"* Bern, 17 November 1990, Martin Matter, P-26. *"Die Geheimarmee, die keine war"* (Baden: Verlag hier + jetzt, 2012), Tribelhorn, 12.

534. Grdovic, *A Leader's Handbook to Unconventional Warfare*.

535. Matter, P-26, 47.

536. Matter, P-26, 153–154.

537. Matter, P-26, 82.

538. Matter, P-26, 50–58.

539. Fluri.

540. Condensed from Leopoldo Nuti, "The Italian 'Stay-Behind' Network–The Origins of Operation 'Gladio,'" *Journal of Strategic Studies*, 30:6, (2007): 955-980, DOI: 10.1080/01402390701676501.

541. Nuti, "The Italian 'Stay-Behind' Network," 955-956.

542. Virgilio Ilari, *Storia militare della prima repubblica, 1943–1993* (Ancona: Casa editrice Nuove Ricerche 1994), 524–25.

543. Nicola. Tranfaglia, *Come nasce la repubblica: La Mafia, il Vaticano e il neofascismonei documenti americani e italiani, 1943–1947* (Milano: Bompiani 2004), 178–88 and 204–10.

544. Ilari, *Il generale col monocolo: Giovanni De Lorenzo, 1907–1973*, 68.

545. Nuti, "The Italian 'Stay-Behind' Network," 959-960.

546. Nuti, "The Italian 'Stay-Behind' Network," 961-962.

547. Nuti, "The Italian 'Stay-Behind' Network," 964.

548. Nuti, "The Italian 'Stay-Behind' Network," 965-966.

549. Nuti, "The Italian 'Stay-Behind' Network," 966-967.

550. Nuti, "The Italian 'Stay-Behind' Network," 967-969.

551. Nuti, "The Italian 'Stay-Behind' Network," 969-970.

552. Nuti, "The Italian 'Stay-Behind' Network," 970.

553. Ten of the original 139 caches were not retrieved in 1973 but in 1990, as they had been hidden in places where their retrieval would require complex demolition work. Nuti, "The Italian 'Stay-Behind' Network," 970.

554. Nuti, "The Italian 'Stay-Behind' Network," 974-975.

555. Nuti, "The Italian 'Stay-Behind' Network," 977-978.

556. Condensed from Olav Riste, "With an Eye to History: The Origins and Development of 'Stay-Behind' in Norway," *Journal of Strategic Studies* 30:6 (2007): 997-1024, DOI: 10.1080/01402390701676527.

557. *Rocambole* was the name of the hero of the French 19th century author Ponson du Terrail, whose long series of books about *Rocambole*—each volume ending with the words 'Men Rocambole var ikke død' (but Rocambole was not dead)—was popular reading in Norway, from Riste; "With an Eye to History," 1004.

558. Riste, "With an Eye to History," 1000. Quoted from Defence Minister Jens Christian Hauge in the post-war plan for the reconstruction of the Norwegian armed forces, which was presented and published by parliament as a White Paper, (1945–6).

559. The doctrine was based on a speech by President Harry S. Truman to a joint session of the U.S. Congress on 12 March 1947. As background, in February of 1947 Great Britain announced that it could no longer afford to economically and militarily aid the nations of Greece and Turkey, which it had been doing since the end of the war. Greece was suffering an internal civil war and Turkey had internal communist problems as well as perceived threats by the Soviet Union regarding the Dardanelles Straights. In the speech, Truman declared that "it must be the policy of the United States to support free peoples who are resisting attempted subjugation by armed minorities or by outside pressures." He requested and received a large assistance package for Greece and Turkey. He also identified the USSR as the place from which all communist activity emanated and which intended to invade or internally subvert other nations. This speech is used as a historical marker for the beginning of the Cold War.

560. Riste, "With an Eye to History," 998-999.

561. Riste, "With an Eye to History," 999.

562. A group within a country at war who are sympathetic to or working with its enemies. The term dates from the Spanish Civil War, when General Mola, leading four columns of troops towards Madrid, declared that he had a fifth column inside the city. Cited from the Oxford Dictionary.

563. Riste, "With an Eye to History," 1001-1002.

564. Riste, "With an Eye to History," 1002-1003.

565. Riste, "With an Eye to History," 1003-1004; In conversation with Olav Riste, Defence Minister Jens Christian Hauge stated that this initiative had not been cleared with his Cabinet colleagues

566. Riste, "With an Eye to History," 1004-1005.

567. Riste, "With an Eye to History," 1006.

568. Riste, "With an Eye to History," 1006.

569. Riste, "With an Eye to History,"1007-1008.

570. Riste, "With an Eye to History," 1008.

571. Riste, "With an Eye to History," 1008-1009.

572. Riste, "With an Eye to History," 1009-1010.

573. Riste, "With an Eye to History," 1020.

574. Riste, "With an Eye to History," 1010-1011.

575. Riste, "With an Eye to History," 1013.

576. Riste, "With an Eye to History," 1013.

577. Riste, "With an Eye to History," 1014.

578. Riste, "With an Eye to History," 1015-1018.

579. Riste, "With an Eye to History," 1020-1024.

580. Uwe Hartmann, *The Evolution of the Hybrid Threat, and Resilience as a Countermeasure* (Rome: NATO Defense College, September 2017), http://www.ndc.nato.int/about/search.php.

581. Hoffman, *Conflict in the 21st Century.*

582. Charles K. Bartles, "Getting Gerasimov Right," *Military Review* (January/February 2016): 30-38.

583. Sir Rupert Smith, *The Utility of Force* (New York: Random House, 2007).

584. Stephen R. Covington, *The Culture of Strategic Thought Behind Russia's Modern Approaches to Warfare*, Belfer Center for Science and International Affairs, Harvard Kennedy School (Cambridge: Harvard Kennedy School, 2016), 13-20; Bartles, "Getting Gerasimov Right," 31.

585. Jamie Shea, "Resilience: A Core Element of Collective Defence," *NATO Review Magazine*, 4, http://www.nato.int/docu/Review/2016/Also-in-2016/nato-defence-cyber-resilience/EN/index.htm.

586. Shea, "Resilience: A Core Element of Collective Defence."

587. Carl von Clausewitz,*On War*, edited and translated by Michael Howard and Peter Paret (Princeton: Princeton University Press, 1984), 89.

588. Sven Biscop, "Hybrid Hysteria," *Security Policy Brief*, no. 64 (June 2015): 3-4, http://aei.pitt.edu/64790/.

589. These samples are based on products provided by the Navanti Group, LLC., Arlington, Virginia.

590. Marek Jan Chodakiewicz, *Between Nazis and Soviets, Occupation Politics in Poland, 1939-1947* (Lanham: Lexington Books, 2004), 1. Though Professor Chodakiewicz' study is a micro-study of one county in Poland, he offers this spectrum as universally applicable occupier driven reactions, because the same Nazi policies were similarly applied throughout Poland, see page 338.

591. Chodakiewicz, *Between Nazis and Soviets*, 319.

592. Chodakiewicz, *Between Nazis and Soviets*, 335.

593. Chodakiewicz, *Between Nazis and Soviets*, 334; See also Kalyvas, *The Logic of Violence in Civil War*, 160.

594. Chodakiewicz, *Between Nazis and Soviets*, 322.

595. Chodakiewicz, *Between Nazis and Soviets*, 320.

596. Molnar, *Undergrounds in Insurgent, Revolutionary and Resistance Warfare*, 107.

597. Chodakiewicz, *Between Nazis and Soviets*, 334.

598. Chodakiewicz, *Between Nazis and Soviets*, 333.

599. *If Crisis or War Comes (Om krisen eller kriget kommer)* (Karlstad: Swedish Civil Contingencies Agency, May 2018), https://www.msb.se/Upload/Forebyggande/Krisberedskap/Krisberedskapsveckan/Fakta%20om%20broschyren%20 Om%20krisen%20eller%20Kriget%20kommer/om-krisen-eller-kriget-kommer---engelska.pdf.

Bibliography

U.S. Department of Defense Joint Publications

United States Department of Defense, Joint Publication 3-05. *Special Operations*. Washington D.C.: Department of Defense, 2014.

United States Department of Defense, Joint Publication 3-08, *Interorganizational Cooperation*. Washington D.C.: Department of Defense, 2017.

United States Department of Defense, Joint Publication 3-13.3, *Operations Security*. Washington D.C.: Department of Defense, 2012.

United States Department of Defense, Joint Publication 3-22. *Foreign Internal Defense*. Washington D.C.: Department of Defense, July 2010.

United States Department of Defense, Joint Publication 3-27, *Homeland Defense*. Washington D.C.: Department of Defense, 2018.

United States Department of Defense, Joint Publication 3-61, *Public Affairs*. Washington D.C.: Department of Defense, 2016.

United States Department of Defense, *DOD Dictionary of Military and Associated Terms* (As of 15 October 2016). Washington D.C.: Department of Defense, 2016.

United States Department of Defense, Joint Doctrine Note 2-13, *Commander's Communication Synchronization*. Washington D.C.: Department of Defense, 2013.

United States Department of Defense, *Law of War Manual*, Washington D.C.: Department of Defense, 2015 (as updated in December 2016)

U.S. Special Operations Command Publications

United States Special Operations Command. *Counter-Unconventional Warfare*. White Paper. 2014.

United States Special Operations Command and United States Joint Forces Command. *Irregular Warfare Joint Operating Concept, v.2.0: Countering Irregular Threats*, 2010.

United States Special Operations Command. *SOF Support to Political Warfare*. White Paper. 2015.

United States Special Operations Command. *The Gray Zone*. White Paper. 2015.

United States Army Special Operations Command. *Unconventional Warfare Pocket Guide*, 2016.

United States Army and Service Publications

Grdovich, Mark. *A Leader's Handbook to Unconventional Warfare*. Fort Bragg: U.S. Army John F. Kennedy Special Warfare Center and School, 2009.

Lee, David, ed. *Law of Armed Conflict Deskbook*. Fifth Edition. Charlottesville: International and Operational Law Department, 2015.

Department of the Army. Army Doctrine Publication (ADP) 6-0, *Mission Command*, Washington D.C.: Headquarters, Department of the Army, 2014.

Department of the Army. FM 3-24, MCWP 3-33.5, *Insurgencies and Countering Insurgencies*, Washington D.C.: Headquarters, Department of the Army, 2014.

United States Army Institute for Military Assistance. *The Underground*, ST 31-202. Fort Bragg: U.S. Army Institute for Military Assistance, 1978.

Department of the Navy. *Operations Security*, NTTP 3-54M/ U.S. Marine Corps Publication MCWP 3-40.9, 2009.

United States Army, *Operations Security*, Regulation 530-1, 2014.

United States Air Force, *Operations Security*, Instruction 10-701, 2011.

Other United States Government Publications

Catherine Dale, Nina Serafino, and Pat Towell, *Organizing the U.S. Government for National Security: Overview of the Interagency Reform Debates*. Washington D.C.: Congressional Research Service, 2008.

United States Department of Homeland Security, *Continuity of Operations Plan Template for Federal Departments and Agencies*. Washington, D.C.: Department of Homeland Security, 2013.

United States Department of Homeland Security, *National Response Framework*. Washington D.C.: Department of Homeland Security, 2008.

Frederick M. Kaiser, *Interagency Collaborative Arrangements and Activities: Types, Rationales, Considerations*. Washington, D.C.: Congressional Research Service, 2011.

NATO Publications

Allied Joint Publication (AJP) 3.10, *Allied Joint Doctrine for Information Operations*. Brussels: NATO, 2009.

Other Publications

Adamsky, Dmitry. *Cross-Domain Coercion: The Current Russian Art of Strategy*. Paris: Institut Francais des Relations Internationales, November 2015. http://www.ifri.org/sites/default/files/atoms/files/pp54adamsky.pdf.

After Action Report, Resistance Seminar, Baltic Defence College, Tartu, Estonia, November 4-6, 2014.

After Action Report, Resistance Seminar, The General Jonas Žemaitis Military Academy of Lithuania, Vilnius, Lithuania, 30 June–2 July 2015.

After Action Report, Resistance Seminar, Riga, Latvia, 223-25 November 2015.

After Action Report, Resistance Seminar, Krakow, Poland, 26-27 April 2016.After Action Report, Resistance Seminar, Swedish Defence University, Stockholm, Sweden, July 26-27, 2016.Asprey, Robert. *War in the Shadows, The Guerilla in History*. New York: William Morrow and Co., 1994.

Arrow, Ruaridh. "Gene Sharp: Author of the Nonviolent Revolution Rulebook." *BBC News*, February 21, 2011. http://www.bbc.co.uk/news/world-middle-east-125228-48.

Barber, Victoria, Andrew Koch, and Kaitlyn Neuberger. "Russian Hybrid Warfare." Capstone Project, Tufts University, The Fletcher School of Law and Diplomacy, June 2017.

Bartles, Charles K. "Getting Gerasimov Right." *Military Review* (January/February 2016): 30-38.

Baxter, Richard R. "So-Called 'Unprivileged Belligerency': Spies, Guerillas, and Saboteurs," *British Yearbook of International Law* 28 (1951).

Bērziņš, Jānis. "Russia's New Generation Warfare in Ukraine: Implications for Latvian Defense Policy." *Policy Paper No. 02*. National Defence Academy of Latvia, Center for Security and Strategic Research. (April 2014).

Beznosiuk, Maksym. "Russia's Military Reform: Adapting to the Realities of Modern Warfare." *New Eastern Europe*. (October 2016). http://neweasterneurope.eu/2016/10/13/russia-s-military-reform-adapting-to-the-realities-of-modern-warfare/.

Biscop, Sven. "Hybrid Hysteria." *Security Policy Brief*, No. 64 (June 2015). http://aei.pitt.edu/64790/.

Bos, Nathan, ed. "Human Factors Consideration of Undergrounds in Insurgencies." In *Assessing Revolutionary and Insurgent Strategies* 2nd ed. Fort Bragg: United States Army Special Operations Command, 2013.

Bramstedt, Ernest K. *Dictatorship and Political Police: The Technique of Control by Fear.* New York: Oxford University Press, 1945.

Calha, Julio Miranda. "Hybrid Warfare: NATO's New Strategic Challenge?" Draft General Report, NATO Parliamentary Assembly, 7 April 2015. http://www.nato-pa.int/documents.

Chekinov, S.G. and S.A. Bogdanov. "The Nature and Content of a New-Generation War." *Military Thought* (October-December 2013): 12–23. http://www.eastviewpress.com/Files/MT_FROM%20THE%20CURRENT%20ISSUE_No.4_2013.pdf

Chodakiewicz, Marek Jan. *Between Nazis and Soviets; Occupation Politics in Poland, 1939-1947.* Lanham: Lexington Books, 2004.

Chodakiewicz, Marek Jan. *Intermarium; The Land Between the Black and Baltic Seas.* New Brunswick NJ: Transaction Publishers, 2016.

Clausewitz, Carl von. *On War.* Edited and translated by Michael Howard and Peter Paret. Princeton: Princeton University Press, 1984.

Cogan, Charles. "'Stay-Behind in France: Much Ado About Nothing?" *Journal of Strategic Studies*, 30:6 (2007): 937-954. DOI: 10.1080/01402390701676493.

Conley, Jerome. "Orange Revolution (Ukraine): 2004-2005." In *Casebook on Insurgency and Revolutionary Warfare, Volume II: 1962-2009*, edited by Charles Crossett. Ft Bragg: United States Army Special Operations Command 2012.

Corum, James S. "A View from Northeastern Europe: The Baltic States and the Russian Regime." *South Central Review* vol. 35, no. 1 (spring 2018): 129-130.

Covington, Stephen R. *The Culture of Strategic Thought behind Russia's Modern Approaches to Warfare.* Belfer Center for Science and International Affairs, Harvard Kennedy School. Cambridge: Harvard Kennedy School, 2016.

Crossett, Chuck and Summer Newton. "Solidarity." In *Casebook on Insurgency and Revolutionary Warfare, Volume II: 1962-2009*, edited by Charles Crossett. Fort Bragg: United States Special Operations Command, 2012.

Crossett, Chuck and Summer Newton. "The Provisional Irish Republican Army: 1969-2001." In *Casebook on Insurgency and Revolutionary warfare, Volume II: 1962-2009*, edited by Charles Crossett. Fort Bragg: United States Special Operations Command, 2012.

Dabkowska-Cichocka, Lena, Dariusz Gawin, Pawel Kowal, Jan Oldakowski, Agnieszka Panecka, Pawel Ukielski, Ewa Ziolkowska eds. *Warsaw Rising Museum Guidebook*. Warsaw: Warsaw Rising Museum, undated.

Dach, H. Von. Total Resistance: *Swiss Army Guide to Guerilla Warfare and Underground Operations*. Boulder: Paladin Press, 1992.

De Allende, Veronica and Alison Vernon, *Coordination in a JTF Formation*. Arlington: Center for Naval Analysis, 2018.

De Brun, Bairbre. *The Road to Peace in Ireland*. Berlin: Konstruktive Konfliktbearbeitung, 2008.

Deksnis, Eduards Bruno. "Latvian Exile Government Proposals." *Journal of the University of Latvia, Law, Riga* no. 9 (2016): 79.

Dinstein, Yoram. *The Conduct of Hostilities Under the Law of International Armed Conflict*. New York: Cambridge University Press, 2004.

Dinstein, Yoram. *The International Law of Belligerent Occupation*. New York: Cambridge University Press, 2009.

Dunlap, Charles J. Jr. "Law and Military Interventions: Preserving Humanitarian Values in 21st Century Conflicts." *Humanitarian Challenges in Military Intervention Conference*. (November 2001).

Dunlap, Charles J. Jr. "Lawfare: A Decisive Element of 21st Century Conflicts?" *Joint Forces Quarterly* 54 (July 2009): 34-39. http://scholarship.law.duke.edu/cgi/viewcontent.cgi?article=6034&context=faculty_scholarship.

Elkin, Mike. "Tunisia Internet Chief Gives Inside Look at Cyber Uprising." *Wired*, 28 January 2011, www.wired.com/dangerroom/2011/01/as-egypt-tightens-its-internet-grip-tunisia-seeks-to-open-up/.

Estonica, Encyclopedia about Estonia. "The post-WW II armed resistance to Soviet power in Estonia." Accessed May 2018. http://www.estonica.org/en/The_post-WW_II_armed_resistance_to_Soviet_power_in_Estonia/.

Fitzsimmons, Michael. *Governance, Identity, and Counterinsurgency: Evidence from Ramadi and Tal Afar*. Carlisle: Strategic Studies Institute and U.S. Army Way College Press, 2013.

Fletcher, George and Jen David Ohlen. *Defending Humanity*. New York: Oxford University Press, 2008.

Fluri, Lucien. "Ehemalige GeheimarmeeP-26." *Solothurner Zeitung*, 14 July 2012. "France Since 1940, Wartime France." Encyclopedia Britannica. https://www.britannica.com/place/France/France-since-1940#ref465491.

Galvach, Zane M., Thomas B Everett, Matthew J. Mesko, Jeffrey V. and Anton V Soltis. "Russian Political Warfare: Origin, Evolution, and Application." thesis, Naval Postgraduate School, 2015. https://calhoun.nps.edu/bitstream/handle/10945/45838/15Jun_Dickey_Everett_Galvach_Mesko_Soltis.pdf?sequence=1&isAllowed=y.

Ganser, Daniele. "The British Secret Service in Neutral Switzerland: An Unfinished Debate on NATO's Cold War Stay-behind Armies." *Intelligence and National Security* 20, no. 4 (December 2005): 553-580.

Gazula, Mohan B. "Cyber Warfare Conflict Analysis and Case Studies." *Cybersecurity Interdisciplinary Systems Laboratory (CISL) Sloan School of Management, Massachusetts Institute of Technology.* Submitted in partial fulfillment of master's degree requirements. May 2017.

Gerasimov, Valery. "The Value of Science in Prediction." *Military-Industrial Kurier*, 27 February 2013. http://inmoscowsshadows.wordpress.com/2014/07/06/the-gerasimov-doctrine-and-russian-non-linear-war/.

Gerasimov, Valery. "The Value of Science Is in the Foresight: New Challenges Demand Rethinking the Forms and Methods of Carrying out Combat Operations." Translated by Robert Coalson. *Military Review*, vol 96, no. 1 (January-February 2016): 23-29. https://www.armyupress.army.mil/Portals/7/military-review/Archives/English/MilitaryReview_20160228_art008.pdf.

Gervais, Bryan. "Hutu-Tutsi Genocides." In *Casebook on Insurgency and Revolutionary Warfare, Volume II: 1962-2009*, edited by Charles Crossett. Fort Bragg: United States Special Operations Command, 2012.

Gildea, Robert. *Fighters in the Shadows, A New History of the French Resistance.* Cambridge: The Belknap Press, 2015.

Gordon, Max. "Lessons From the Front: A Case Study of Russian Cyber Warfare." Submitted in partial fulfillment of Masters Degree requirements. Air Command and Staff College, Air University. Maxwell AFB, Alabama. December 2015. http://www.dtic.mil/dtic/tr/fulltext/u2/1040762.pdf

Grdovic, Mark. *A Leader's Handbook to Unconventional Warfare.* Fort Bragg: U.S. Army John F. Kennedy Special Warfare Center and School, 2009.

Gross, Feliks. *The Seizure of Political Power in a Century of Revolutions.* New York: Philosophical Library, 1958.

Guisnel, Jean. "Chevenement balance le glaive." *Liberation*, 13 November 1990.

Hahn, Erin H., ed. *Special Topics in Irregular Warfare: Understanding Resistance, Assessing Revolutionary and Insurgent Strategies.* Fort Bragg: United States Army Special Operations Command, 2018.

Hartmann, Uwe. "The Evolution of the Hybrid Threat, and Resilience as a Countermeasure." NATO Defense College, Research Division no. 139 (September 2017). http://www.ndc.nato.int/about/search.php.

Herem, Martin. "The Strategy and Activity of the Forest Brothers: 1947-1950." *Combatting Terrorism Exchange* vol. 3, no.3 (August 2013).

Hoffman, Frank G. *Conflict in the 21st Century: The Rise of Hybrid Wars.* Arlington: Potomac Institute for Policy Studies, 2007. http://www.potomacinstitute.org/images/stories/publications/potomac_hybridwar_0108.pdf.

Hoffman, Frank and James N. Mattis. "Future Wars: The Rise of Hybrid Wars." *Proceedings* vol. 131/11/1,233, (November 2005).

Hoffman, Frank G. "Hybrid Warfare and Challenges." *Joint Forces Quarterly* vol. 52, no. 1 (2009): 34-39.

If Crisis or War Comes (Om krisen eller kriget kommer) Karlstad: Swedish Civil Contingencies Agency (MSB) May 2018. https://www.msb.se/Upload/Forebyggande/Krisberedskap/Krisberedskapsveckan/Fakta%20om%20broschyren%20Om%20krisen%20eller%20Kriget%20kommer/om-krisen-eller-kriget-kommer---engelska.pdf.

International Conferences (The Hague), *Hague Convention (IV) Respecting the Laws and Customs of War on Land and Its Annex: Regulations Concerning the Laws and Customs of War on Land*, 18 October 1907.

Jackson, Julian. *France: The Dark Years, 1940-1944*. New York: Oxford University Press, 2001.

Johnson, Robert. "Hybrid War and its Countermeasures: A Critique of the Literature." *Small Wars and Insurgencies* vol. 29, no. 1 (January 2018): 141-163.

Jones, Benjamin. *Eisenhower's Guerillas, The Jedburghs, the Marguis, and the Liberation of France*. New York: Oxford University Press, 2016.

Jureidini, Paul A., Norman A La Charite, Bert H. Cooper, and William A. Lybrand. *Casebook on Insurgency and Revolutionary Warfare, Volume I: 1933-1962*. Revised Edition. Fort Bragg: United States Army Special Operations Command, 2013.

Kalyvas, Stathis N. *The Logic of Violence in Civil War*. New York: Cambridge University Press, 2006.

Karolis, Aleksa, ed. *Prepare to Survive Emergencies and War: A Cheerful Take on Serious Recommendations*. Vilnius: Ministry of National Defence, 2015.

Kaszeta, Daniel J. "Lithuanian Resistance to Foreign Occupation 1940-1952." *Lituanus Lithuanian Quarterly Journal of Arts and Sciences* vol. 43, no. 3 (fall 1988).

Keats, John. *They Fought Alone*. Philadelphia: Lippincott, 1963.

Kerry, John. Secretary of State, Opening Statement Before the Senate Committee on Foreign Relations. *National Security and Foreign Policy Priorities in the FY 2015 International Affairs Budget* 113th Cong., 2d sess., April 8, 2014.

Kersaudy, Francois. *Churchill and De Gaulle*. London: Collins, 1981.

Kilcullen, David. *Out of the Mountains*. Oxford: Oxford University Press, 2015

Klein, Margarete. "Russia's Military: On the Rise?" Transatlantic Academy, 2015-16 Paper Series, no. 2 (March 2016).

Kramer, Franklin D. and Lauren M. Speranza. *Meeting the Russian Hybrid Challenge; A Comprehensive Strategic Framework*. Washington D.C.: Atlantic Council, Brent Scowcroft Center on International Security, May 2017.

Kuder, Edward M. "The Moros in the Philippines." *The Far Eastern Quarterly*, vol. 4, no. 2, 1945, pp. 119–126. JSTOR, www.jstor.org/stable/2048961.

Kuodvte, Dalia and Rokas Tracevskis. *The Unknown War; Armed Anti-Soviet Resistance in Lithuania in 1944-1953*. Vilnius: Genocide and Resistance Research Center of Lithuania, 2013.

Kwitny, Jonathon. "The CIA's Secret Armies in Europe." *The Nation*, 6 April 1992.

Larr, Mart. *War in the Woods: Estonia's Struggle for Survival, 1944-1956*. Washington D.C.: Compass Press, 1992.

Leonhard, Robert, ed. *Assessing Revolutionary and Insurgent Strategies: Undergrounds in Insurgent, Revolutionary, and Resistance Warfare*, 2nd Edition. Fort Bragg: United States Army Special Operations Command, 2013.

Leskys, Vylius M., "Forest Brothers, 1945: The Culmination of the Lithuanian Partisan Movement." thesis, United States Marine Corps Command and Staff College, 2009.

Levinson, Charles and Matt Bradley. "Egypt's Regime on the Brink." *Wall Street Journal Online*, 29 January 2011. http://online.wsj.com/article/SB10001424052748703956604576109323492986438.html?mod=WSJ World LeadStory.

Liang, Qiao and Xiangsui, Wang. *Unrestricted Warfare*. Translated by Ian Strauss. Brattleboro: Echo Point Books and Media, 2015.

Lowe, Keith. Savage Continent, *Europe in the Aftermath of World War II*. New York: St. Martin's Press, 2012.

Maigre, Merle. "Nothing New in Hybrid Warfare: The Estonian Experience and Recommendations for NATO." The German Marshall Fund of the United States, Foreign Policy Program, Policy Brief (February 2015). http://www.gmfus.org/publications/nothing-new-hybrid-warfare-estonian-experience-and-recommendations-nato.

Mantovani, Mauro. "Der 'Volksaufstand': Vorstellungen und Vorbereitungen der Schweiz im 19. umd 20. Jh." *Power Military Revue de Schweizer Armee*, r.1 (2012): 52-60.

Matthews, Matt M. *We Were Caught Unprepared: The 2006 Hezbollah-Israeli War*. Fort Leavenworth: Combat Studies Institute Press, 2008.

May, Ernest, ed. *American Cold War Strategy: Interpreting NSC 68*. Boston: Bedford Books of St. Martin's Press, 1993.

McCulloh, Timothy and Richard Johnson. *Hybrid Warfare*. MacDill Air Force Base: Joint Special Operations University Press, 2013.

McDermott, Roger. "Russia Activates New Defense Management Center." *Eurasia Daily Monitor*, vol. 11, Issue 196 (November 2014). https://jamestown.org/program/russia-activates-new-defense-management-center/.

McGowan, Sam. "Guerilla War on Luzon During WWII." *Warfare History Network*, 12 August 2015. http://warfarehistorynetwork.com/daily/wwii/guerrilla-war-on-luzon-during-world-war-ii/.

Melzer, Nils. *Interpretive Guidance on the Notion of Direct Participation in Hostilities Under International Humanitarian Law*. Geneva: International Committee of the Red Cross, 2009.

Miller, Robert A. and Daniel T. Kuehl. "Cyberspace and the 'First Battle' in 21st -Century War." Defense Horizons, no. 68, September 2009, National Defense University. https://ndupress.ndu.edu/Media/News/Article/1006254/cyberspace-and-the-first-battle-in-21st-century-war/.

Molnar, Anthony R. *Undergrounds in Insurgent, Revolutionary and Resistance Warfare*. Washington, D.C.: The American University, 1963.

Mosquera, Andrés B. Muñoz and Sascha-Dominik Bachmann. "Understanding Lawfare in a Hybrid Warfare Context." *NATO Legal Gazette*, Issue 37, (October 2016): 20-37.

Murphy, David E., Sergei A. Kondrashev and George Bailey. *Battleground Berlin: CIA vs. KGB in the Cold War*. New Haven: Yale University Press, 1997.

Murray, Williamson and Peter R. Mansoor. *Hybrid Warfare: Fighting Complex Opponents from the Ancient World to the Present*. New York: Cambridge University Press, 2012.

Neiburgs, Uldis. "Latvia during World War II," http://okupacijasmuzejs.lv/en/history/nazi-occupation/latvia-during-world-war-ii.

Nemeth, William J. "Future Wars and Chechnya: A Case for Hybrid Warfare." MA thesis, Naval Postgraduate School, 2002. http://calhoun.nps.edu/bitstream/handle/10945/5865/02Jun_Nemeth.pdf?sequence=1.

Nix, Margen and Dru Daubon. "Kosovo Liberation Army (KLA), 1996-1999." In *Casebook on Insurgency and Revolutionary Warfare, Volume II: 1962-2009*, edited by Charles Crossett. Fort Bragg: United States Special Operations Command, 2013.

Nonviolent Civic Action, A Study Guide Series on Peace and Conflict. Washington D.C.: United States Institute of Peace, 2009.

Nowak, Chester M. "The Polish Resistance Movement in the Second World War." *Bridgewater Review* vol. 4, issue 1 (1986): 4-7. http://vc.bridgew.edu/br_rev/vol4/iss1/6.

Nowak, Jan. *Courier From Warsaw*. Detroit: Wayne State University Press, 1982.

Nuland, Victoria. Assistant Secretary, Bureau of European and Eurasian Affairs, Statement before the Senate Committee on Foreign Relation. *Ukraine: Countering Russian Intervention and Supporting Democratic State* 113th Cong., 2d sess., 6 May 2014.

Nuti, Leopoldo. "The Italian 'Stay-Behind' Network – The origins of operation 'Gladio,'" *Journal of Strategic Studies, 30:6*, (2007): 955-980, DOI: 10.1080/01402390701676501.

Pajaujis-Javis, Joseph. *Soviet Genocide in Lithuania*. New York: Maryland Books Inc., 1980.

Petersen, Roger D. *Resistance and Rebellion, Lessons from Eastern Europe*. Cambridge: Cambridge University Press, 2001.

Pictat, Jean, ed. *Commentary, Convention Relative to the Treatment of Prisoners of War*. Geneva: International Committee of the Red Cross, 1960.

Poli, Fulvio. "An Asymmetrical Symmetry: How Convention has Become Innovative Military Thought." Master of Strategic Studies thesis, Carlisle: U.S. Army War College, 2010.

Prazmowska, Anita Jean. "Anticipation of Civil War: The Polish Government-in-Exile and the Threat Posed by the Communist Movement during the Second World War." *Journal of Contemporary History* vol. 48, no. 4 (October 2013): 717-741. http://www.jstor.org/stable/24671829.

Punga, Olavi. "Estonia's Forest Brothers in 1941: Goals, Capabilities, and Outcomes." *Combating Terrorism* Exchange vol. 3, no. 3 (August 2013) https://globalecco.org/estonias-forest-brothers-in-1941-goals-capabilities-and-outcomes.

Racz, Andras. *Russia's Hybrid War in Ukraine; Breaking the Enemy's Ability to Resist*. Helsinki: The Finnish Institute of International Affairs, FIIA Report 43, 16 June 2015.

Remeikis, Thomas. *Opposition to Soviet Rule in Lithuania, 1945-1980*. Chicago: Institute of Lithuanian Studies Press, 1980.

Riste, Olav and Berit Nokleby. *Norway 1940-45: The Resistance Movement*. Otta: Arthur Vanous Co., 1970.

Robertson, Nic, Antonia Mortenen, Elizabeth Robers and Woj Treszczynski. "Lithuania Issues Manual on What to do if Russia Invades." *CNN Online*, 28 October 2016. http://edition.cnn.com/2016/10/28/europe/lithuania-war-manual/index.html.

Salonius-Pasternak, Charly. "An effective antidote: The four components that make Finland more resilient to hybrid campaigns." The Finnish Institute of International Affairs (FIIA), October 3, 2017. https://storage.googleapis.com/upi-live/2017/10/comment19_finland.pdf.

Sawyer, John E. "The Reestablishment of the Republic in France: The De Gaulle Era, 1944-1945." *Political Science Quarterly* vol. 62, no. 3 (September 1974): 354-367. http://www.jstor.org/stable/2144294.

Schmidt, Andreas. "The Estonian Cyberattacks." (2013). https://www.researchgate.net/publication/2644 18820_The_Estonian_Cyberattacks.

Schmidt, Larry S. "American Involvement in the Filipino Resistance Movement on Mindanao During the Japanese Occupation, 1942-1945." thesis, United States Army Command and General Staff College, 1982.

Sharp, Gene. *Sharp's Dictionary of Power and Struggle: Language of Civil Resistance in Conflicts.* New York: Oxford University Press, 2011.

Shea, Jamie. "Resilience: A Core Element of Collective Defence." *NATO Review Magazine*, 30 March 2016. http://www.nato.int/docu/Review/2016/Also-in-2016/nato-defence-cyber-resilience/EN/index.htm.

Smith, Sir Rupert. *The Utility of Force.* New York: Random House, 2007.

Snegovaya, Maria. *Putin's Information Warfare in Ukraine: Soviet Origins of Russia's Hybrid Warfare.* Washington D.C.: Institute for the Study of War, September 2015.

Sontheimer, Michael. "When We Finish, Nobody is Left Alive." *Spiegel Online*, 27 May 2011. http://www.spiegel.de/international/europe/germany-s-wwii-occupation-of-poland-when-we-finish-nobody-is-left-alive-a-759095-3.html.

Spector, Ronald H. *Eagle Against the Sun, the American War with Japan.* Norwalk: Easton Press, 1985.

Stein, George J. "Total Defense: A Comparative Overview of the Security Policies in Switzerland and Austria." *Defense Analysis* vol.6, no. 1 (1990): 17-33.

Steinberg, David Joel. *Philippine Collaboration in World War II.* Ann Arbor: University of Michigan Press, 1967.

Stringer, Kevin D. "Building a Stay-Behind Resistance Organization; The Case of Cold War Switzerland Against the Soviet Union." *Joint Forces Quarterly*, 85 (2nd Quarter 2017): 109-114.

Solis, Gary D. *The Law of Armed Conflict: International Humanitarian Law in War.* New York: Cambridge University Press, 2012.

Swedish Defence Commission Report, "Resilience- The Total Defense Concept and the Development of Civil Defence 2021-2025." 20 December 2017. https://www.government.se/4afeb9/globalassets/government/dokument/forsvarsdepartementet/resilience---report-summary---20171220ny.pdf.

"Sweden's Defence Policy, 2016 to 2020." Government Offices of Sweden. 1 June 2016. https://www.government.se/globalassets/government/dokument/forsvarsdepartementet/sweden_defence_policy_2016_to_2020.

"Sweden Mobilises entire Home Guard for First Time since 1975." *The Local*, 6 June 2018. https://www.thelocal.se/20180606/sweden-mobilises-entire-home-guard-for-first-time-since-1975.

Tamkin, Emily. "Ten Years After the Landmark Attack on Estonia, Is the World Better Prepared for Cyber Threats?" foreignpolicy.com, 27 April 2017. https://foreignpolicy.com/2017/04/27/10-years-after-the-landmark-attack-on-estonia-is-the-world-better-prepared-for-cyber-threats/.

The Telegraph. "Nazi fake banknote 'part of plan to ruin British economy.'" *The Telegraph*, September 29, 2010. https://www.telegraph.co.uk/history/world-war-two/8029844/Nazi-fake-banknote-part-of-plan-to-ruin-British-economy.html.

The United Nations War Crimes Commission, *Law Reports of Trials of War Criminals*, vol. VIII. London: His Majesty's Stationery Office, 1948.

Tribelhorn, Marc. "Terror-Rezepte aus der Schweiz." *Neue Zürcher Zeitung*, 26 July 2013.

Tse-Tung, Mao. *On Guerrilla Warfare*. New York: Frederick A. Praeger, Inc., 1965.

Trueman, C N. "The Polish Resistance." The History Learning Site. Accessed 12 January 2018. historylearningsite.co.uk.

United States Institute of Peace. *Nonviolent Civic Action: A Study Guide Series on Peace and Conflict*. Washington D.C.: Endowment of the United States Institute of Peace, 2009.

Vagts, Detlev F., and Theodor Meron. "So-Called 'Unprivileged Belligerency': Spies, Guerrillas, and Saboteurs." *Humanizing the Laws of War: Selected Writings of Richard Baxter* (September 2013). DOI:10.1093/acprof:oso/9780199680252.003.0003.

Venter, Al J. *War Dog: Fighting Other People's Wars*. Drexel Hill: Casemate, 2003.

Voelzke, Marlies ed., *War After War, Armed anti-Soviet Resistance in Lithuania, 1944-1953*. Vilnius: The Museum of Genocide Victims of the Genocide and Resistance Research Centre of Lithuania, n.d.

Vojtiskova, Vladislava, Vit Novotny, Hubertus Schmid-Schmidsfelden and Kristina Potapova. *The Bear in Sheep's Clothing: Russia's Government-Funded Organizations in the EU*. Brussels: Wilfried Martens Centre for European Studies, 2016. https://www.martenscentre.eu/sites/default/files/publication-files/russia-gongos_0.pdf.

Volkmann, Russell W. *We Remained*. New York: Norton, 1954.

Von Dach, H. *Total Resistance: Swiss Army Guide to Guerrilla Warfare and Underground Operations*. Boulder: Paladin Press, 1992.

Warner, Michael ed. *CIA Cold War Records: The CIA under Harry Truman* doc. 30. Washington D.C.: CIA History Staff, Center for the Study of Intelligence, 1993.

Wright, Gordon. "Reflections on the French Resistance (1940-1944)." *Political Science Quarterly*, vol. 77, no. 3 (September 1962): 336-349. http://www.jstor.org/stable/2146309.

Yuk-wai, Li. "The Chinese Resistance Movement in the Philippines During the Japanese Occupation." *Journal of Southeast Asian Studies* vol.23 issue 2 (September 1992): 308.

Zimmerman, Jon. "The 'V for Victory' Campaign." *DefenseMediaNetwork*, July 24, 2011. https://www.defensemedianetwork.com/stories/the-v-for-victory-campaign/.

Commentaries

Pictet, Jean ed., *Commentary, Convention Relative to the Treatment of Prisoners of War*. Geneva: International Committee of the Red Cross, 1960.

Pictet, Jean ed., *Commentary, Convention Relative to the Protection of Civilians in Time of War*. Geneva: International Committee of the Red Cross, 1958.

Statements And Speeches

Nuland, Victoria. "Ukraine: Countering Russian Intervention and Supporting Democratic States." Statement before the Senate Committee on Foreign Relations. 113th Cong., 2d sess., May 6, 2014.

Kerry, John. "National Security and Foreign Policy Priorities in the FY 2015 International Affairs Budget." Opening statement before the Senate Committee on Foreign Relations. 113th Cong., 2d sess., 8 April 2014.

"Wales Summit Declaration." Issued by the Heads of State and Government participating in the meeting of the North Atlantic Council of the North Atlantic Treaty Organization (NATO) in Wales, 5 September 2014.

www.ingramcontent.com/pod-product-compliance
Lightning Source LLC
Chambersburg PA
CBHW052109020426

42335CB00021B/2694